P9-DNV-038
UNH LIBRARY

6.30

QUANTITATIVE ELECTRON-PROBE MICROANALYSIS

QUANTITATIVE ELECTRON-PROBE MICROANALYSIS

Editors:
V. D. SCOTT and G. LOVE
School of Materials Science
University of Bath

ELLIS HORWOOD LIMITED
Publishers · Chichester

Halsted Press: a division of
JOHN WILEY & SONS
New York · Brisbane · Chichester · Ontario

First published in 1983 by

ELLIS HORWOOD LIMITED
Market Cross House, Cooper Street, Chichester, West Sussex, PO19 1EB, England

The publisher's colophon is reproduced from James Gillison's drawing of the ancient Market Cross, Chichester.

Distributors:

Australia, New Zealand, South-east Asia:
Jacaranda-Wiley Ltd., Jacaranda Press,
JOHN WILEY & SONS INC.,
G.P.O. Box 859, Brisbane, Queensland 40001, Australia

Canada:
JOHN WILEY & SONS CANADA LIMITED
22 Worcester Road, Rexdale, Ontario, Canada.

Europe, Africa:
JOHN WILEY & SONS LIMITED
Baffins Lane, Chichester, West Sussex, England.

North and South America and the rest of the world:
Halsted Press: a division of
JOHN WILEY & SONS
605 Third Avenue, New York, N.Y. 10016, U.S.A.

© 1983 V.D. Scott and G. Love/Ellis Horwood Ltd.

British Library Cataloguing in Publication Data
Quantitative electron probe microanalysis.
1. Electron probe microanalysis
I. Scott, V.D. II. Love, G.
543'.0812 QD98.E4

Library of Congress Card No. 83-18366

ISBN 0-85312-514-7 (Ellis Horwood Ltd. – Library Edn.)
ISBN 0-85312-672-0 (Ellis Horwood Ltd. – Student Edn.)
ISBN 0-470-27510-3 (Halsted Press)

Typeset in Press Roman by Ellis Horwood Ltd.
Printed in Great Britain by Unwin Brothers of Woking.

COPYRIGHT NOTICE –

All Rights Reserved. No part of this publication may be reproduced, stored in a retrieval system, or transmitted, in any form or by any means, electronic, mechanical, photocopying, recording or otherwise, without the permission of Ellis Horwood Limited, Market Cross House, Cooper Street, Chichester, West Sussex, England.

Table of Contents

Preface

This book owes its origin to a course designed to accommodate scientists with differing backgrounds and experience working in industry, government and universities. The scientific programme considered both the acquisition of accurate data from the instrument and the application of corrections to convert the results into quantitative information. Since nowadays users rarely need to be involved with the development of equipment, only those aspects of instrument design were included which were fundamental to reliable measurements.

The book follows the same pattern. After a general introduction to the subject, Chapter 1, the physical basis of x-ray production in bulk specimens by a beam of energetic electrons is developed in Chapter 2. Methods of measuring and analysing the x-ray emission spectra are dealt with in Chapters 3, 4 and 5 and here, for the convenience of workers interested in one specific technique, descriptions of wavelength-dispersive spectrometry (WDS) and energy-dispersive spectrometry (EDS) are given separately. An important factor in quantitative microanalysis concerns the quality of output data from the instrument, and criteria such as selection of experimental conditions and specimen requirements are discussed in Chapter 6. The correction factors -- atomic number, absorption and fluorescence - are treated in separate chapters (7, 8 and 9), each of which discusses various analytical expressions which have been proposed for calculating them. Correction programmes combining the three factors are analysed critically in Chapter 10 and the merits and deficiencies of the respective analytical expressions are considered. The following chapter describes methods of iteration involved in the corrections as well as computer programs for carrying out the calculations; also in Chapter 11 reference is made to alternative correction procedures such as the use of alpha factors and the 'no-standards technique'. Chapter 12 is devoted to the use of the Monte Carlo method in microanalysis studies since, although not generally utilised for routine quantitative work, it has proved its value for developing and evaluating correction factors. The final chapter in the book deals with cases where corrections developed for bulk specimens may no longer be applicable, as in the analysis of thin films and particles; such studies are proving to be an important growth area in the field, especially now that x-ray facilities are often fitted to transmission electron microscopes.

Doubtless we can expect in the future to see further advances made in extracting microanalysis data rapidly with the help of sophisticated computer control and in presenting quantitative information in a variety of forms (such as concentration maps) which may be easily assimilated by the scientist or engineer without recourse to a detailed understanding of the various physical processes involved. Nevertheless there will undoubtedly be occasions where such processed information may be misinterpreted and where some appreciation of the fundamentals of the technique is needed to identify the likely cause of the difficulty. This is another reason why this book should prove invaluable to scientists depending upon electron-probe microanalysis for answers to their various problems.

Finally, we would like to take this opportunity to express our thanks to Professor Tom Mulvey and Drs Stephen Reed and Mike Cox for their collaboration and encouragement in initiating the Summer School and for their respective contributions to this book.

V. D. SCOTT and
G. LOVE

School of Materials Science,
University of Bath

Contributors

Dr M. G. C. COX — Bryant & May Ltd.
Garston, Liverpool, UK

Professor T. MULVEY — Department of Physics
University of Aston in Birmingham
Birmingham, UK

Dr S. J. B. REED — Department of Earth Sciences
University of Cambridge
Cambridge, UK

1

Development of electron-probe microanalysis – an historical perspective

T. MULVEY

In the electron-probe microanalyser, a pre-selected small area of a solid specimen is bombarded with electrons. The resulting emission includes backscattered primary electrons, low energy photo-electrons and Auger electrons, together with characteristic x-ray emission superimposed on a background of continuous x-radiation (*Bremsstrahlung*). In suitably transparent crystal specimens, optical photo-emission (cathodoluminescence) spectra can also be observed whilst in semiconductors, electron-induced currents in the specimen may be measured and imaged to give analytical information. If the samples are sufficiently thin then analysis of the energy of transmitted electrons (energy loss spectrometry) will also yield data on the chemical composition of the specimen.

In electron-probe x-ray microanalysis attention is concentrated on the emitted x-ray spectrum from the specimen which is recorded by means of a crystal spectrometer or an energy-dispersive detector. In a solid target the x-ray spectrum originates from a volume of a few cubic micrometers, compatible with the resolution of an optical microscope. In a thin target, on the other hand, reduced electron scattering causes this volume to be decreased (to that occupied by some 10^4 atoms) so that an electron microscope is needed to direct the probe onto the appropriate point for analysis.

The accurate recording of x-ray spectra from solid targets or thin slices of material is now carried out routinely. A more difficult problem, however, is deducing from the observed x-ray spectrum the composition of the specimen to an accuracy of about 1%. The practical solution of this problem is the central concern of the present book.

In this chapter the subject is approached from an historical viewpoint, detailing not only the early approaches to quantitative analysis but also those major instrumental developments which led up to the electron-probe micro-analyser as we know it today.

1.1 BACKSCATTERED ELECTRONS, X-RAYS AND THE ELECTRON-PROBE ANALYSER

The first electron-probe analyser was probably that built by Starke (1898) in Berlin. A beam of electrons (strength I_p) produced in a cathode ray tube was directed at a suitably positioned specimen and the net current (I_s) entering the specimen was measured. The backscattered fraction $\eta = (I_p - I_s)/I_p$ was found to vary with the atomic weight of the target as shown in Fig. 1.1 in which the analysis of a brass sample is included. The method was not developed further as an analytical technique, since it is not sufficiently specific for element identification. Nevertheless it is still used in the preliminary analysis of a sample.

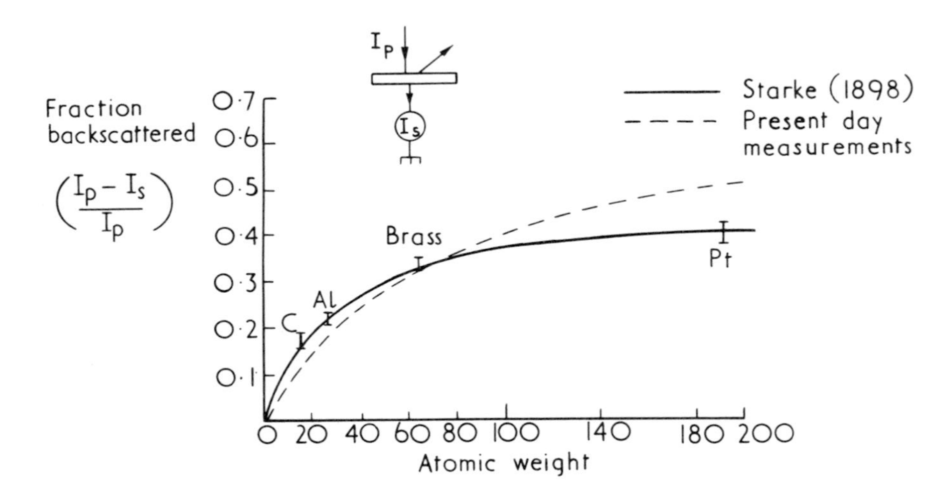

Fig. 1.1 – Electron-probe backscattering analysis (Starke, 1898).

The term 'characteristic x-rays' was used by Barkla and Sadler (1909) during their studies of x-ray emissions from pure elements subjected to irradiation by a beam of x-rays. Kaye (1909) was the first to show that similar characteristic rays are emitted when an element is bombarded by cathode rays. His experimental electron-probe x-ray analyser is shown in Fig. 1.2. A beam of 28keV electrons, produced by the demountable cathode ray tube (A) irradiated a specimen (B) mounted on a carriage which ran on rails. The carriage and hence the specimens mounted on it could be moved successively under the electron beam by means of an external magnet which attracted a block of iron (C) fastened to the carriage. Characteristic x-rays emitted by the specimen were detected with an ionisation chamber (D) and identified by successively interposing absorbing screens (E) of various materials of selected thickness between specimen and detector. Different areas of the specimen could be examined by suitably deflecting the electron beam using an external permanent magnet – the first hint of a magnetic scanning system?

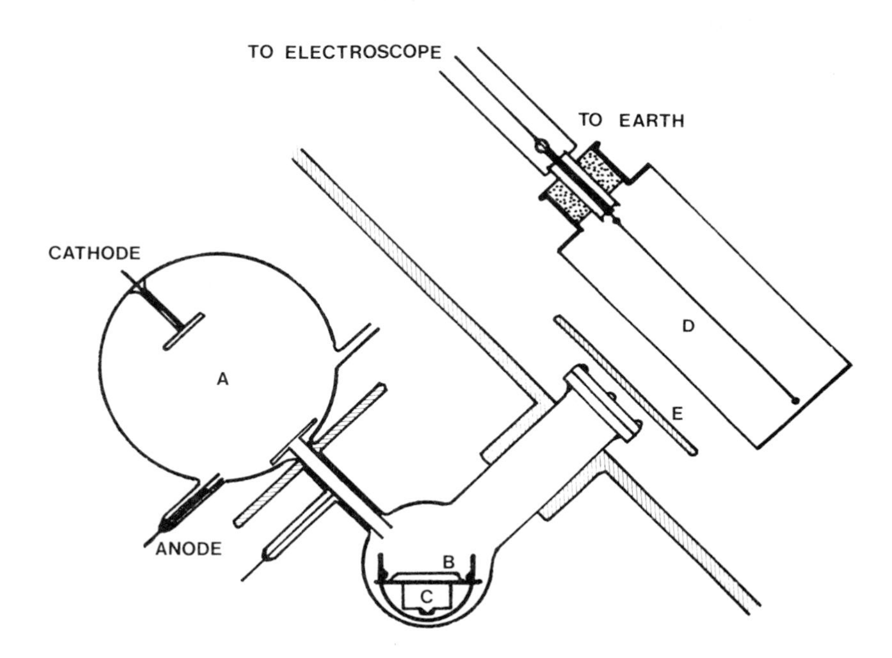

Fig. 1.2 – Electron-probe x-ray analyser (Kaye, 1909).

At this period the physical origin of characteristic rays was not clear. Barkla and Sadler (1909) favoured the view that incident electrons produced non-characteristic (fluorescence) radiation in the target, which in turn generated characteristic radiation. Others believed that electrons could produce characteristic radiation directly. Among the latter was Beatty (1912, 1913) who demonstrated that characteristic radiation is predominantly caused by direct electron excitation. His experimental technique was a model of ingenuity in avoiding experimental errors and also provided an insight into the subsequent development of the theory of fluorescence corrections in x-ray microanalysis. He showed that from a target bombarded with electrons there emerged two radiations (a) independent radiation (*Bremsstrahlung*)† which was always present and (b) characteristic radiation which occurred only when the cathode rays were

† Beatty coined the term independent radiation (*Bremsstrahlung*) to designate that x-radiation which occurs when a target is bombarded with electrons of energy too low to stimulate characteristic radiation.

sufficiently fast. Beatty argued correctly that a target of a few micrometres thickness would stop the electrons but hardly affect the passage of the x-rays giving rise to fluorescence effects. It was found that the x-ray output from a copper target of 5μm thickness was about as large as one of 2mm thickness indicating that the x-rays were produced very near the surface of the target. By covering a thick copper target with an aluminium foil of 3μm thickness he was able to isolate and measure the characteristic copper radiation produced solely by fluorescence in the target since the electrons themselves were unable to reach the copper. Knowing from previous experiments that the continuous background from aluminium was about a third of that from copper, Beatty estimated that the characteristic radiation produced by direct electron excitation in copper was at least an order of magnitude greater than that produced by fluorescence. Direct experimental confirmation that the effective x-ray source in electron-probe analysis is within a few micrometres of the surface was obtained using the arrangement shown in Fig. 1.3. An electron beam irradiated an inclined copper target 10μm in thickness and x-rays emitted by the target were detected using two ionisation chambers (A and B) positioned on either side of the target. A second copper foil, also 10μm thick, was placed between the target and ionisation chamber B so that if the effective x-ray source is at the target surface equal x-ray attenuation would occur in each target. Experimentally no difference in intensity was observed in the two detectors. Beatty therefore concluded that

(a) Cathode rays were able to produce characteristic x-rays directly in atoms, and the effective depth of x-ray production was less than 10μm.
(b) The atomic mechanisms which accounted for the independent radiation and for the characteristic radiations were not connected with each other.

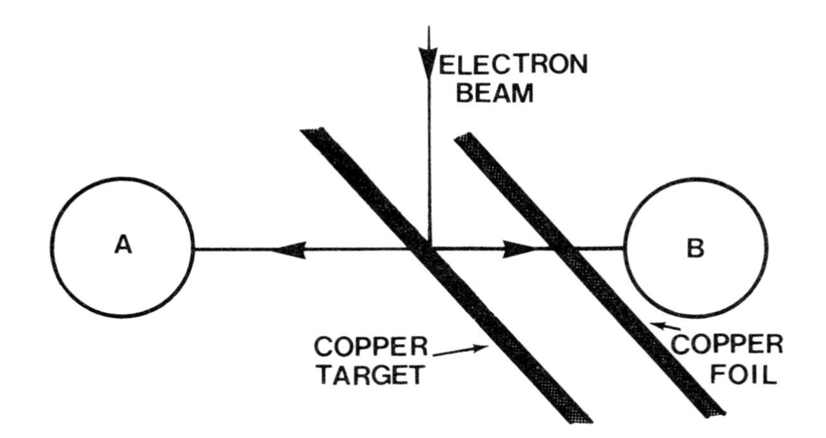

Fig. 1.3 – Experiment to show that the effective x-ray source in electron-probe analysis is close to the surface (Beatty, 1912, 1913).

The first conclusion was crucial for electron-probe analysis because it meant that microanalysis of a volume of the order of a few cubic micrometres could be envisaged, whereas with x-ray fluorescence analysis the corresponding volume would be several orders of magnitude greater. The second conclusion was important because analysis by means of characteristic x-rays would be specific and well defined.

1.1.1 Analysis of x-rays – the crystal spectrometer

In 1912 von Laue proposed, and Friedrich and Knipping verified, that a crystal could be used as a three-dimensional diffraction grating for x-rays. These ideas were taken up by W. H. and W. L. Bragg (1913) who conceived and built the first x-ray crystal spectrometer. Their interpretation of von Laue's results in terms of optical reflections from certain crystal planes was expressed as

$$n\lambda = 2d \sin\theta \ ,$$

where λ is the x-ray wavelength, n is an integer, θ is the angle of incidence of the x-rays and d is the spacing of the appropriate reflecting planes. Fig. 1.4 shows the arrangement of their spectrometer, an inspiration being to place the slit, crystal, and detector on a circle. As the crystal was rotated, radiation of a given wavelength always passed through the exit slit into the detector enabling all parts of the crystal and all parts of the target to contribute *successively* to the output. The modification of this arrangement needed to attain *simultaneous*

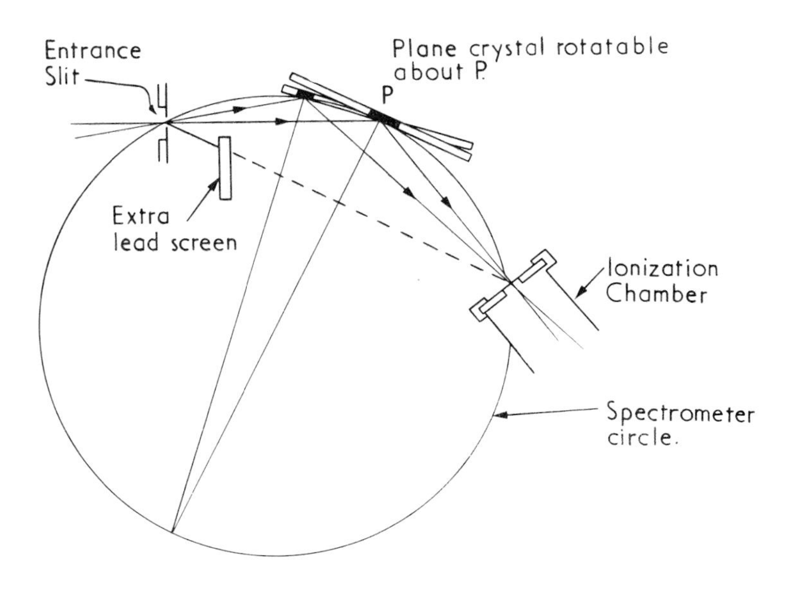

Fig. 1.4 – Rotating crystal x-ray spectrometer of W. H. and W. L. Bragg (1913).

focusing from all parts of the crystal was to take many years to arrive at and was to surprise everybody by the simplicity of the final solution.

Moseley (1913), however, had to manage without such refinements. He was not primarily interested in how the characteristic x-rays were produced but rather

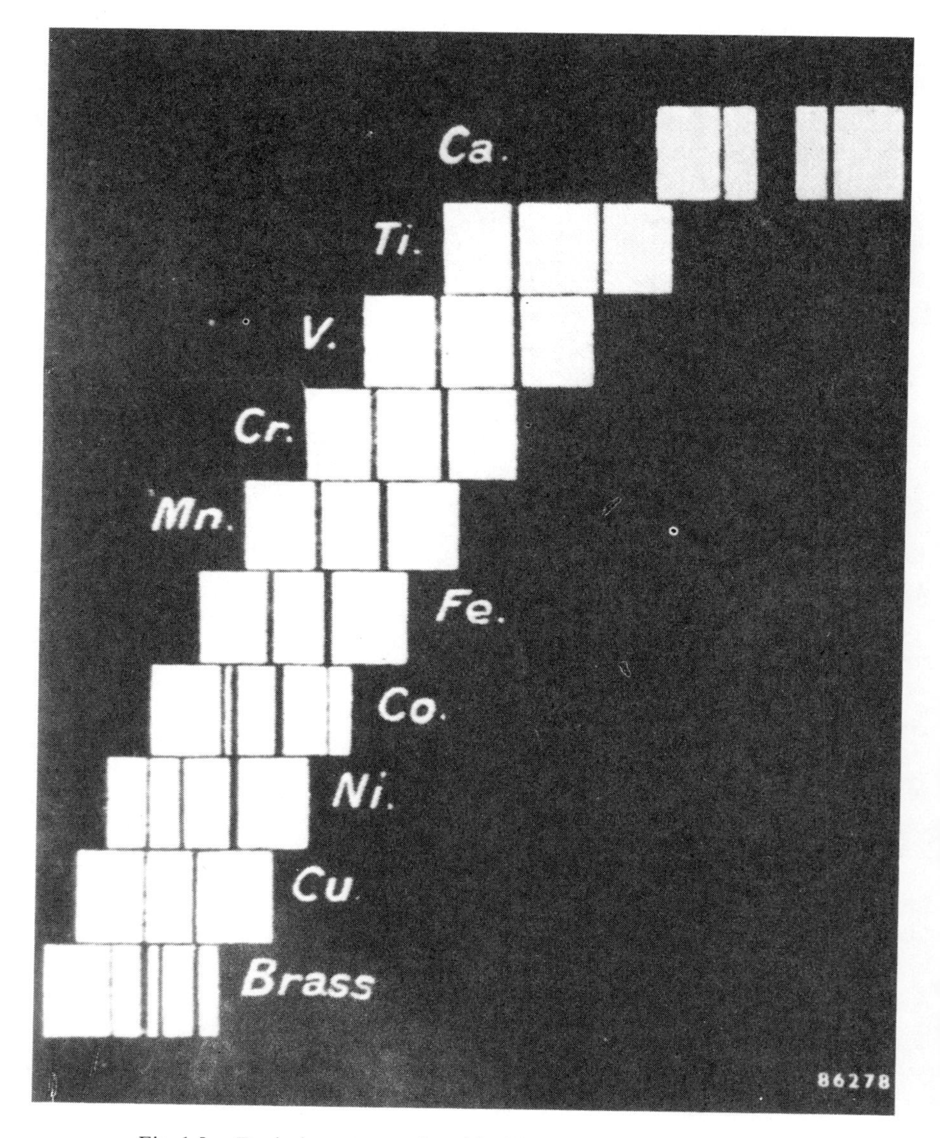

Fig. 1.5 – Typical spectra produced by Moseley's electron-probe x-ray analyser (Moseley, 1914); note impurity lines in Co and Ni spectra.

in their wavelengths and hence their energies. He returned to Kaye's design of a demountable x-ray tube and, remarkably for the time, constructed an x-ray spectrometer capable of covering the elements zinc to calcium; later its range was extended to aluminium (Moseley, 1914). He decided against an ionisation chamber as an x-ray detector and used instead the less sensitive photographic plate since this could record several spectral lines simultaneously. A wide entrance slit was employed to give adequate intensity from a fixed, flat crystal of potassium ferrocyanide. After recording one section of the spectrum, the crystal and photographic plate were moved to predetermined positions and the next (overlapping) section recorded. When this spectrometer was being tested at the University of Manchester, Moseley was at first unable to locate the spectral lines but fortunately was able to persuade W. H. Bragg from the nearby University of Leeds to help. Bragg, although primarily a mathematician, was a gifted instrument designer and soon found the lines. Figure 1.5 shows typical spectra obtained on this instrument including the first x-ray analysis of a technological material (brass) by an electron probe. However, Moseley's apparatus was unsuitable for analytical work because it was difficult to control the electron beam and his x-ray spectrometer was far too insensitive.

In the next few years considerable effort was expended in many countries to design better x-ray spectrometers. Maurice de Broglie, the brother of Louis de Broglie, devised a curved crystal spectrometer (de Broglie, 1914) which, although not focusing, produced a complete spectrum on a photographic plate without the complication of having to move the crystal.

Gouy (1916) approached the problem of spectrometer focusing from the viewpoint of geometrical optics, subject to the additional requirement of satisfying Bragg's Law, and put forward the arrangement shown in Fig. 1.6. Here a hollow cone of x-rays from a point source is focused into a point image of monochromatic x-rays. The reflection takes place along a narrow strip of the crystal (shaded) giving approximate paraxial focusing. Gouy did not try out these ideas in practice but they were to be used later in x-ray spectroscopy.

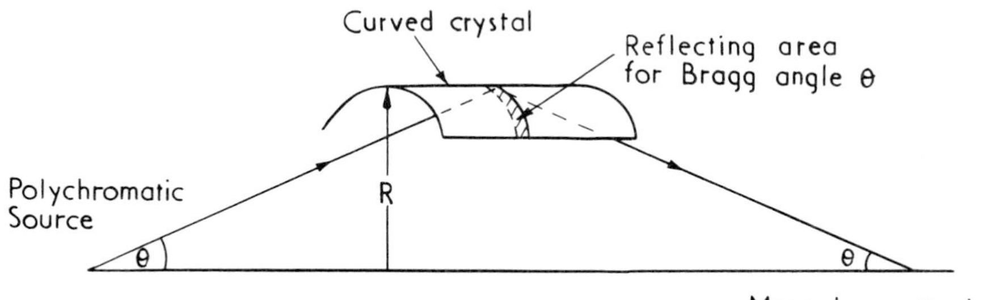

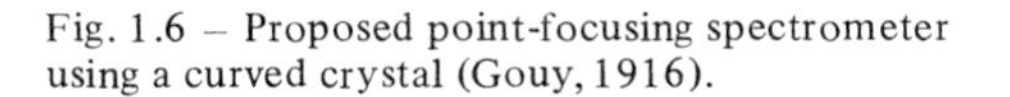
Fig. 1.6 – Proposed point-focusing spectrometer using a curved crystal (Gouy, 1916).

The quest for a focusing x-ray spectrometer using curved crystals continued, in spite of Wagner's pronouncement that no surface shape existed for which the double condition would be satisfied, namely that the angle of incidence must equal the angle of reflection and Bragg's Law must hold (Wagner, 1917). Darbord (1922) indicated an approximate solution, pointing out that attention should be concentrated on the atomic planes rather than on the shape of the crystal. By bending the crystal as shown in Fig. 1.7 several points in the crystal would satisfy the reflection conditions. However, Darbord did not pursue his ideas to their logical conclusion but approximate focusing, at least, could be realised for x-rays.

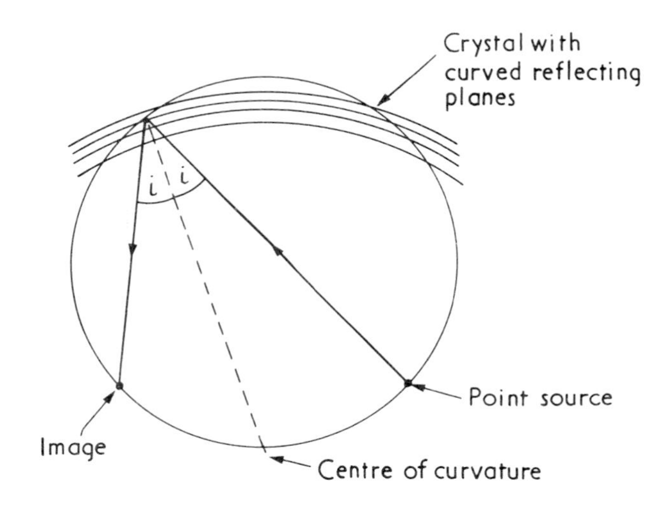

Fig. 1.7 – Suggestion for a curved-crystal focusing spectrometer (Darbord, 1922).

Johann (1931) and Cauchois (1932) proposed and built approximate focusing systems using thin crystals bent to twice the radius of curvature of the Rowland circle. This method is suitable for crystals such as mica which cannot readily be ground. In Johann's spectrometer the crystal was used in reflection, in that of Cauchois it was used in transmission.

A better solution was given by Johansson (1932, 1933) who bent the crystal to **twice** the radius of curvature of the spectrometer circle and ground away the inner surface of the crystal to a radius equal to that of the spectrometer circle. This ensured that the angle of incidence equalled the angle of reflection and that the Bragg angle remained constant over the entire surface of the crystal. It also resulted in practically perfect focusing and because the crystal size was not unduly restricted a high intensity could be obtained (see section 3.1).

The comparatively high sensitivity of the Johann, Johansson and Cauchois spectrometers was indispensable to the eventual development of a practical microanalyser since it enabled probe currents of less than 10^{-7} A to be usefully employed in the early instruments. However, it was still not possible to produce the necessary fine electron probe of adequate intensity at the specimen and this had to wait until a basic understanding of electron optics had been achieved.

1.1.2 Electron optics and early electron microscopes

Busch (1926, 1927) showed mathematically that a beam of electrons could be brought to focus by a magnetic coil in much the same way as a beam of light is focused by a glass lens. A similar result would also be expected for an arrangement of charged electrodes of axial symmetry. Hence began the science and art of electron optics. With limited experimental apparatus Busch was unable to confirm his theoretical prediction unambiguously but his findings greatly stimulated engineers and physicists to set about focusing a beam of electrons in a cathode ray oscillograph. In particular, Knoll and Ruska (1932a and 1932b) devised and produced the first electron microscope, with the modest magnification of 13 times, which led to the development of the high resolution transmission instrument. Shortly afterwards Knoll (1935) invented the scanning electron microscope, later to be of crucial importance in electron-probe microanalysis.

Knoll's instrument, shown schematically in Fig. 1.8, was a very advanced concept for its time. The specimen was scanned in television fashion by a focused electron probe. Local variations in specimen topography or chemical composition caused corresponding variations in the number of backscattered and secondary electrons leaving the target. The net specimen current was amplified and caused

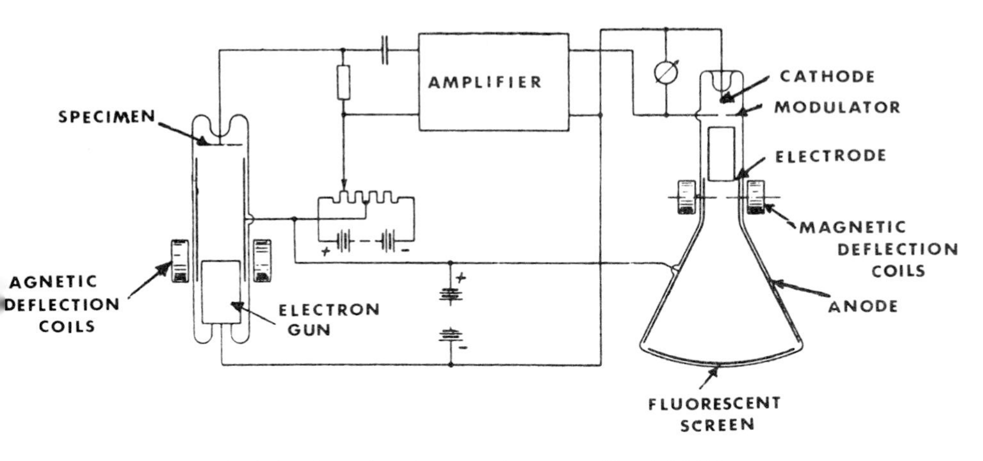

Fig. 1.8 – Scanning electron microscope (Knoll, 1935).

to modulate a cathode ray display tube scanned in synchronism with the electron probe. The resulting image that appeared on the display tube was remarkably similar to that observed in an optical microscope, even though the image forming mechanisms have little in common. The image magnification is simply the ratio of the length of the line scan on the display tube to that on the specimen. In this early instrument with its fairly large probe size (probably about 100 μm since high resolution was not the principal aim), the foundations of scanning electron microscopy were laid.

During the war period 1939–1945 considerable development of electron optics took place in Germany, mainly in connection with transmission electron microscopy. Of particular importance was Boersch's (1939) technique of forming fine electron probes ~2.5 nm in diameter which were used to project a highly magnified (shadow) image of a small object onto a fluorescent screen. Such an instrument would clearly be useful in electron-probe microanalysis.

Ruthemann (1941, 1942) in the Physikalisches Institut of the Technische Hochschule at Danzig, at the instigation of Professor Kossel, investigated the 'characteristic' electron energy losses experienced by a beam of 7.5 keV electrons in passing through a thin film of collodion (100 nm thick). He analysed the velocity and hence the energy of the transmitted electrons and showed that some of these electrons had suffered an energy loss corresponding to the excitation of K radiation in carbon (291.2 eV), oxygen (543 eV) and nitrogen (392 eV). Such a technique is suited for the transmission electron microscope, where the thin film specimen minimises the effects of inelastic scattering, and is particularly useful for the analysis of light elements such as carbon, oxygen and nitrogen which are difficult to analyse in an x-ray spectrometer.

1.1.3 RCA electron-probe microanalyser

Hillier (1943) and Hillier and Baker (1944) at the RCA Laboratories at Princeton, New Jersey, ingeniously combined an electron projection microscope and energy-loss spectrometer in the arrangement shown schematically in Fig. 1.9. The instrument relied heavily on electron-optical techniques already developed by Hillier and others for the RCA electron microscope. An electron source was demagnified by two magnetic lenses to form a probe of 20 nm or so in diameter at the specimen, which took the form of a thin film some tens of nanometres thick of the type used in transmission electron microscopy. A magnified shadow image of the specimen, projected onto the fluorescent screen in the way previously devised by Boersch, enabled any part of it to be positioned under the probe and studied by viewing the fluorescent screen through an optical microscope of moderate magnification. A third lens could be switched in to focus the electrons on to the slit of a 180° magnetic velocity spectrometer mounted above the lens system and the resulting energy spectrum was recorded on a photographic plate. With this instrument, spectra were obtained in which the presence of carbon,

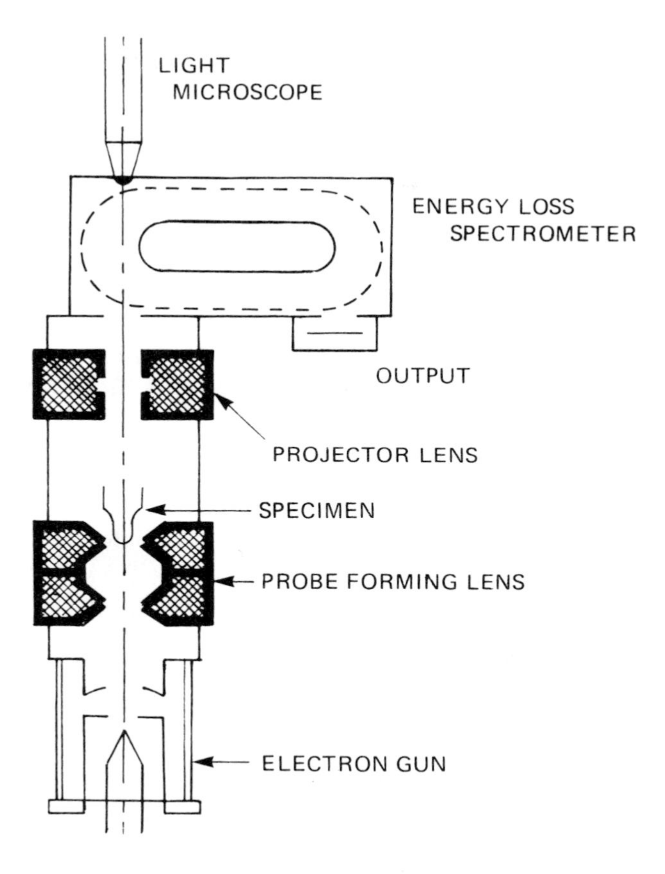

Fig. 1.9 – RCA electron-probe microanalyser (Hillier and Baker, 1944).

nitrogen and oxygen K radiation in a collodion film was clearly demonstrated by discontinuities in the spectral curve. The sensitivity of the technique was high and about 10^{-13}g of these light elements could be detected provided that a probe of about 100 nm was used; even the analysis of beryllium was possible. No attempt was made to observe the characteristic x-ray emission directly and neither was the instrument developed commercially. However, Hillier did take out a patent on the electron-probe x-ray microanalyser and in Patent Law is, therefore, the inventor although he apparently did not proceed with its actual construction. The schematic diagram included in the patent (Hillier, 1947) shows an electron probe-forming system, similar to that used in the RCA instrument, together with an x-ray spectrometer incorporating a flat crystal and photographic plate strongly reminiscent of Moseley's design.

1.1.4 Castaing's electron-probe x-ray microanalyser

It would appear that, because of post-war difficulties in scientific communication, news of the Hillier Patent had not reached Castaing and Guinier in France in 1947. In January of that year Raymond Castaing had joined the research staff of the Materials Department of the Office National de Rechèrches Aéronautiques (ONERA) and became involved in the setting up of an electron microscope laboratory for metallurgical and materials research. In 1948 during the course of an investigation into the properties of copper–aluminium alloys Professor Guinier, who was associated with the project, asked Castaing about the possibility of making a point by point analysis of a metal sample by bombarding it with electrons and measuring the characteristic x-ray emission. The idea was to analyse at least qualitatively areas of some hundreds of angstrom units in diameter although it was realised that the counting rates would be low, perhaps a few pulses a minute. It was a tall order but by the beginning of 1949 Castaing had succeeded in producing an electron probe of about 1 μm in diameter with a current approaching 10^{-8} A when everything was working properly.

In July 1949 at the first European Conference on electron microscopy in Delft, Castaing described this experimental electron-probe x-ray microanalyser (Castaing and Guinier, 1949, 1950). The instrument was in fact a commercial (TSF) electrostatic electron microscope modified for use as an electron shadow microscope. A fine probe of $\sim$1 μm in diameter was formed by two electrostatic lenses and the shadow image of a transparent specimen, or the edge of an opaque one, was projected onto a fluorescent screen to facilitate location of a probe onto the specimen. The specimen could be moved by a mechanical stage and some movement of the electron spot could be obtained by an electrostatic deflector placed between the first and second lenses. There was no means of viewing a solid sample but the specimen micrometer controls provided movement along a more or less predetermined path. The equipment differed from that described in the Hillier Patent since x-rays emitted from the specimen were detected by a Geiger counter secured to the outside of the microscope body, an arrangement reminiscent of Kaye's apparatus. A Geiger counter cannot, of course, separate radiations of different wavelengths, but copper and aluminium in a copper–aluminium alloy could be distinguished one from another by the greater total x-ray output from copper, an effect previously noted by Beatty (1912). Graphite was distinguished in cast iron by the same method. In 1950, the fitting of a Johansson focusing spectrometer incorporating a quartz crystal and Geiger counter enabled the possibility of quantitative analysis to be investigated. A schematic diagram of Castaing's instrument is shown in Fig. 1.10. The small plane mirror inclined at 45° to the specimen permitted an optical microscope to be inserted in the vacuum; the specimen could then be observed and manipulated while under the electron probe. The point of impact of the electron probe could be detected either by the fluorescence of impurities on the specimen surface or by the carbon contamination (an important specimen marker in the early days of microanalysis). The concentration of a given element was obtained by setting the crystal spectrometer to the appropriate spectral line and comparing the counting rate from a selected point on the specimen with that of a standard under the same irradiation conditions. This was truly the beginning of the electron-probe x-ray microanalysis as we know it today.

The foundations of the entire technique were displayed in a remarkable Ph.D. thesis (Castaing, 1951), which is a classic on the subject. The thesis set out the fundamental principles of the method and its experimental realisation as a metallurgical tool. A theoretical basis was established for the corrections needed to take fluorescence effects into account and experiments were also conducted to determine the variation of the characteristic x-ray intensity as a function of electron penetration into the specimen and standard. These studies provided a firm foundation for the quantitative analysis of alloys and compounds and Castaing, although not the inventor in Patent Law, may be rightly regarded as the one who established, single-handed, the technique of electron-probe microanalysis.

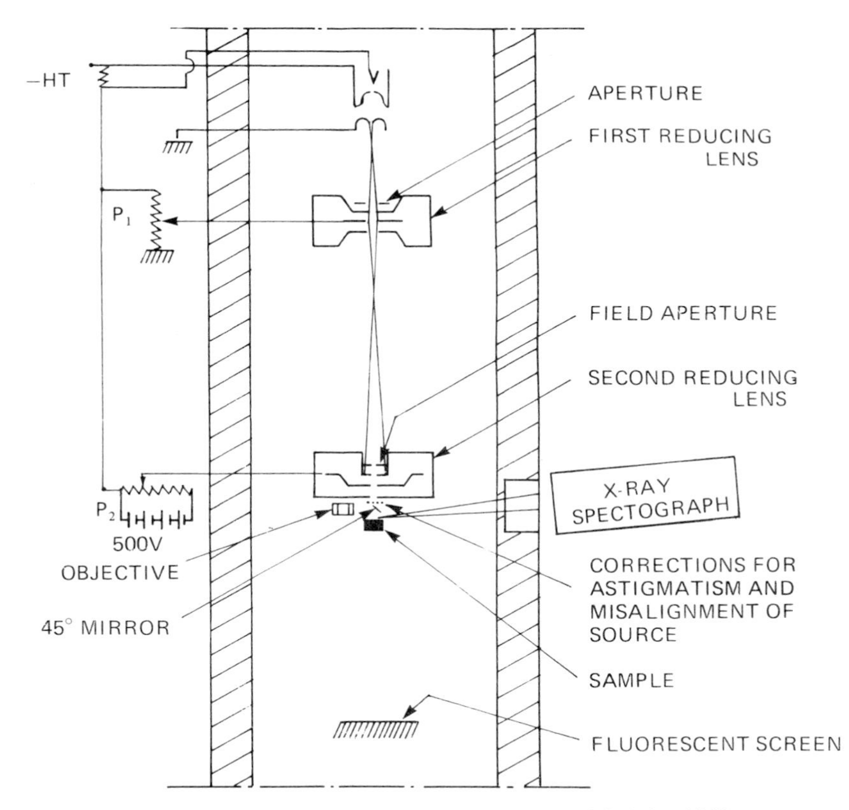

Fig. 1.10 – Electron-probe x-ray microanalyser (Castaing and Guinier, 1949).

1.1.5 Scanning electron-probe microanalyser

In 1953 Peter Duncumb, a Ph.D. student, joined the team of Dr Cosslett at the Cavendish Laboratory, Cambridge, in order to investigate the possibility of replacing the optical microscope by a scanning electron microscope. The experimental instrument (Fig. 1.11) was completed in 1956. An RCA model B electron microscope column was modified by inserting a new final probe-forming lens in which x-rays emitted from the specimen could pass through a side tube in order to reach an energy-dispersive x-ray detector or a crystal spectrometer. Scanning coils placed in front of the probe-forming lens allowed the electron beam to be scanned across the specimen. The x-ray signal was then caused to modulate the brightness of a cathode ray tube scanned in synchronism with the electron probe. The resulting 'x-ray image' therefore gave a visual picture of the distribution of a chosen element in the specimen. In addition, the electron probe could be positioned on the specimen more precisely than was possible with the optical microscope. The development made feasible the transfer of electron-probe microanalysis into the electron microscope so that high resolution analysis of thin specimens became a future possibility.

1.1.6 Electron microscope/microanalyser

The technical problems of forming a fine electron probe of diameter less than 0.1 μm on a specimen in the transmission electron microscope were elegantly solved by Cooke and Duncumb (1969) who supplemented the conventional illuminating system of an electron microscope by fitting an iron-free mini-lens of the le Poole (1964) type immediately in front of the objective lens. This greatly reduced probe size (from 1 μm to some 0.05 μm) at the specimen also permitted improved selected area diffraction patterns to be obtained. In addition the x-ray spectrum from the selected area was recorded by two fully focusing spectrometers mounted on the electron optical column. This design was later produced commercially in the form of the EMMA 4 electron microscope/microanalyser. Because of the high accelerating voltage, and hence increased electron-optical brightness of the electron gun, acceptable counting rates were obtainable on precipitates and inclusions in the size range of 0.1 μm or less. In such an instrument, the combination of high resolution image, selected area diffraction pattern and x-ray spectral output have proved of inestimable value in characterising with a high degree of certainty important microstructural features such as precipitates and grain boundaries.

Subsequent instrumental developments by manufacturers in different parts of the world have enabled micro-areas as small as 2 nm in diameter in thin specimens to be analysed in the transmission microscope. This has been greatly facilitated by the development of lithium-drifted silicon detectors as energy-dispersive x-ray analysers (see Chapter 4). A detailed survey of recent instrumental developments including the increasing use of computer control is, however, beyond the scope of the present book. Suffice it to say that the EMMA instrument

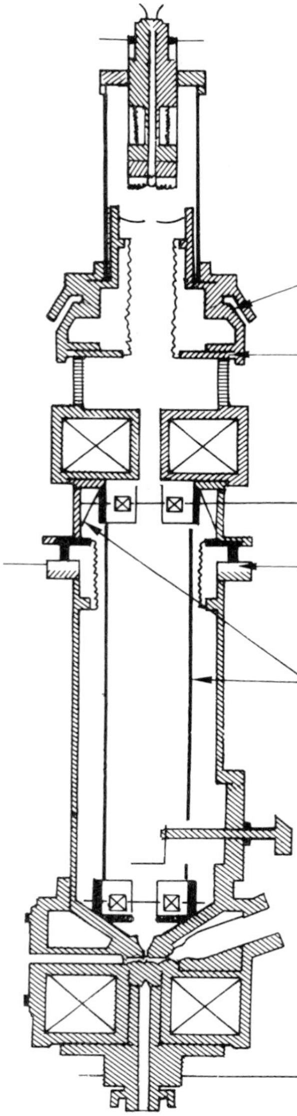

Fig. 1.11 – Experimental scanning x-ray projection microscope/microanalyser (Duncumb and Cosslett, 1956).

and the recently developed analytical scanning transmission electron microscope have brought to full realisation in a remarkably short time the original daring concept of Castaing and Guinier, namely that of analysing by x-ray spectroscopy the finest metallurgical details that are visible in the transmission electron microscope.

1.2 THE APPROACH TO QUANTITATIVE ELECTRON-PROBE MICROANALYSIS

1.2.1 Castaing's treatment

In order to carry out quantitative electron-probe microanalysis, Castaing (1951) introduced the concept of measuring the ratio (k_A) of the characteristic x-ray emission from the specimen (I_{spec}) to that (I_{stnd}) from a pure standard, both measurements being recorded under identical analysis conditions. He then proposed that to a first approximation

$$k_A = c_A \ , \tag{1.1}$$

where c_A is the mass concentration of the analysed element (A). This proposition was based upon consideration of the number of ionisations (dn) produced in a particular electronic shell by a single electron travelling a distance dx in the target. Let us examine the parameters involved in calculating dn for a pure element standard. The number of ionisations produced will be directly proportional to the number of atoms in the layer of thickness dx and inversely proportional to the area over which these atoms are distributed. Thus

$$(dn)_{stnd} = Q\left[\frac{N\rho}{A}a\,dx\right]\frac{1}{a} \ , \tag{1.2}$$

where the bracketed term is the number of atoms in a layer of area a and thickness dx; N is Avogadro's number, ρ is the density and A is the atomic weight. The term Q is known as the ionisation cross-section and has units of area. Its value is a function of the electronic shell of the element and also of the energy of the electron, Q tending to increase as the electron energy falls because the electron will remain close to an atom for a greater length of time. Equation (1.2) may be rewritten

$$(dn)_{stnd} = \frac{N\rho}{A}\frac{Q}{dE/dx}dE \ .$$

If Williams' law (Williams, 1932) is used to describe the deceleration ($-dE/dx$) of the electron

$$\frac{-dE}{dx} = \text{const.}\left(\frac{c}{v}\right)^{1.4}\rho \ ,$$

where c is the velocity of light and v the electron velocity, then

$$(\mathrm{d}n)_{\mathrm{stnd}} = -\frac{N}{A}\left(\frac{v}{c}\right)^{1.4} \text{const.}\, Q\mathrm{d}E \ .$$

Castaing made the assumption that the electron was not backscattered from the target so that the total number of ionisations produced by the electron over the whole of its trajectory is

$$n_{\mathrm{stnd}} = -\int_{E_0}^{E_c} \frac{N}{A}\left(\frac{v}{c}\right)^{1.4} \text{const.}\, Q\mathrm{d}E \ , \qquad (1.3)$$

where E_0 is the energy of the incident electron and E_c the minimum energy required to produce ionisations of the electronic shell of interest, referred to hereafter as the critical excitation energy. Both Q and v are functions of electron energy so equation (1.3) may be rewritten

$$n_{\mathrm{stnd}} = -\frac{\text{const.}}{A}\int_{E_0}^{E_c} f_{\mathrm{A}}(E)\mathrm{d}E \ ,$$

where the function $f_{\mathrm{A}}(E)$ depends solely upon the characteristics of the pure element A. In an analogous manner the number of ionisations of element A produced by an electron travelling in a multi-element specimen containing A in mass concentration c_{A} is

$$n_{\mathrm{spec}} = -c_{\mathrm{A}}\frac{\text{const.}}{A}\int_{E_0}^{E_c} f_{\mathrm{A}}(E)\mathrm{d}E \ .$$

Thus $n_{\mathrm{spec}}/n_{\mathrm{stnd}} = c_{\mathrm{A}}$ and, if the effects of x-ray absorption are ignored for the present, n is directly proportional to the x-ray intensity, I.
Hence $n_{\mathrm{spec}}/n_{\mathrm{stnd}} = I_{\mathrm{spec}}/I_{\mathrm{stnd}} = k = c_{\mathrm{A}}$ which is Castaing's first approximation.

On the other hand if Webster's law (Webster *et al.*, 1933) is obeyed,

$$\frac{-\mathrm{d}E}{\mathrm{d}x} = \text{const.}\,\rho\left(\frac{c}{v}\right)^{1.4}\frac{Z}{A}$$

and a similar calculation of n_{spec} and n_{stnd} is performed, one obtains

$$\frac{I_{\mathrm{spec}}}{I_{\mathrm{stnd}}} = \frac{c_{\mathrm{A}}(Z_{\mathrm{A}}/A_{\mathrm{A}})}{\sum c_{\mathrm{i}}(Z_{\mathrm{i}}/A_{\mathrm{i}})} \ ,$$

where c_i, Z_i and A_i refer to the ith element in the sample.

However, there is no compelling reason to believe that this is a better approximation, since electron backscattering effects are not taken into account in either of the calculations. Castaing therefore proposed that an empirical coefficient, α, should be assigned to each element in the specimen so that

$$\frac{I_{\text{spec}}}{I_{\text{stnd}}} = \frac{\alpha_A c_A}{\sum \alpha_i c_i} .$$

This is often known as Castaing's second approximation. It should be stressed that neither of these treatments takes account of absorption or fluorescence effects.

However, in spite of the fact that x-rays are generated within a few micrometres of the surface, x-ray absorption effects in the sample and standard can be severe. Furthermore, absorption in the sample will usually be much greater than that in the standard, the reason for this being that a pure element is comparatively transparent to its own radiation, whereas a multi-component alloy is not, a point which we shall return to later. Nevertheless in the early measurements on materials such as brass, in which x-ray absorption effects are small and similar in specimen and standard, the linear relationship between k_A and c_A given by equation (1.1) seemed to hold. It proved surprisingly difficult to test the relationship experimentally for other materials since it is also necessary to make allowance for characteristic x-rays generated in the specimen by other x-rays (fluorescence), either by characteristic x-rays from heavier elements present in the specimen or even from the high-energy components of the continuous background spectrum. In his thesis Castaing developed the theory for calculating the characteristic fluorescence correction arising from K x-rays exciting other K-radiations, an approach which has formed the basis for subsequent work in this area (see Chapter 9), and also carried out experiments to determine the magnitude of the continuum fluorescence correction.

In order to estimate absorption effects, Castaing decided to determine the way in which the intensity, $\phi(\rho z)$, of characteristic x-rays was distributed with mass depth, ρz, in the target. Although difficult and time consuming, Castaing and Descamps (1955) succeeded in experimentally determining $\phi(\rho z)$ distributions for copper, gold, and aluminium at an accelerating voltage of 29 kV (see section 2.3). From such curves (see, for example, Fig. 2.11) the absorption factor, $f(\chi)$, can be calculated as described in section 2.5. (A more detailed discussion of this approach is given in section 8.1.1) The resulting $f(\chi)$ curves were for many years the only readily available means of calculating absorption effects in microanalysers which therefore were constrained to operate at an accelerating voltage of 29 kV.

1.2.2 Philibert's calculation of the absorption factor

In view of the above difficulties Philibert (1963) attempted to deduce analytically

$\phi(\rho z)$ curves for a wide range of materials and accelerating voltage. He made use of Lenard's exponential law of attenuation of electron density in the target (Lenard and Becker, 1927) and Bothe's law of multiple electron scattering (Bothe, 1929) neither of which is strictly applicable. However, by an ingeniously chosen set of assumptions on the nature of electron backscattering and by a gross simplification of the ionisation cross-section for characteristic x-ray production, he was able to synthesise an analytical expression which gave a general shape resembling that of the measured $\phi(\rho z)$ curves. This expression had adjustable constants which allowed it to be matched to experimental $\phi(\rho z)$ curves. It was then a simple matter, after making some further approximations, to deduce an expression for $f(\chi)$. The Philibert treatment of absorption corrections is discussed in detail in section 8.2.

1.2.3 Discovery of the atomic number effect

Philibert's $\phi(\rho z)$ curves, although useful for deducing absorption effects, were completely inadequate for determining whether or not Castaing's first approximation was valid and an independent calculation was needed using a different approach. This was provided by Archard's electron diffusion and backscattering model (Archard, 1961). In this method a parallel beam of normally incident electrons was assumed to proceed in a straight line until it reached the mean depth of diffusion for the material in question. After that the electrons moved in straight lines with equal probability in every direction. This model enabled the backscattering coefficient to be calculated with reasonable accuracy. The range of each group of electrons was calculated by Bethe's law (Bethe, 1930), a more well-founded expression than that of Williams or of Webster. It was soon realised by Archard and Mulvey (1963) that this model could be used to investigate the 'atomic number' effect, that is the possible failure of Castaing's first approximation. Although the Archard model turned out to be too crude for accurate quantitative analysis, it revealed without any doubt that the 'atomic number' effect was a serious problem and that only in exceptional cases could it be neglected.

Experimental confirmation of this came at about the same time from microanalysis experiments on 'nuclear' materials by Scott and Ranzetta (1961–2), and later by Kirianenko *et al.* (1963) which showed that when samples contained elements widely different in atomic number, application of only absorption and fluorescence corrections was not enough. One of the first attempts to produce an 'atomic number' correction was by Poole and Thomas (1961–2), who showed it could be considered in two parts, an electron backscatter factor and an electron stopping power factor. These effects tend to compensate for one another, that is in a heavier element more electrons are lost from the sample by scattering but those that remain in the target are more effective at producing x-rays. This explains why the atomic number effect went unnoticed in earlier work on systems such as brass.

Further confirmation of the existence of the atomic number effect came from the Monte Carlo calculations of Green (1963a). In this method the solid specimen was effectively replaced by a succession of thin foils whose electron scattering properties had been determined experimentally. The electron in its passage through the system was arranged to strike each foil at a randomly chosen angle, and backscattering coefficients, $\phi(\rho z)$ curves and $f(\chi)$ curves were obtained with considerable precision.

A few years later, analytical expressions were introduced by Philibert and Tixier (1968a, 1968b) and by Duncumb and Reed (1968) which, as with the Poole and Thomas method, treated the atomic number correction in two parts. These corrections could also be applied separately from the absorption and fluorescence factors and so the matrix or 'ZAF' correction procedure (see section 2.7) became the method most commonly adopted for quantitative analysis.

1.2.4 Concluding remarks

ZAF corrections for quantitative analysis are now applied routinely and accuracies claimed are often better than 1%. There are, however, cases where large discrepancies arise especially when dealing with highly absorbing systems. As regards the atomic number correction either the method proposed by Philibert and Tixier (1968a, 1968b) or by Duncumb and Reed (1968) would appear to give acceptable results; the physical basis of these methods is discussed in Chapter 7. The characteristic fluorescence correction most frequently employed is that given by Reed (1965) which is based upon Castaing's approach and this would also appear to be satisfactory especially since the correction is generally the smallest of the three. Concerning the absorption correction the simpler Philibert equation, developed in 1963 and referred to in this book as the simplified Philibert model, is in most common use today. Since, however, the absorption correction is almost always the largest of the three corrections any failure of the model may produce significant discrepancies in the final corrected data and, consequently, most research on developing new corrections has concentrated on the absorption factor. A full description of a number of proposed absorption corrections is given in Chapter 8 and their performance is assessed in Chapter 10.

We have referred in the previous section to the use of Monte Carlo calculations for studying electron–solid interactions and for obtaining electron backscattering data and x-ray distributions, that is $\phi(\rho z)$ curves, in different target materials. A more detailed discussion of this subject is given in Chapter 12. Monte Carlo methods are, however, time consuming and so are not usually used directly in practical quantitative microanalysis. Nevertheless they have become a powerful means of providing essential data for calculating atomic number and absorption effects and for evaluating models used in practical correction procedures. Monte Carlo calculations have also proved useful in quantitative studies on thin films and small particles where only a small part of the incident electron energy may be given up to the specimen and, in view of the increasing interest in this field, methods for obtaining quantitative data on such specimens are included in this book (see Chapter 13).

2

The physical basis of quantitative analysis

V. D. SCOTT

As mentioned in the preceding chapter, quantitative microprobe analysis is most commonly carried out by comparing the intensity (that is the number of x-ray photons per second) of characteristic x-ray emission from element of interest (A) in the specimen with that from a standard of known composition, often for convenience the pure element A. Keeping the analysis conditions (probe voltage, current, x-ray detector efficiency) the same enables an x-ray intensity ratio (k_A) to be obtained which to a first approximation equals the weight (or mass) concentration (c_A) of A in the specimen. In only exceptional cases, however, does k_A give exactly c_A and generally corrections have to be applied which take into account the differences in electron scattering, x-ray generation and x-ray emission between specimen and standard.

The physical basis of quantitative analysis is the subject of this chapter. It outlines the origin of x-ray emission spectra, introduces the principles involved in the generation of x-rays in a solid by a beam of electrons, and describes the way in which x-ray production is affected by target composition and electron energy. The treatment shows that the correction factor required for quantitative microanalysis can be considered in three parts – an atomic number, an absorption and a fluorescence component – and that they may be applied separately to obtain the final result. This approach is adopted by most workers in the field and is sometimes referred to as the 'matrix' or 'ZAF' correction method. Later in the book (Chapters 7, 8 and 9) detailed descriptions are given of the various models which have been proposed to take the three effects into account.

2.1 ELECTRON INTERACTIONS IN SOLIDS

The interactions which occur between incident electrons and target atoms may be divided into elastic and inelastic processes.

Elastic events arise from electron encounters with the relatively massive atomic nuclei where, because the nuclear energy levels are widely spaced (~MeV) compared with electron energies used in microprobe analysis (tens of keV), there is little possibility of energy interchange between electron and nucleus. However, strong Coulombic attraction deflects the much lighter electron, the magnitude of the deflection (β) being determined by the distance of closest approach (p), electron energy (E) and atomic number (Z) according to

$$\cot \beta/2 = \text{const.} \frac{E}{Z} p \; .$$

The reader may recognise that this equation is based, essentially, upon the Rutherford model for scattering of alpha particles. Typically the electron scattering angle is $\sim 5^\circ$ but it may range from 0° to 180°. Hence one or more of these elastic scattering events may result in an incident electron being back-scattered from the surface of the target.

Inelastic scattering results from interactions between the primary electron and orbital electrons of the atom. Large scattering angles are here unlikely owing to the diffuseness of the electron 'cloud' and the relative mass parity but, since the differences between energy levels of the atom are less than the energy of the primary electron, energy transfer readily occurs. Hence inelastic scattering causes the primary electron to slow down as it progresses through the target until, eventually, its energy reaches the Fermi level of the target material and the electron flows from specimen to earth. The electron loses energy in a discontinuous fashion and the magnitude of each loss may be equated with the energy required to excite an electron in the target. If the energy of the electron is greater than that of the K electron level of the atom then ionisation of any of the shells (K, L, M etc.) is involved in the energy-loss process. However, as the electron energy decreases, fewer and fewer shells can contribute to its slowing down. Whilst in reality inelastic scattering is a discontinuous process, it may be treated as continuous if the energy lost at each interaction is small compared with the electron energy. Using a continuous slowing-down model Bethe and Ashkin (1953) were able to deduce that

$$\frac{-\mathrm{d}E}{\mathrm{d}\rho s} = 78\,500 \frac{Z}{AE} \ln(1.166\, E/J)$$

where $-\mathrm{d}E/\mathrm{d}\rho s$ is the stopping power, that is the rate of loss of energy per unit mass path length transversed. J is the mean ionisation energy of the target atoms and represents the average energy lost at each inelastic collision in the target; its magnitude is given approximately by $J = 13.5Z$ (in eV). Bethe's energy-loss expression is the one most commonly used in electron-probe microanalysis (section 7.1).

The region over which incident electrons diffuse within the solid and deposit their energy is termed the interaction volume, and with knowledge of the spatial distribution of electrons as a function of energy, it becomes possible to establish the corresponding distribution of generated x-rays. As we shall see this information is needed, not only for developing corrections in quantitative analysis, but also for understanding factors which affect the spatial resolution of the technique (section 2.8). First, however, the production of x-ray spectra needs to be considered.

2.2 X-RAY EMISSION SPECTRA

The x-ray emission spectrum generated by a beam of energetic electrons (~ one to tens of keV typically) when it impinges on a target consists of a background of x-rays (continuum x-rays) which extends up to a limiting energy corresponding to the incident electron energy, together with superimposed peaks at discrete energies (characteristic x-rays) which are a function of the target composition. The relationship between the energy of characteristic x-ray emission lines and atomic number of the target was first demonstrated by Moseley (1913, 1914) who found that x-ray frequency, ν, and atomic number, Z, could be related by the equation

$$\nu = 2.48 \times 10^{15} (Z-1)^2 \quad .$$

This equation refers to $K\alpha$ series wavelengths, although similar relationships are found to occur for L and M spectra when these are generated. Fig. 2.1 shows, schematically, x-ray spectra produced from a selection of elements spanning most

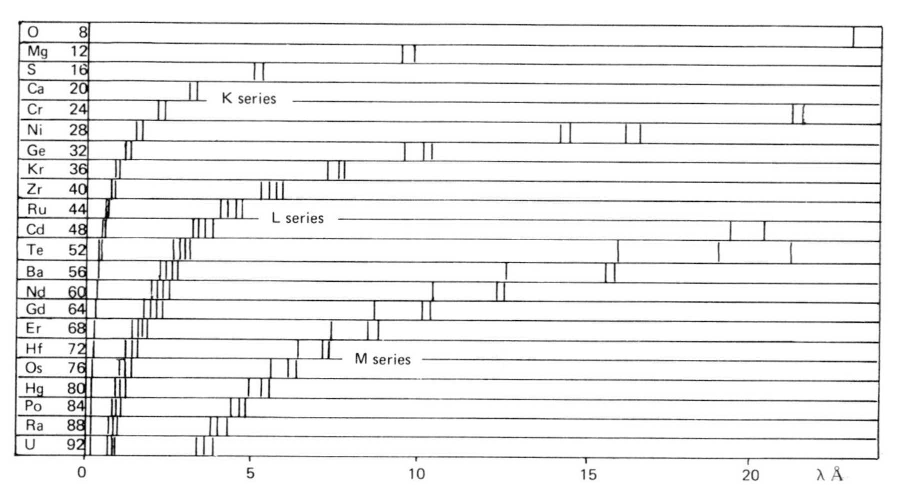

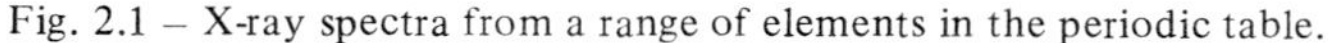

Fig. 2.1 – X-ray spectra from a range of elements in the periodic table.

of the periodic table. Moseley's results provided striking confirmation of the quantum theory and indicated, from the high energies involved, that x-ray generation must be associated with events occurring deep within the atom. A detailed description of the principles governing the production of x-ray spectra is to be found in most undergraduate textbooks of physics and we shall, therefore, confine ourselves to a simplified account of those aspects which are fundamental to the present treatment of quantitative electron-probe microanalysis.

2.2.1 Characteristic x-ray spectra

The systematic manner in which the energies of the characteristic x-ray lines from different elements vary with atomic number is readily explained by considering the energy states in which a particular atom can exist.

With the hydrogen atom, its single electron has a binding energy of 13.6 eV. As the atomic number, Z, increases so the added electrons build up the number of orbitals surrounding the nucleus, starting with electrons closest to the nucleus. The first orbital in an atom to be filled (K shell) is the most tightly bound and has the lowest (largest negative) potential energy. This can accept two electrons, the third electron (when $Z \geqslant 3$) starting the L orbital. Eight electrons can be accommodated in the L energy level but the Pauli exclusion principle dictates that they cannot all occupy the same orbital and so sub-orbitals occur with small energy differences between them. With $Z > 10$ (neon), the electrons start to fill M orbitals and the eighteen electrons which comprise a completed M energy level must be distributed among five sub-orbitals. By the time we reach lutetium ($Z = 71$) in the periodic table, seven N sub-orbitals have been filled. The O and P energy levels are never fully occupied, however, the Q levels starting to accept electrons at $Z = 87$. The binding energy of the K electron progressively increases (approximately with Z^2) and for uranium ($Z = 92$) is 116 keV. The binding energy for L electrons increases from 3 eV for lithium to ~20 keV for uranium while, for M electrons, it increases from a few eV for sodium up to ~4 keV for uranium. The customary way of showing these energy level relationships is by means of a diagram such as that illustrated in Fig. 2.2. This represents an atom of atomic number >29 with filled orbitals up to and including the M level and with electrons in the valence band.

When a K electron is ejected from the atom, for example by an incident electron, the atom becomes ionised and goes to a higher energy state, the K state. For the atom to regain equilibrium, the vacancy in the K orbital must be reoccupied by an electron from one of the outer levels. Suppose this electron is obtained from the L_{III} sub-orbital. Then according to energy conservation laws a quantum of radiation will be emitted with a discrete energy corresponding to the difference in energy between the levels (or states) involved, that is $h\nu = E_K - E_{L_{III}}$ (see Fig. 2.3). For elements of higher atomic number than lithium, the emitted photon has an energy in the x-ray region and is denoted a $K\alpha_1$ x-ray. The atom is still in the ionised state, just as if it were an L_{III} electron which had

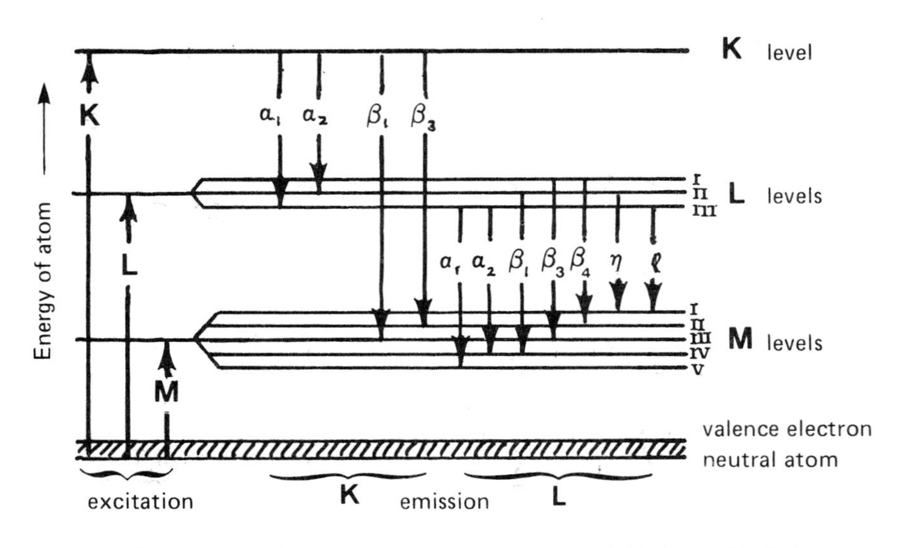

Fig. 2.2 – Schematic diagram showing common x-ray emission lines with their designation for an element with atomic number Z, where $29 < Z < 37$.

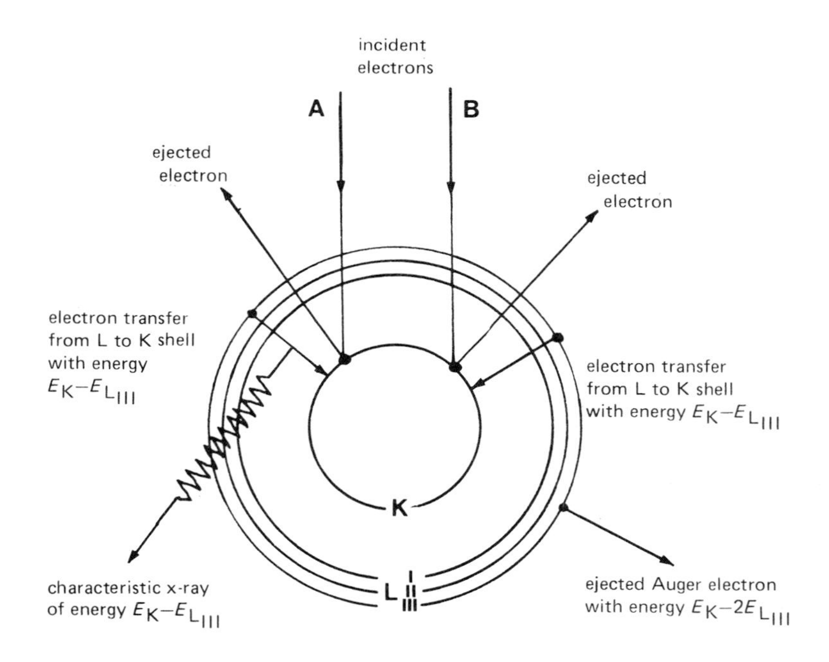

Fig. 2.3 – Schematic diagram showing emission of characteristic x-ray by electron A and emission of Auger electron by electron B.

been initially ejected by the incident electron and, consequently, the L_{III} sub-orbital could be refilled with an M electron. The radiation emitted is then designated an L radiation, the convention adopted for denoting x-rays being that the first letter identifies the initial excited state. If the M_V sub-orbital supplies the electron, the radiation is denoted $L\alpha_1$, if M_{IV} then the radiation is labelled $L\alpha_2$. Alternatively, the original vacancy in the K orbital may be filled directly by an M electron, giving a $K\beta$ x-ray. This will, of course, have a higher energy than $K\alpha$ x-rays since $E_M < E_L$. Not all transitions between available energy levels are permitted and those which are allowed are governed by selection rules provided by quantum theory principles. Some of the more common emission lines together with their designation are given in Fig. 2.2.

Only some inner shell ionisations result in the emission of an x-ray and the energy available when an electron drops into the initial vacancy may be used to eject another electron from an inner energy level, for example a K vacancy can be filled by an L_{III} electron and a second L_{III} electron emitted (see Fig. 2.3). This is known as an Auger process and the ejected Auger electron has a kinetic energy $E_K - 2E_{L_{III}}$. The atom is left in a doubly ionised state. Since the energy of Auger electrons is related to the electron energy levels of an atom, they too may be used for spectrochemical analysis, although Auger electron analysis will not be discussed further in this volume.

The probability of an x-ray rather than an Auger electron being produced is the fluorescence yield. With heavier elements, x-ray production is more likely but the probability decreases with decrease in atomic number, see Fig. 2.4. For example, the fluorescence yield of the K shell (ω_K) decreases from ~0.8 for elements above $Z = 40$, to ~0.1 for $Z = 16$ (see Fink *et al.*, 1966); in the light

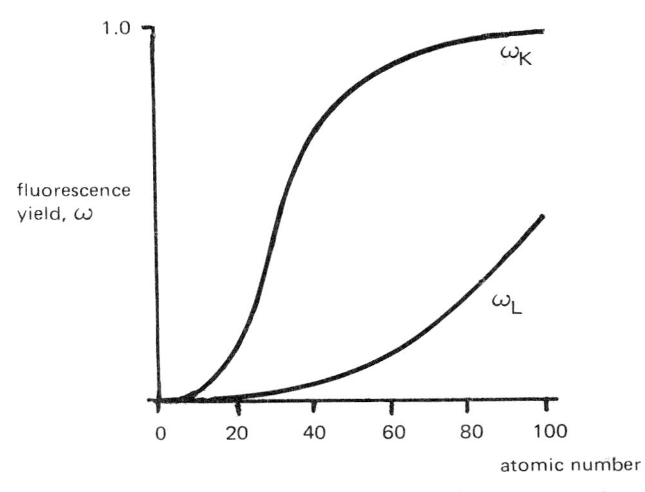

Fig. 2.4 – Variation of fluorescence yield with atomic number.

element region fluorescence yields are very low, ~0.001 for carbon x-rays. A number of relationships between ω_K and atomic number, Z, have been proposed such as $\omega_K = Z^4/(a + Z^4)$ (Burhop, 1955) where a is a constant. Less extensive data are available on fluorescence yields for L and M x-radiations.

The relative intensities of generated characteristic x-ray lines may be deduced, in principle, from the population of the respective sub-orbitals (see Bambynek *et al.*, 1972); for example, a ratio of 2:1 in the population of the L_{III} and L_{II} sub-orbitals results in the $K\alpha_1$ line being twice as intense as the $K\alpha_2$. The relative intensities of the lines change with atomic number as the probability of particular transitions occurring changes. Hence the ratio of $K\alpha_1$ and $K\beta_1$ lines is ~10 for aluminium (Z = 13) but drops to ~5 for copper (Z = 29). Data are available (White and Johnson, 1970) on line intensities, commonly expressed by taking the intensity of the strongest line in a series as unity, but may be somewhat unreliable. Principal lines in selected series from a few elements are shown schematically in Fig. 2.5.

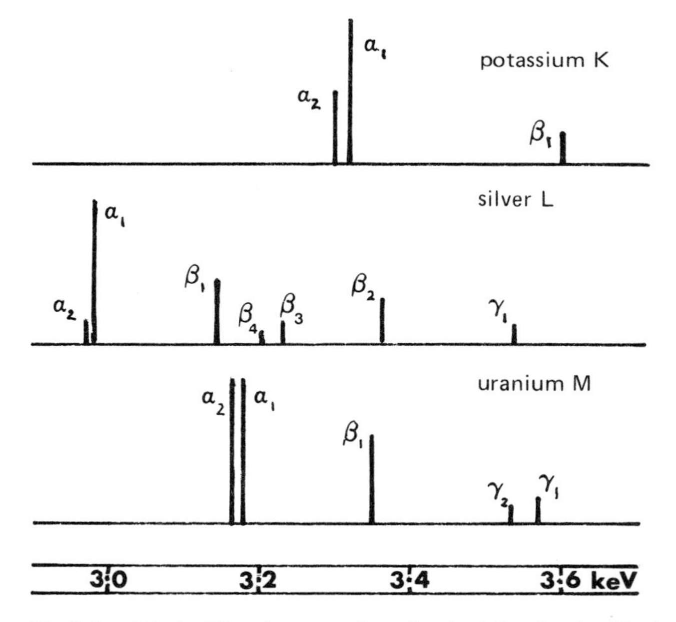

Fig. 2.5 – Principal lines in some selected series (after Reed, 1975b).

It is important to note that the energy required to produce an ionised state will always be greater than the energy of the associated emitted radiations, that is $E_K > E_{K\alpha}$. This is because the initial vacancy is created by ejecting an electron completely from the atom, a process which requires an energy (the critical excitation energy) greater than that involved in any subsequent electron transition. Also since the excitation energy relates to the shell being initially ionised, it is the same for both Kα and Kβ emissions despite their difference in energy. This means that for atoms containing M electrons Kα x-rays cannot be generated without producing Kβ x-rays. Nevertheless, reducing the energy of incident (exciting) radiation to prevent K emissions may still allow L x-ray generation, etc.

2.2.2 Satellite lines

Although most of the characteristic x-ray emission can be explained on the basis of transitions allowed by the selection rules as described above, weak lines may appear which do not fit into this pattern. They occur as satellites close to one of the principal lines and can be separately resolved only with a high resolution x-ray spectrometer. Their production has been explained by assuming that an atom may be doubly ionised by an incident radiation, for example vacancies being created in both K and L orbitals. (The two ionisations have to occur virtually simultaneously since the lifetime of an excited state is very short, $\sim 10^{-14}$ s.) If now the KL state is changed to an LL state by an electron transition, the energy released results in the emission of a single photon. Its energy will be slightly higher than that associated with an ordinary L–K transition owing to the fact that the original extra vacancy would have reduced the degree of screening of the nucleus by the electrons and thus increased their binding energy. Hence on high energy side of the K$\alpha_1\alpha_2$ peak, satellites appear at energies associated with the L sub-orbitals involved in the double ionisations (Fig. 2.6). The

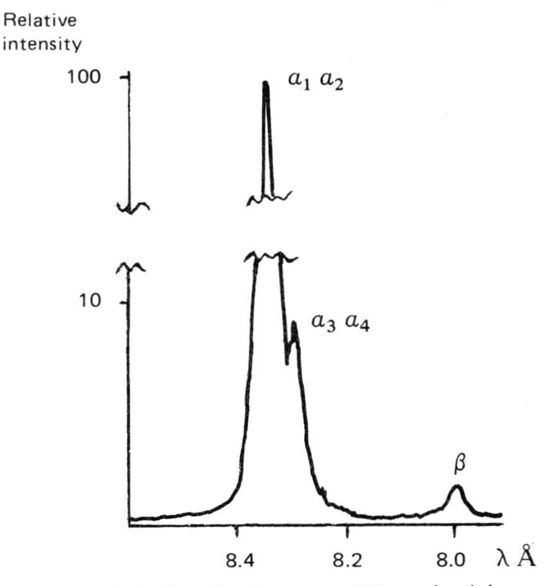

Fig. 2.6 – Satellite lines (α_3, α_4) from aluminium.

satellites are relatively more intense for lighter elements because the lifetime of an excited state is longer and the probability of double ionisation is higher. In the case of aluminium Kα emissions, for example, approximately 10% of the total intensity is contained in the satellite lines (White and Johnson, 1970), all of which lie within ~20 eV of the Al Kα_1 line.

2.2.3 Soft x-ray spectra

In the foregoing we have considered x-rays generated as a result of electron transitions involving inner orbitals of the atom, x-rays which have discrete energies and give sharp emission lines. However, x-rays may be produced as a result of transitions associated with the outer orbitals containing valence electrons. These electrons are not unique to a particular atom due to bond formation and, consequently, some may leave the vicinity of one atom in favour of its neighbours. This then alters the overlap of the valence energy states and leads to a decrease in x-ray intensity from those atoms which lose electrons (Nagel, 1970). Moreover, since the inner levels are relatively discrete compared with the valence band, the transitions that occur between them will sensitively reflect the energy states of the valence electrons. Such changes in the soft x-ray emission spectra that result from chemical bonding effects are well documented, see for example the reviews by Baun (1969) and Fabian *et al.* (1971). Generally, the differences between the pure element and its compounds are more pronounced in insulators than in conducting materials because the valence electron binding energies increase in progressing from metallic, through covalent, to ionic bonding and the energy of the soft x-ray emission line decreases correspondingly. As well as changes in the Kα spectra from the light elements ($Z < 10$), where the L shell involved in K–L transitions is incomplete, chemical bonding effects may be observed in other series such as the L spectra from transition-metal elements ($21 \leqslant Z \leqslant 28$) where the M orbitals contain the valence electrons. In Fig. 2.7(a) are shown the oxygen K emission bands from different oxides (Love *et al.*, 1974a), while Fig. 2.7(b) illustrates changes in the Ti L x-ray emission from titanium oxide as a function of chemical composition (Fischer, 1970). Undoubtedly, analysis of soft x-ray emission spectra can give important additional structural information but it appears not to be fully exploited in electron-probe micro-analysis. Perhaps this is partly due to the increasing use of energy-dispersive spectrometers for x-ray analysis and this system (Chapter 4) has not sufficient energy resolution to detect such changes. On the other hand where the x-ray emissions are analysed using the wavelength-dispersive spectrometer fitted with a high resolution crystal or diffraction grating (section 3.2) quantitative analysis may then be complicated by these chemical bonding effects (section 6.2).

2.2.4 The continuous x-ray spectrum

As stated earlier, the continuous x-ray spectrum consists of a background which extends up to an energy corresponding to the energy of the incident electron. The

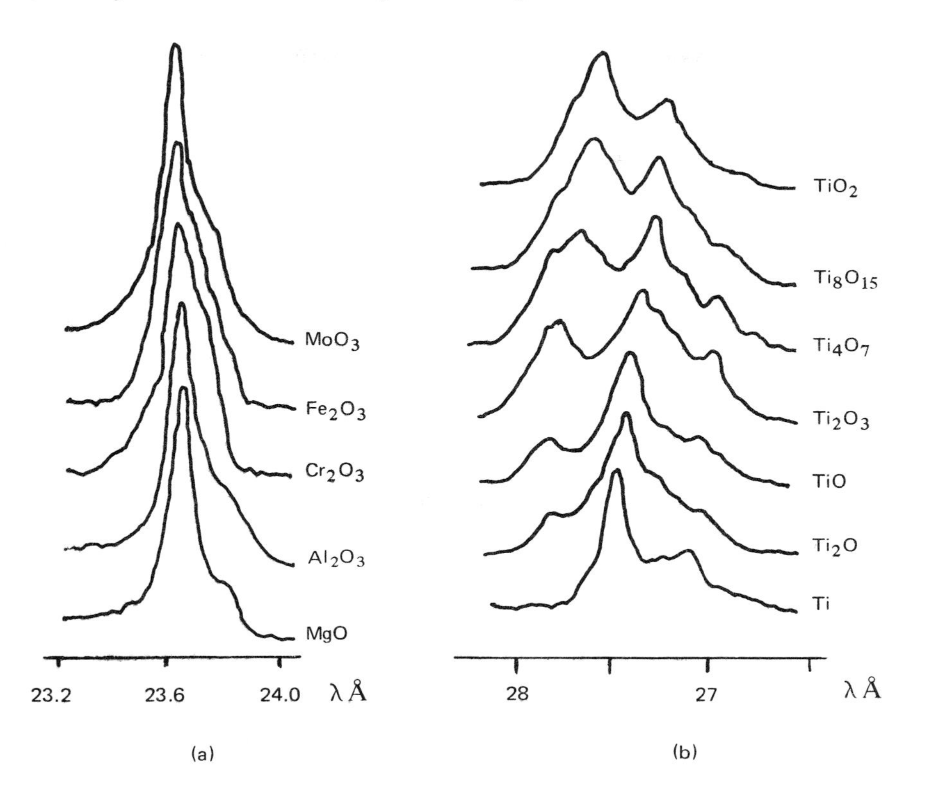

Fig. 2.7 – (a) Oxygen emission from a range of oxides (after Love *et al.*, 1974a). (b) Titanium L emission from titanium and its oxides (after Fischer, 1970).

classical view of continuum x-ray production is of a process involving the rapid slowing down of electrons in the vicinity of the nuclear field; the x-ray of highest energy (at the Duane–Hunt limit) is produced when the entire energy of an incident electron is converted into a radiation in one single collision. Hence measurement of this cut-off value of the x-ray spectrum is a simple means of establishing the true electron accelerating voltage in electron-probe micro-analysis. The probability of such a single interaction occurring is, however, very low and the likelihood is that only part of the electron energy is given up in any deceleration event. Furthermore, since the electrons rapidly lose energy as they enter the target, x-rays of maximum energy will be emitted only from the surface. The energy distribution of the x-ray continuum from a thin surface layer has the form shown in Fig. 2.8(a) (Nicholas, 1929), where the distribution is plotted in arbitrary units on a frequency scale and ν_0 corresponds to the Duane–Hunt limit. This relationship may be expressed by $I_\nu \, d\nu$ = constant for

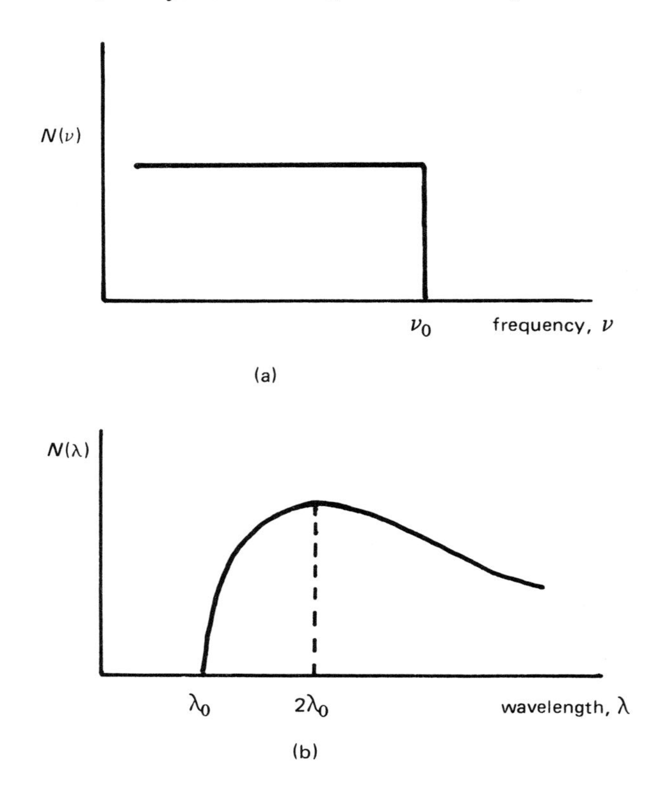

Fig. 2.8 – (a) Energy distribution of x-ray continuum from a thin layer.
(b) Energy distribution of continuum radiation from a solid target.

values of ν less than the incident electron energy ν_0, where I_ν is the energy density of radiation generated in the energy interval ν and $(\nu + d\nu)$. For a thick target Kramers (1923) applied the Thomson–Whiddington law (Whiddington, 1912) and showed that $I_\nu\, d\nu = a\,Z(\nu_0 - \nu)\,d\nu$, where a is a constant. Since $E = h\nu$, the equation may be rewritten as $I_E\, dE = bZ(E_0 - E)dE$. Now the number of photons $N(E)dE$ within the energy interval E and $(E + dE)$ is given by $I_E\, dE/E$ and hence

$$N(E)dE = bZ\frac{E_0 - E}{E}\,dE\ .$$

This is the form of Kramers' equation customarily used in electron-probe micro-analysis (see, for example, section 5.3.1); b is known as Kramers's constant and has a value close to 2×10^{-9} photons $s^{-1} eV^{-1}$ electron^{-1}, although Green and Cosslett (1968) found some variation with atomic number.

The equation may be used to derive the generated photon distribution with wavelength (λ) as follows. Since $E = hc/\lambda$,

$$\mathrm{d}E = -hc\,d\lambda/\lambda^2 \ ,$$

and

$$N(\lambda)\mathrm{d}\lambda = bZ\left(\frac{1}{\lambda_0} - \frac{1}{\lambda}\right)\lambda\left(-hc\,\frac{\mathrm{d}\lambda}{\lambda^2}\right)$$

$$= hcbZ\left(\frac{1}{\lambda^2} - \frac{1}{\lambda\lambda_0}\right)\mathrm{d}\lambda \ .$$

A plot of $N(\lambda)$ is given in Fig. 2.8(b) and it may be seen that the number of photons generated rises rapidly from zero at λ_0 as the wavelength increases to reach a maximum value. The position of the maximum, found by putting $[\mathrm{d}N(\lambda)]/\mathrm{d}\lambda = 0$, is at $\lambda = 2\lambda_0$. The intensity of the continuum actually emitted from the target will be reduced, owing to absorption. The longer the wavelength of the photon the more likely it will be absorbed and, consequently, there will be a shift in the maximum towards shorter wavelengths, very approximately to where $\lambda = 1.5\,\lambda_0$. A typical distribution of emitted continuum x-rays is illustrated in Fig. 2.9(a) for a molybdenum target; the characteristic lines have been omitted from the drawing. It follows that the relative intensity of the continuous x-ray spectrum and the position of the intensity maximum, as well as the short wavelength limit, are affected by the energy of the incident electrons, although the general shape of the curve is little changed. The continuous spectrum is affected too by the atomic number of the target, although now the change is restricted simply to an overall increase in intensity with increase in atomic number (Fig. 2.9(b)).

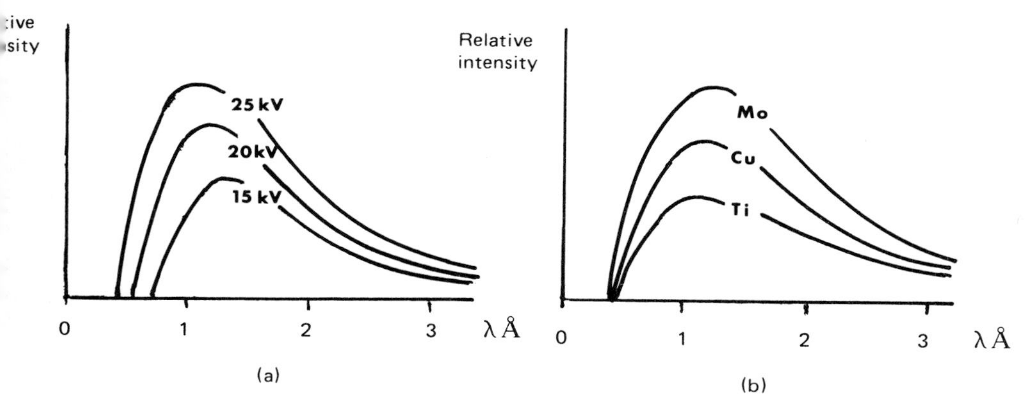

Fig. 2.9 – (a) Energy distribution of continuum radiation as a function of electron accelerating voltage, molybdenum target. (b) Energy distribution of continuum radiation for different target materials, 25 keV electrons.

The continuum is generally a nuisance as far as electron-probe microanalysis is concerned, since the lower the ratio of characteristic line intensity to background upon which it lies, the lower the precision of measurement. The problems of background subtraction become greater when using energy-dispersive (ED) spectrometers because the poorer resolution of the system (see Chapter 5) results in statistical broadening of the peak with a consequent decrease in recorded peak-to-background ratios. One method of carrying out background subtractions in ED spectra is to calculate the intensity of the continuum using a Kramers-type equation with the appropriate factors introduced to allow for absorption in the specimen and the detector. The same equation is used for calculating fluorescence enhancement effects caused by the continuous spectrum (section 9.6).

2.3 DISTRIBUTION WITH DEPTH OF GENERATED X-RAYS

Thus far we have described the nature and origin of x-ray emission spectra which are generated by the electrons but rather more information than this is required since, before the x-rays can be measured in the spectrometer, they must find their way out of the sample. The chance of x-rays escaping will depend upon the degree of x-ray absorption they may suffer, and this is a function of the absorption properties of the material and the amount of material through which they must pass. While information on the former may be fairly readily acquired, the second may be more difficult to estimate since it will depend upon the depth below the sample surface that a particular x-ray is produced. In fact x-rays will be generated over a range of depths and we need to know about this distribution before any satisfactory adjustment for absorption losses in the material can be made.

One way of establishing the distribution of x-rays with depth in a target is the tracer method described by Castaing and Descamps (1955). It involves first depositing a thin film of element B onto a solid substrate of element A and then measuring the x-ray yield from the film under electron bombardment. The thin film, or tracer, is then covered by successively thicker layers of substrate material (Fig. 2.10) and the x-ray intensity from the film measured as a function of its depth below the surface. Element B is chosen such that it has similar electron scattering properties to element A, the substrate material, otherwise the behaviour of the electrons in the sample, and the subsequent generation of x-rays, will not be properly representative of A. For example, when measuring the x-ray distribution in copper, zinc is selected as the tracer element. The thickness of the tracer is sufficient to give a detectable x-ray signal but not so thick that it significantly alters scattering within the target. The measurements are then compared with the x-ray intensity obtained under the same analysis conditions from an isolated layer of element B of identical thickness, for convenience supported on a film of say collodion. The reference layer enables data to be

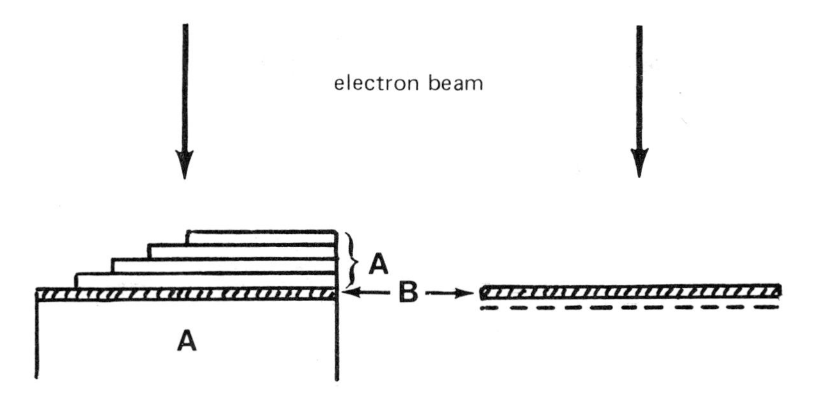

Fig. 2.10 – Specimens used in tracer method for determining depth distribution of generated x-rays.

obtained which are independent of x-ray spectrometer, etc. The x-ray generation profile is then derived from the x-ray emission measurements by making adjustments for x-ray absorption losses, a relatively straightforward calculation since the depth of the tracer below the surface for each measurement is known precisely from accurate determinations of coating thickness. A typical distribution of generated x-rays, plotted as a function of mass depth ρz, is shown in Fig. 2.11; the units on the vertical axis are based upon taking the x-ray emission from the isolated film as unity. Such x-ray depth distributions are termed $\phi(\rho z)$ curves

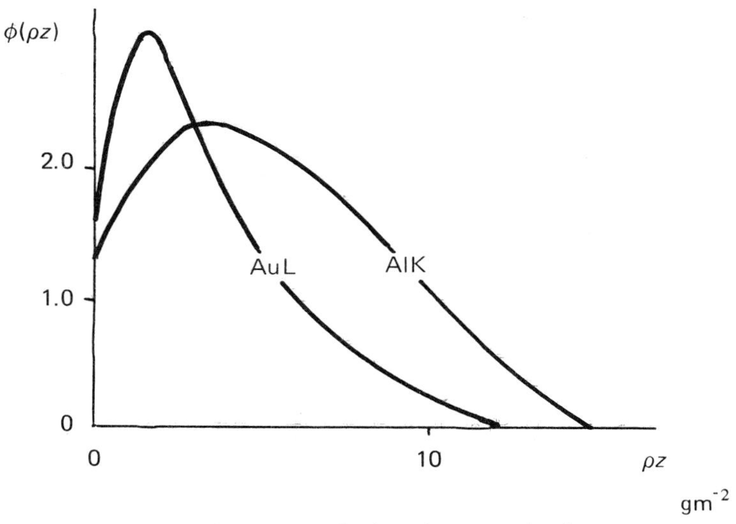

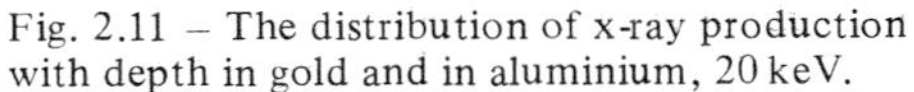
Fig. 2.11 – The distribution of x-ray production with depth in gold and in aluminium, 20 keV.

and the shape depends upon the energy of the incident electron beam and the atomic number of the target; included in Fig. 2.11 are two $\phi(\rho z)$ curves which illustrate the point. From these data it may be seen that:

(i) x-ray production in the immediate surface regions of the solid sample is greater than that from the isolated thin film, that is $\phi(0) > 1$. This is because some electrons are backscattered from underlying regions of the solid specimen, giving them a second opportunity to generate x-rays;

(ii) there is an initial rise in the curves. This is due to a progressive increase in electron scattering as the electrons penetrate the target which, in turn, increases their path length through each elemental layer, $\Delta\rho z$, and increases the probability of x-ray production. An additional factor contributing to the rise in the curve is the increase in efficiency of ionisation as the energy (E) of the electron in the target drops to a value approximately twice the critical ionisation energy (E_c). This change in ionisation efficiency is illustrated in Fig. 2.12, which shows the ionisation cross-section per atom (Q) plotted against overvoltage ratio, U, where $U = E/E_c$;

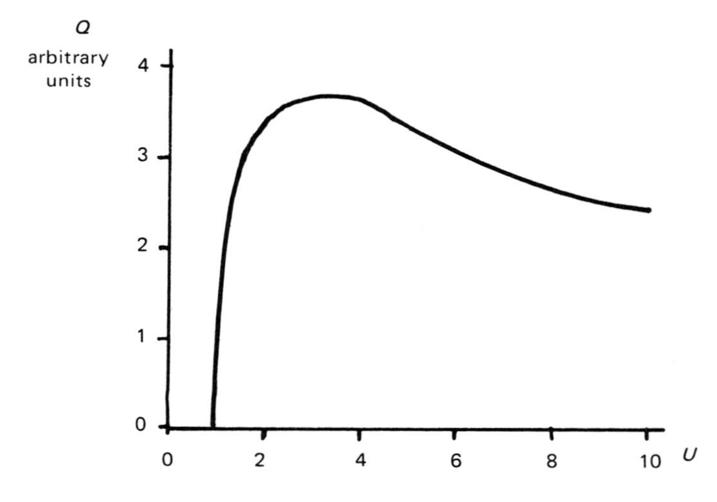

Fig. 2.12 – K shell ionisation cross-section (Q) as a function of overvoltage ratio (U).

(iii) at greater depths in the target the x-ray production begins to decrease, firstly because there will be fewer electrons reaching these levels and, secondly, because many of those which do arrive will have reduced energy ($<2E_c$) and be less effective or incapable of generating x-rays.

Comparison of the $\phi(\rho z)$ data from a gold target with that obtained from aluminium indicates that the increased electron scattering from a sample containing heavier elements results in a higher $\phi(0)$ value and a smaller mass depth at the x-ray intensity maximum. Changing the incident electron energy would also alter the $\phi(\rho z)$ curve; the higher the electron energy the more deeply are x-rays generated in the target.

The use of the tracer method in deriving x-ray depth distributions is, because of practical difficulties, limited to simple pure element targets whereas in electron-probe microanalysis information is needed on specimens which contain a mixture of different elements. Furthermore, tracer studies are difficult or impossible to carry out on highly absorbing systems, at low energies, and where the energy is close to the critical excitation potential. An alternative approach is to use computer simulations to establish the position and energy of many individual electrons as they travel through the target. This is determined by the probable occurrence of both elastic and inelastic interactions with constituent atoms which is, in turn, affected by the energy of the electron and the composition of the target. Such calculations have been carried out using Monte Carlo methods (Chapter 12) which incorporate appropriate expressions for electron scattering. Fig. 2.13(a) shows data obtained with a simplified Monte Carlo approach (Curgenven and Duncumb, 1971) for 20 keV electrons incident on a gold target. It may be seen that the overall effect of a succession of scattering events is to produce generally random movement of electrons as they reach the end of their path. It should also be noted that many of the incident electrons are back-scattered out of the target and carry away energy which otherwise would have generated x-rays. To obtain the x-ray distribution, the probability of inner shell ionisation is calculated using an equation of the type Q = const. $(1/U) \ln U$ (Bethe, 1930; Green and Cosslett, 1961). Finally, the probability of the ionisation resulting in emission of an x-ray may be deduced by incorporating the fluorescence yield term, ω.

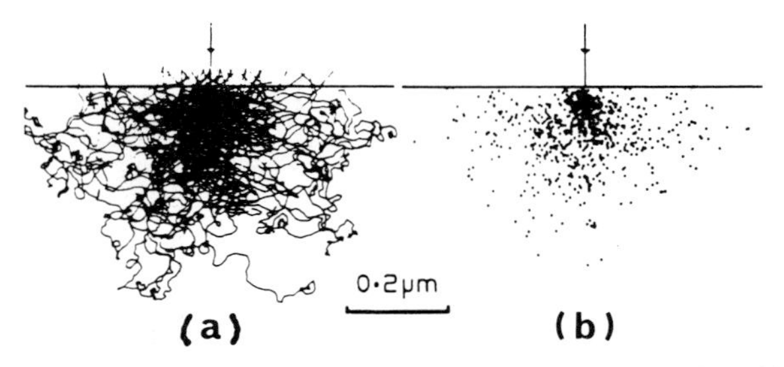

Fig. 2.13 – (a) Electron trajectories and (b) x-ray distribution in gold at 20 keV computed by Monte Carlo method (Curgenven and Duncumb, 1971; courtesy Tube Investments Res. Ltd.).

Fig. 2.13(b) illustrates the x-ray distribution in the gold target produced by the electrons depicted in Fig. 2.13(a). A similar pair of diagrams for an aluminium target, Figs. 2.14(a) and 2.14(b), clearly show the difference in the distribution of electrons and of generated x-rays in the two materials.

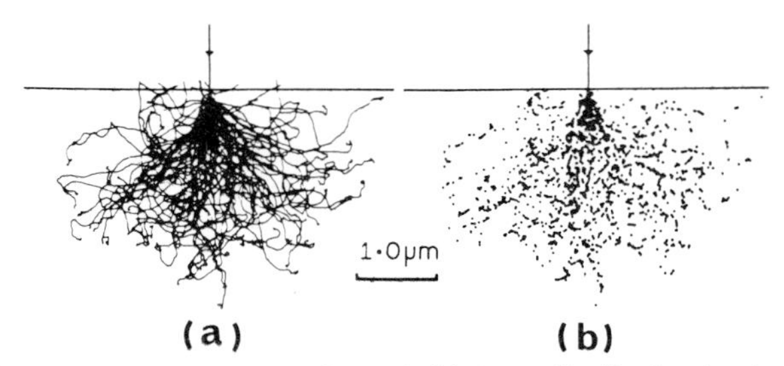

Fig. 2.14 – (a) Electron trajectories and (b) x-ray distribution in aluminium at 20 keV computed by Monte Carlo method (Curgenven and Duncumb, 1971; courtesy Tube Investments Res. Ltd.).

Once the validity of such a theoretical approach to obtaining $\phi(\rho z)$ curves has been established, by for example checking calculations against available tracer measurements on pure elements and/or electron backscattering data, the Monte Carlo model may be used to generate new information covering a wide range of experimental conditions and then to derive analytical expressions for $\phi(\rho z)$.

However, in quantitative electron-probe microanalysis, measurements are made of *emitted* x-ray intensities not generated x-ray intensities and these will always be lower owing to the absorption which occurs as the x-rays travel to the surface. Before we can proceed further with quantitative electron-probe microanalysis some discussion of x-ray absorption is therefore required.

2.4 X-RAY ABSORPTION

X-rays may suffer absorption owing to a number of processes:

(i) Compton scattering, which is an inelastic collision with an atom causing the x-ray energy to be reduced;
(ii) Rayleigh scattering, where there is no exchange of energy with the atom (diffraction effects occur by this process);
(iii) the photoelectric effect, where the x-ray photon is completely absorbed by the atom and an electron is ejected.

The most important of these as far as microanalysis is concerned is the photoelectric effect; Rayleigh scattering is important only when considering absorption of collimated beams while Compton scattering is significant only for energies above those of interest here.

The probability of an x-ray photon being absorbed to produce a photoelectron is a function of the photon energy. If this is below the energy of the electron in its orbital, it will not be absorbed. If the photon energy equals the electron binding energy the probability of capture will be highest. Hence it follows that the absorption properties of a given material when plotted as a function of energy of absorbed radiation will give a curve containing sharp discontinuities (Fig. 2.15).

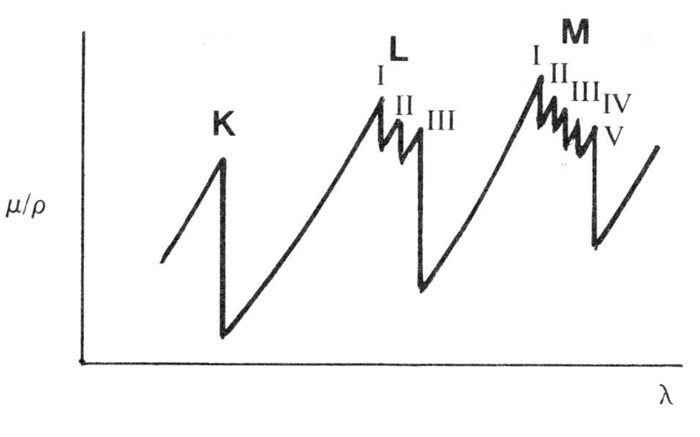

Fig. 2.15 – Mass absorption coefficient of a given material as a function of wavelength of absorbed radiation.

Each step in the curve, termed an absorption edge, will correspond to a particular energy level, or excited state of the atom, and may be designated K, L_I, L_{II}, etc. as defined in section 2.2. Hence the measured position of absorption edges may give information on energy levels for the atom, in the same way that characteristic x-ray emission spectra do. It also follows that if we were to measure the energies of the emitted photoelectrons from a given material, information concerning its chemical composition could be obtained, an approach which is exploited in the technique of x-ray photoelectron spectroscopy. In electron-probe microanalysis, absorption edge effects can be seen as differences in the continuum background level either side of a characteristic peak, a feature which has to be taken into account (section 5.2) when estimating peak intensities.

The energy of the atom after emission of a photoelectron will, as in the case of excitation by an incident electron, be reduced by an electron from an outer level dropping into the vacancy. Hence emission of a characteristic x-ray may follow. This process, whereby x-rays produce characteristic spectra, is termed x-ray fluorescence, a method of chemical analysis much older than electron-probe microanalysis. However, x-ray fluorescence effects are relevant also to the latter (section 2.6 and Chapter 9).

The amount of absorption, dI, experienced by a monochromatic beam of x-rays of intensity, I, when passing through a material of thickness dx is given by $dI/I = -\mu dx$, where μ is the linear coefficient of absorption. Integrating the expression leads to $I = I_0 \exp(-\mu x)$, where I_0 is the incident x-ray intensity. However, it is more useful to use mass absorption coefficients since these are essentially independent of the physical and chemical state of the atom, that is $I = I_0 \exp(-(\mu/\rho)\rho x)$, where ρ is density and ρx refers to mass thickness. Mass absorption coefficients for most elements have been tabulated for a range of x-ray energies, customarily in units of $cm^2 g^{-1}$ (Heinrich, 1966a) or barns $atom^{-1}$ (Bracewell and Viegele, 1971).

Mass absorption coefficients for compound specimens are calculated by averaging the values for the constituent elements according to their respective mass concentrations (c). We may justify the procedure as follows. Consider an x-ray beam of intensity I_0 (Fig. 2.16) passing first through a slab of material consisting of element A and of mass thickness $(\rho x)^A$. The transmitted intensity (I^A) will be $I^A = I_0 \exp\{-(\mu/\rho)^A(\rho x)^A\}$.

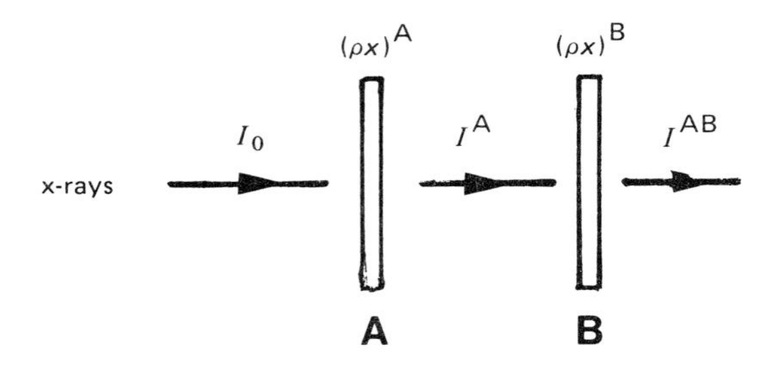

Fig. 2.16 – Mass absorption coefficients for compound specimens are calculated by mass averaging μ/ρ for the constituent elements.

After passing through a second slab of material, this time of element B and of mass thickness $(\rho x)^B$, the x-ray intensity I^{AB} will be given by

$$\begin{aligned} I^{AB} &= I^A \exp\{-(\mu/\rho)^B(\rho x)^B\} \\ &= I_0 \exp\{-(\mu/\rho)^A(\rho x)^A - (\mu/\rho)^B(\rho x)^B\}. \end{aligned}$$

Now $(\rho x)^A = c_A . (\rho x)^{\text{total}}$, etc., and thus

$$\begin{aligned} I^{AB} &= I_0 \exp\{-(\mu/\rho)^A\, c_A(\rho x)^{\text{total}} - (\mu/\rho)^B\, c_B(\rho x)^{\text{total}}\} \\ &= I_0 \exp\{-(\rho x)^{\text{total}}\, [c_A(\mu/\rho)^A + c_A(\mu/\rho)^B]\} \end{aligned}$$

or, more generally,

$$I = I_0 \exp\{-(\rho x)^{\text{total}} \sum c_i(\mu/\rho)_i\}\ .$$

Thus μ/ρ for compound specimens is given by

$$\sum c_i(\mu/\rho)_i.$$

For example, consider a specimen of $CuAl_2$ containing weight fractions (c_i) of 0.46 copper and 0.54 aluminium.

Now

$$(\mu/\rho)^{Cu}_{CuK} = 54 \quad \text{and} \quad (\mu/\rho)^{Al}_{CuK} = 50$$

and hence

$$(\mu/\rho)^{\text{spec}}_{CuK} = 0.46 \times 54 + 0.54 \times 50 = 52\ .$$

Similarly,

$$(\mu/\rho)^{\text{Al}}_{\text{AlK}} = 386 \quad \text{and} \quad (\mu/\rho)^{\text{Cu}}_{\text{AlK}} = 5377$$

and

$$(\mu/\rho)^{\text{spec}}_{\text{AlK}} = 0.46 \times 5377 + 0.54 \times 386 = 2682 \; .$$

The dependence of mass absorption coefficient upon x-ray energy and atomic number of the absorber has been given by an equation of the form $\mu/\rho = b\lambda^3 Z^4$ (Kramers, 1923), where λ is the x-ray wavelength and b is a constant with a different value either side of the absorption discontinuity; for K shell excitations μ/ρ may differ by up to a factor of ten either side of the edge. Usually, however, the expression $\mu/\rho = \text{const.}\,\lambda^n$ (Leroux, 1961) is preferred and here the value for n is dependent upon atomic number and x-ray energy; Heinrich uses such an equation to produce his tables of mass absorption coefficients. In correction programmes for quantitative electron-probe microanalysis it is probably more convenient to utilise equations for μ/ρ, since otherwise a substantial amount of computer storage is occupied with large numbers of μ/ρ values. A more detailed discussion of mass absorption coefficient data is given in section 8.7.

2.5 INTENSITY OF EMITTED X-RAYS

We have now seen that the intensity of characteristic x-radiation emerging from a sample is strongly affected by sample composition, the probability of a generated x-ray being emitted being given by $I = I_0 \exp\{-(\mu/\rho)\rho x\}$. The distance x denotes the x-ray path length in the sample and with reference to the geometrical arrangement in an electron-probe microanalyser (Fig. 2.17) it can be replaced by z cosec ψ, where ψ is the x-ray take-off angle, that is, $I = I_0 \exp\{-(\mu/\rho)\rho z \operatorname{cosec} \psi\}$.

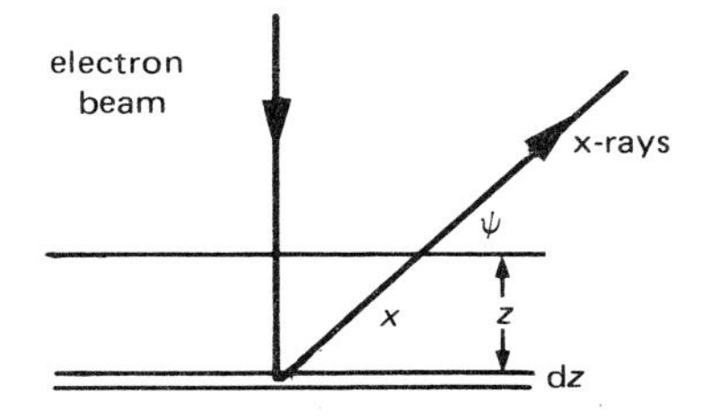

Fig. 2.17 – Geometrical arrangement with electron beam at normal incidence; ψ is x-ray take-off angle.

Consequently if we consider the x-ray intensity generated in an element $\mathrm{d}\rho z$ at depth z below the surface, as given by $\phi(\rho z)\mathrm{d}\rho z$, the fraction emerging at angle ψ to the surface is

$$\mathrm{d}I = \phi(\rho z) \exp\{-(\mu/\rho)\rho z \operatorname{cosec} \psi\}\mathrm{d}\rho z$$

and the total intensity emerging at this angle becomes

$$I = \int_0^\infty \phi(\rho z) \exp\{-(\mu/\rho)\rho z \operatorname{cosec} \psi\} \, \mathrm{d}\rho z \quad .$$

Since the total generated intensity is given by $\int_0^\infty \phi(\rho z)\mathrm{d}\rho z$, the absorption factor may be defined as

$$\frac{\int_0^\infty \phi(\rho z) \exp\{-(\mu/\rho)\rho z \operatorname{cosec} \psi\} \, \mathrm{d}\rho z}{\int_0^\infty \phi(\rho z)\mathrm{d}\rho z} \, ,$$

a ratio denoted conventionally by $f(\chi)$, where $\chi = (\mu/\rho) \operatorname{cosec} \psi$.

The depth distribution of emitted x-rays may be represented in graphical form by superimposing an absorption curve (an exponential profile whose steepness relates to the μ/ρ value), see Fig. 2.18, upon the generated x-ray intensity distribution (the $\phi(\rho z)$ curve). The shaded area under the composite curve gives the total x-rays emitted at an angle ψ to the surface. From this construction it may be seen that the greater the degree of x-ray absorption in a particular specimen, the more the x-ray emission is confined to its surface regions.

We now have to consider the additional contribution that x-ray fluorescence effects may make to the intensity of characteristic x-ray emission.

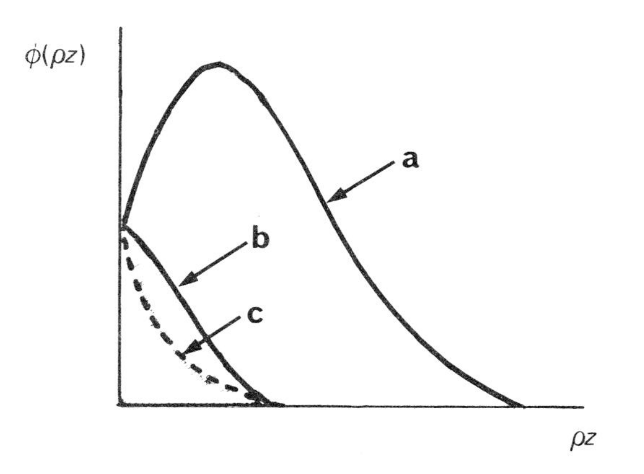

Fig. 2.18 – Distribution with depth of generated x-rays (curve a) and emitted x-rays (curve b); curve c represents the absorption profile.

2.6 X-RAY FLUORESCENCE

So far we have discussed the production of x-rays in the target by the electron beam directly (primary x-rays) and then deduced what fraction of these will eventually emerge from the surface after absorption in the specimen. However, the generated x-rays may in turn produce secondary x-rays (fluorescence) from within the target and this factor must be taken into account. For fluorescence to occur, the radiation giving rise to it must, of course, have an energy greater than the critical excitation energy of the radiation to be fluoresced. This is illustrated in Fig. 2.19 where both the Kα and Kβ lines of element B are capable of fluorescing the K x-ray lines of element A; this process is termed 'characteristic fluorescence'. In addition, part of the x-ray continuum will have sufficient energy (the shaded region where $E > E_c^A$) to cause fluorescence of K x-ray emission from A, this process being referred to as 'continuum fluorescence'. Corrections for the two effects are made separately (Chapter 9).

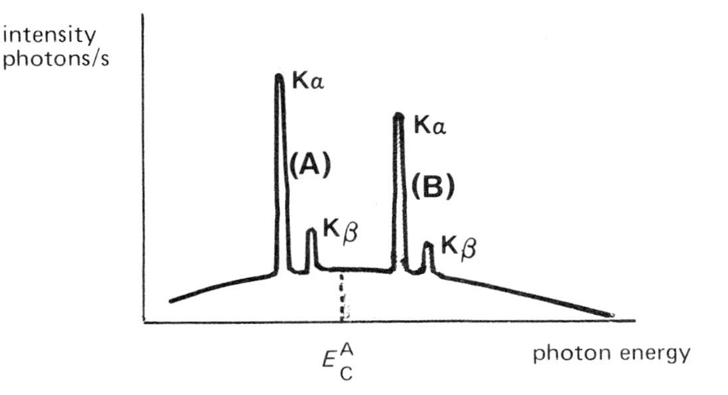

Fig. 2.19 – X-ray emission spectrum for binary specimen AB.

The magnitude of the fluorescence effect is greatest when the primary (electron-excited) x-ray energy does not greatly exceed the relevant critical excitation energy. For example, excitation of FeKα radiation ($E_c = 7.110$ keV) can be more readily produced by NiKα x-rays (7.471 keV) than by CuKα (8.040 keV) or ZnKα (8.630 keV) x-rays. As indicated in section 2.4, x-ray fluorescence and x-ray absorption effects have the same origin and a study of the absorption curve (Fig. 2.15) can give a guide to the degree of fluorescence that might be expected in a particular system. In fact, the upper and lower values of μ/ρ at the absorption edge discontinuity provide an important parameter, the 'absorption jump ratio', which is used in the calculation of fluorescence corrections.

The volume of material within which fluorescence is excited is generally much larger than that associated with the primary x-radiation since the primary x-rays may travel further than electrons into the specimen. This has some important consequences.

Firstly, it means that the distribution with depth of fluorescence will be different from the distribution of electron-excited x-rays (that is the $\phi(\rho z)$ curve). Consequently, when deriving a fluorescence correction an appropriate absorption correction term must be included at this stage of the calculation; it is not sufficient to subtract the contribution due to fluorescence generation and then multiply the total x-ray emission by the $f(\chi)$ absorption factor. In practice the fluorescence correction factor is applied in the form $(1 + \gamma + \delta)$ where γ is the ratio of the intensity of fluorescence emission to primary characteristic x-ray emission, and δ is the corresponding ratio for the continuum fluorescence contribution.

Secondly, since the x-ray spectrometer cannot distinguish between primary and secondary x-rays of the same energy, it is possible to get misleading results when analysing near an interface where x-rays may be collected inadvertently from the other side of the boundary; the effect is discussed further in section 9.7.

Fortunately, however, the intensity of fluorescence generation is, in almost all cases, very much less than that of primary generation and any uncertainties due to 'boundary' fluorescence or in the calculation of fluorescence corrections do not usually create insurmountable problems for the electron-probe microanalyst.

2.7 THE 'ZAF' OR 'MATRIX' APPROACH TO QUANTITATIVE ANALYSIS

In order to obtain quantitative electron-probe microanalysis data, the measured intensity of a particular characteristic x-ray line from the specimen is compared with that from a reference standard of known composition. This approach, first proposed by Castaing (1951) in his pioneering work on the subject, remains unchanged today although 'no standards' methods have been suggested for particular applications (sections 11.4 and 13.2.6). Hence by keeping instrumental settings (kV, beam current, x-ray spectrometer) constant while the x-ray intensity readings are being taken, such factors as spectrometer efficiency are eliminated from the calculation and, apart from applying any necessary dead-time corrections associated with the x-ray detection system (section 3.6), we have to consider only the *different* electron and x-ray behaviour in specimen and standard respectively. Suppose then we consider a binary specimen containing elements A and B where the mass concentration of A is to be measured by reference to a standard consisting of the pure element A. The intensity of the primary x-ray emission is given by

$$I = \phi(\Delta\rho z) \int_0^\infty \phi(\rho z) \exp(-\chi \rho z) \mathrm{d}\rho z \quad ,$$

where $\phi(\Delta\rho z)$ corresponds to the emission from an isolated thin film of mass thickness $\Delta\rho z$.

The total emitted x-ray intensity including any fluorescence contributions is then

$$I = \phi(\Delta\rho z)\int_0^\infty \phi(\rho z)\exp(-\chi\rho z)\mathrm{d}\rho z(1+\gamma+\delta)$$

$$= \phi(\Delta\rho z)\int_0^\infty \phi(\rho z)\mathrm{d}\rho z\, f(\chi)(1+\gamma+\delta)\ ,$$

where

$$f(\chi) = \frac{\int_0^\infty \phi(\rho z)\exp(-\chi\rho z)\mathrm{d}\rho z}{\int_0^\infty \phi(\rho z)\mathrm{d}\rho z}\ . \tag{2.1}$$

The ratio of x-ray intensities (A radiation) emitted from specimen AB and standard A is given by

$$\frac{I_A^{AB}}{I_A^A} = \frac{\phi(\Delta\rho z)_A^{AB}}{\phi(\Delta\rho z)_A^A}\ \frac{\int_0^\infty \left[\phi(\rho z)\mathrm{d}\rho z\right]_A^{AB}}{\int_0^\infty \left[\phi(\rho z)\mathrm{d}\rho z\right]_A^A}\ \frac{f(\chi)_A^{AB}}{f(\chi)_A^A}\ \frac{(1+\gamma+\delta)_A^{AB}}{(1+\gamma+\delta)_A^A}\ .$$

Now it may be shown as follows that

$$\frac{\phi(\Delta\rho z)_A^{AB}}{\phi(\Delta\rho z)_A^A} = c_A\ ,$$

where c_A is the mass concentration of element A in specimen AB. Consider an isolated thin film of element AB of mass thickness $\mathrm{d}\rho z$. Then the number of A atoms per unit area is $(Nc_A/A)\mathrm{d}\rho z$, where N is Avogadro's number. The number of ionisations produced by a given flux of electrons will be proportional to $Q(Nc_A/A)\mathrm{d}\rho z$ and the number of x-rays subsequently generated will be proportional to $Q\omega(Nc_A/A)\mathrm{d}\rho z$. Hence the intensity ratio of x-rays from isolated thin films of AB and A of the same mass thickness is given by $\{Q\omega(Nc_A/A)\mathrm{d}\rho z\,/\,Q\omega(N/A)\mathrm{d}\rho z\} = c_A$.

The integral terms correspond to the areas under the respective $\phi(\rho z)$ curves and their ratio is referred to as the atomic number correction factor (**'Z'**). The $f(\chi)$ ratio may be replaced by **'A'** to denote the absorption correction factor while **'F'** may be used to represent the ratio of the fluorescence terms, that is, $I_A^{AB}/I_A^A = c_A$ **'Z' 'A' 'F'**, and it may be seen that each of the correction factors for the specimen AB/standard A can be treated separately.

Finally, it should be mentioned that since the correction factors are dependent upon specimen composition and this is unknown, an iterative procedure has to be used in the calculations, the true composition being obtained by a series of successive approximations (section 11.1).

2.8 SPATIAL DISTRIBUTION OF X-RAYS

We have seen how a succession of scattering events in the target leads to a random movement of electrons as they reach the end of their path which, in turn, causes a volume of material to be excited whose lateral dimensions are much larger than the width of the incident beam. Such spread of electrons in the target hence determines spatial resolution in electron-probe microanalysis and is a factor to be taken into account when deducing quantititative chemical compositions of small microstructural features. Indeed, where the dimensions of the feature are much smaller than the electron range, the application of conventional ZAF corrections (see Chapter 7 to 10) will be completely inappropriate and alternative methods must be sought to obtain reliable quantitative data. Some techniques proposed for the analysis of thin films and particles, where spatial resolution is critical, are given in Chapter 13. Difficulties can arise when dealing also with small particles of a second phase embedded in a matrix where electron excitation of the surrounding matrix can confuse the results. The problem of analysing such constituents will be exacerbated if any elements present in the particle are contained in appreciable concentration within the matrix and it may become altogether impossible to carry out a meaningful analysis. This situation may be partly alleviated by removing the particles with, say, an extraction replica technique and examining them in isolation but the problem still remains of interpreting quantitatively the x-ray intensity measurements.

An indication of the actual volume of material producing x-rays may be obtained from curves such as those illustrated in Fig. 2.11 which give the x-ray distribution with depth in two different target materials at 20 kV electron accelerating voltage. Parameters which determine the depth profile of x-rays are incident electron energy (E_0), critical excitation energy (E_c), atomic number (Z) and the weight fraction of constituent elements which make up the specimen. Analytical expressions are available indicating how the maximum x-ray depth, z_r, depends upon these factors and Castaing (1960) has suggested using

$$z_r = 0.033\,(E_0^{1.7} - E_c^{1.7})\frac{1}{\rho}\frac{A}{Z}\,\mu\text{m} \ ,$$

where A is atomic weight and ρ is density. Some proposed equations are expressed in units of mass thickness and if it is assumed that A/Z is approximately constant, then ρz_r becomes simply a function of E_0 and E_c.

The same parameters control the lateral distribution of generated x-rays and similar expressions may therefore be deduced to give lateral resolution, adding of course the finite size of the electron probe impinging upon the specimen surface. The shape of the x-ray source is not, however, hemispherical with the centre at the specimen surface but tends to be pear-shaped. The lower the atomic number of the specimen and the higher the electron energy, the more deeply the electrons will penetrate before random diffusion sets in. This description of electron trajectories is essentially similar to that developed in the simple electron scattering model of Archard and Mulvey (1963) where it is assumed that all electrons reach a certain depth and then spread uniformly in all directions from this point. The distribution of electron trajectories and the shape and size of the x-ray source have been calculated more rigorously using the Monte Carlo methods described in Chapter 12. In the two cases illustrated in Figs. 2.13(b) and 2.14(b), for example, it may be seen that the shapes of the x-ray source can be regarded as approximate spheres with part cut off by the specimen surface. For 20 keV electrons incident upon gold the shape is almost hemispherical and the lateral width of the x-ray source is approximately twice the maximum depth (z_r), whereas with aluminium the x-ray source is almost spherical and the lateral spread is approximately the same as the depth.

Since it is the distribution of *emitted* x-rays which determines resolution, the effect of x-ray absorption in the target material needs also to be considered. As illustrated in Fig. 2.18 absorption will reduce the effective range for x-ray emission by a factor dependent upon the mass absorption coefficient and the x-ray take-off angle. The higher the mass absorption coefficient for the x-radiation being measured, the more the emitted x-rays will be confined to those generated in the surface regions of the specimen. This is the principle upon which is based the thin-film model of Duncumb and Melford (1966b) for quantitative analysis of the light elements (section 8.4.4), where large mass absorption coefficients are commonplace. The lateral extent of the x-ray source will also be affected by the absorption factor in a less obvious way. Perhaps the point may be demonstrated most simply by considering a high absorber where the majority of x-rays emitted come from the immediate surface regions. If the shape of the volume of generated x-rays is approximately hemispherical, as in Fig. 2.13(b), the net result will be little change in the lateral dimensions of the x-ray source. If however, the case illustrated in Fig. 2.14(b) is considered, where the generated x-ray volume is a sphere with one pole tangential to the specimen surface, the effective lateral dimensions of the volume being analysed become smaller as the x-ray emission is confined more and more to the surface regions.

From the Monte Carlo calculations of Bishop (1965) for copper irradiated with a fine beam of 29 keV electrons, Reed (1966) estimated the effective spatial resolution to be 0.7 μm and gave the following general expression for deducing the lateral extent of the x-ray distribution:

$$d = \frac{0.077}{\rho} (E_0^{1.5} - E_c^{1.5}) .$$

Based upon the assumption that the shape of the outer part of the lateral x-ray distribution is exponential (as with the depth distribution function), it was then shown that the volume of material producing 99% of the generated x-rays was approximately three times the value for d. For example, Reed estimates d for iron at 20 keV to be 0.8 μm to give a spatial resolution of 2.4 μm. A nomogram from Reed's article which gives x-ray spatial resolution as a function of E_0, E_c and ρ is reproduced in Fig. 2.20.

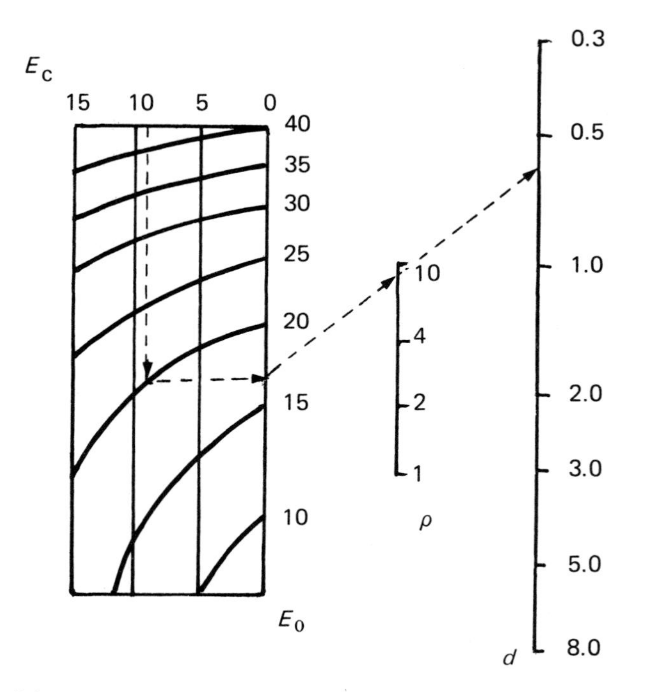

Fig. 2.20 – Nomogram for spatial resolution d (μm) as a function of electron energy E_0 (keV), critical excitation energy E_c (keV) and density ρ ($g cm^{-3}$) (after Reed, 1966); the dotted line gives the spatial resolution for copper at 20keV.

The above discussion has considered the production of primary x-radiation, those x-rays generated directly by the electron beam, but not the possibility of a fluorescence contribution to the size of the x-ray source. As mentioned in section 2.6 fluorescence excitation may take place at appreciably greater distances from the electron irradiated region and can increase significantly the volume of material which produces x-rays (section 9.7).

3

Wavelength-dispersive x-ray spectrometers

V. D. SCOTT

Analysis of the characteristic x-rays emitted from the specimen involves the accurate measurement of their x-ray wavelengths (or energies) and intensities. There are, essentially, two commercially available systems for carrying out such measurements.

The first method utilises an x-ray spectrometer in which the x-ray emission is dispersed by an analysing crystal, via Bragg diffraction from its crystal lattice planes. The diffracted x-rays are then recorded in an x-ray detector, usually a gas proportional counter, and by plotting x-ray intensity against Bragg angle the x-ray spectrum is obtained. This is usually referred to as 'wavelength-dispersive' (WD) spectrometry, since it is essentially the wavelength, λ, of the characteristic x-ray lines which is being measured.

With the second method the x-ray detector is positioned before the dispersing system to collect a sample of all, or most, of the distribution of emitted x-ray energies. Dispersion of the x-ray signal from the detector then takes place by electronic processing using pulse-height analysis equipment, the measured height of a pulse being related to the energy of the incoming x-ray photon. Hence the technique is termed 'energy-dispersive' (ED) spectrometry. The method requires an x-ray detector not only with proportional characteristics but also with good x-ray resolution, and earlier work exploring this principle (Dolby, 1959, 1963) met with limited success because it had to rely upon the performance of available gas proportional counters. Interest was, however, revived with the development of lithium-drifted silicon detectors with their better x-ray energy resolution, and ED systems fitted with these have now become extremely popular for electron-probe microanalysis.

The performance of an x-ray spectrometer system is dependent, of course, upon the efficiency of the analyser in dispersing the x-ray emission and the efficiency of the detector in collecting and recording the number of x-rays which arrive. Since the design and characteristics of WD and ED systems are distinctly different, it should not be surprising that their performance also differs and we shall see later (section 6.5) how they may each fulfil an important role in electron-probe microanalysis and complement one another in range of application. A description of ED methods is given in Chapters 4 and 5; here we shall be concerned with WD spectrometry, treating in turn its main features – the geometry of spectrometer focusing, analysing crystals, x-ray detectors and, finally, the associated electronic circuitry.

3.1 GEOMETRY OF CRYSTAL SPECTROMETERS

The simplest type of spectrometer consists of a flat crystal placed in the x-ray beam and an x-ray counter positioned such that it collects x-rays reflected from a low index plane of the crystal according to the Bragg equation, $n\lambda = 2d \sin \theta$, where d is the interplanar spacing and n is the order of reflection. By rotating both crystal and counter through a range of Bragg angle (θ), the latter at twice the angular speed of the former, a spectral distribution of the x-ray emission may be plotted. However, with the small x-ray source sizes involved in microanalysis, the angle of incidence would equal the Bragg angle over only a small part of a flat crystal (Fig. 3.1) which makes it inefficient. Consequently the analysing crystal is curved to increase its useful working area while maintaining a constant Bragg angle. Furthermore, to ensure that a high proportion of x-rays entering the spectrometer also enters the counter after diffraction by the crystal, a focusing system is used.

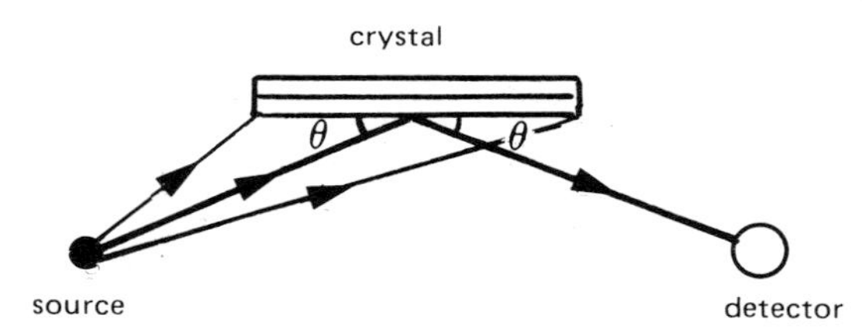

Fig. 3.1 – Flat crystal showing that only part is at the Bragg angle (θ).

Ideally, the geometrical arrangement should be such that the x-ray source, crystal and x-ray detector all lie on the circumference of a circle, the 'Rowland circle' radius R, and the crystal bent to radius $2R$ and its surface ground to radius R so that the elements of the crystal lie exactly on the circle. These conditions are illustrated in Fig. 3.2 and are known as Johansson focusing (Johansson, 1932). It may readily be shown that x-rays from the source S travelling in the plane of the Rowland circle will converge at the detector D after reflection from each part of the crystal ACB at the Bragg angle θ. Assume an x-ray of wavelength λ from S is reflected at C at the Bragg angle. Then $\angle$SCM =

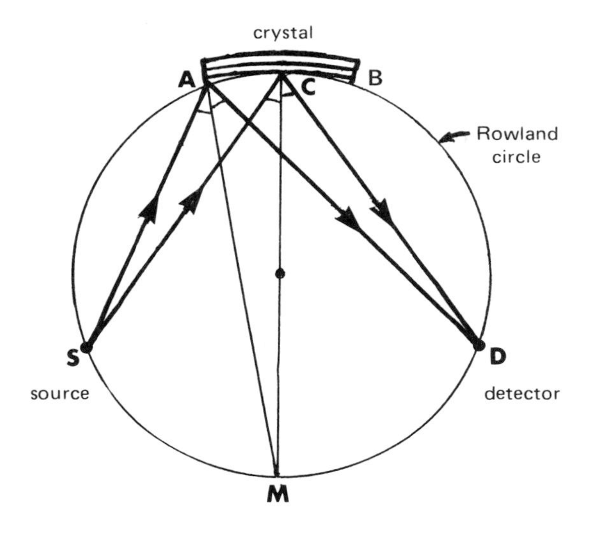

Fig. 3.2 – Johansson focusing arrangement with curved and ground crystal.

$\angle$MCD = 90° − θ, where CM is a diameter of the Rowland circle. Now $\angle$SAM = $\angle$SCM = 90° − θ, and an x-ray of wavelength λ from S which impinges on the crystal at A will also be reflected at the Bragg angle. This x-ray then travels along AD, since $\angle$MAD = $\angle$MCD = 90° − θ. Similarly, all other x-rays of wavelength λ coming from S will converge at D after reflection at the Bragg angle. The Johansson arrangement does not, however, give perfect focusing for all x-rays from S which are incident on the crystal since those not travelling exactly in the plane of the Rowland circle will not all strike the crystal at the exact Bragg angle. Optimum focusing would require therefore a doubly curved crystal (Birks, 1963) if maximum use is to be made of the crystal size, but this geometry is rarely adopted especially since the correct radius of curvature is dependent upon the particular x-ray wavelength to be studied.

In commercial instruments for practical reasons few crystals have their surfaces ground. The crystal is merely curved which means that much of its surface does not lie accurately on the Rowland circle. This is known as Johann focusing (Johann, 1931). The departure from the Bragg angle becomes greater the farther away from the centre the x-ray strikes the crystal, but some improvement in resolution can be achieved by placing apertures in the system to reduce the width of the x-ray beam. With good quality crystals, the $K\alpha_1$ and $K\alpha_2$ x-ray doublet may be separated in many cases. This would correspond to a $\Delta\lambda/\lambda$ better than 2×10^{-3} for Cr $K\alpha_1$ (2.2897 Å) and Cr $K\alpha_2$ (2.2936 Å) wavelengths (LiF crystal used). The improvement in wavelength resolution by introducing apertures would, of course, be at the expense of x-ray intensity.

The choice of Rowland circle radius is governed by the curvature to which a crystal material may be deformed, how close the crystal can approach the specimen position, and by the necessity of keeping the spectrometer and its vacuum enclosure to a convenient size. Radii within the range 100 mm to 250 mm are usual.

Another feature of most microanalyser spectrometers is that they present a constant x-ray take-off angle to the specimen, an arrangement which is convenient for quantitative work. The crystal now moves along a line away from the x-ray source as the Bragg angle is increased, and rotates about its own centre so that the distance from the x-ray source is $2R \sin \theta$ (Fig. 3.3); the detector follows a complicated, non-circular path to stay on the Rowland circle. It is termed a linear spectrometer (Sandstrom, 1952) since the x-ray wavelength is linearly proportional to the source–crystal distance.

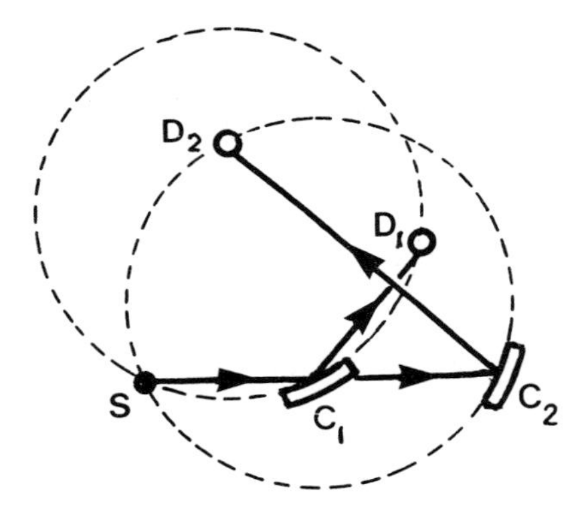

Fig. 3.3 – Linear spectrometer; $C_1 D_1$ low Bragg angle, $C_2 D_2$ high Bragg angle.

Any displacement of the x-ray source away from the Rowland circle, as may arise from changing the position of the specimen or deflecting the electron probe, can result in a reduction of the recorded x-ray intensity. The possible magnitude of the effect is illustrated in Fig. 3.4 which gives data obtained when changing the height of a specimen of titanium. It should be mentioned that the measurements were made on an instrument where the plane containing the Rowland circle also contains the axis of the electron optical column (both were in fact vertical) and the x-ray take-off angle was 35°. Hence in microanalyser designs of this type, correct adjustment of the position of specimen and standard is fairly critical in quantitative work. This may be achieved using a high quality optical microscope which has a short depth of field, adjusting the height of the sample to bring it into sharp focus each time. Alternatively, it may be necessary to redetermine the peak position on each sample. Defocusing effects ($\Delta\theta$) due to specimen height errors (Δz) become less pronounced with higher take-off angle

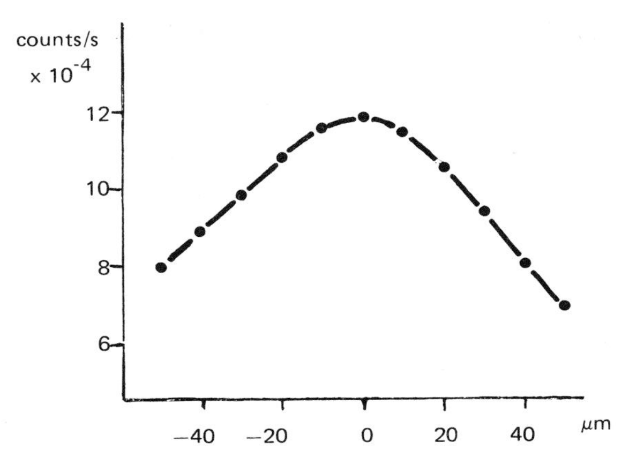

Fig. 3.4 – Change in recorded x-ray counts as specimen height is varied; titanium Kα, PET crystal.

instruments since they are proportional to the cosine of the take-off angle (ψ), that is, from Fig. 3.5 the defocusing is given by $\Delta l/l$ which equals $(\Delta z/l)\cos\psi$ where l is approximately constant for each spectrometer setting. Returning to Fig. 3.2 it may be deduced that any displacement of the x-ray source perpendicular to the plane of the Rowland circle, provided that the movement is not excessive, would cause little defocusing. Thus for any spectrometer geometry there will always be one direction which is fairly insensitive to positional errors, a feature evident in many scanning x-ray pictures (Fig. 3.6). If the production of good quality low magnification (below $\times 500$) x-ray pictures without the defocusing effect is important then either hybrid scanning (CAMECA MS46) where the specimen is moved in the 'defocusing direction' rather than the electron beam so that the x-ray source remains on the focusing circle, or crystal rocking (CAMECA Camebax) where the effective Rowland circle is displaced to keep the

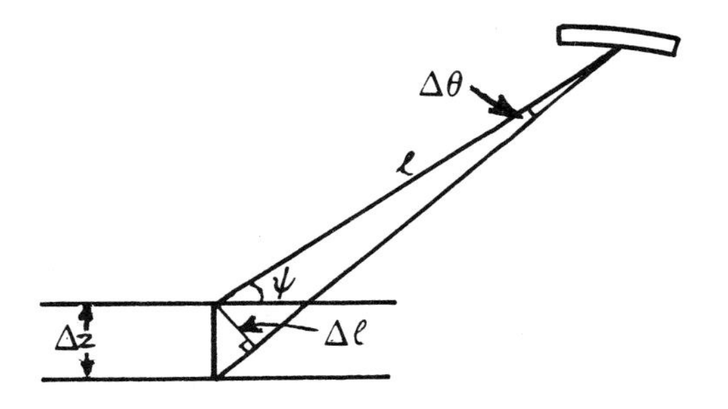

Fig. 3.5 – Defocusing due to specimen height change is a function of x-ray take-off angle, ψ.

Fig. 3.6 – Defocusing effect shown in scanning x-ray picture; titanium Kα, PET crystal.

moving x-ray source on its circumference, may be adopted. It also follows that spectrometers which have the plane of their Rowland circle inclined to the electron beam are less sensitive to variations in specimen height, although they tend to occupy more space around the electron column than the former type of mounting.

Many designs of x-ray spectrometer have been described in the literature (see, for example, Mulvey, 1964; Blokhin, 1965) including semi-focusing systems in which compactness rather than performance has been the prime consideration. A semi-focusing arrangement, which has the merit that it requires a much simpler mechanism than fully-focusing types, has the crystal at a fixed distance from the specimen (Long and Cosslett, 1957). For a given radius of curvature of the crystal, correct focusing is then achieved at only one x-ray wavelength, although to some extent the defocusing problem may be remedied by fitting several interchangeable crystals of different curvature. Since accurate focusing is no longer involved, the x-ray detector can be moved away from the Rowland circle and positioned much closer to the crystal. However, despite its compactness the design is not adopted nowadays.

Although it was mentioned earlier that spectrometers designed to accept x-rays at high take-off angles were less sensitive to variations in specimen height, the main reason for adopting such geometry is to decrease the effective absorption path length in the specimen for generated x-rays. This is particularly beneficial when dealing with the analysis of specimens containing heavily absorbing elements as in light element microanalysis. Hence with the advent of the mini lens many manufacturers have moved towards designs with higher x-ray take-off angles.

Another feature which characterises modern equipment is the increasing number of spectrometers fitted to an electron-probe microanalyser. This permits simultaneous multi-element analysis and enables optimisation of sensitivity and wavelength resolution for each characteristic x-ray over a wide range of elements. However, this approach does create operational problems, particularly when Bragg angles are frequently changed and spectrometers have to be very accurately set each time. Servo systems are therefore becoming common and, when linked with a computer, they can control the spectrometer so that it seeks peaks in the x-ray spectrum and measures their intensities, then estimates the relevant background levels. Indeed, programs are now available which enable WD systems to acquire analysis data almost as quickly as ED methods. An alternative approach to multi-element analysis (an option on the ARL SEMQ instrument) is to fit several simple fixed Bragg-angle spectrometers, each one optimised for the analysis of a single element.

3.2 ANALYSING CRYSTALS

X-ray wavelengths of interest in electron-probe microanalysis range from below 1Å (U L$\alpha \approx$ 0.91Å) to above 100Å (Be K$\alpha \approx$ 113Å). Hence from the Bragg law, $n\lambda = 2d \sin\theta$, it may be seen that a range of d-spacings is required to cover the wavelength range. In fact most spectrometers can function satisfactorily only between $\theta \sim 12°$ (below which angle much x-radiation may enter the x-ray detector directly and produce a high background) and $\theta \sim 60°$ (above which there may be physical constraints connected with the fitting of the spectrometer to the electron column). This reduces the useful range of a particular crystal to wavelengths between ~0.4dÅ and ~1.7dÅ, which means that at least four analysing crystals of different d-spacings need to be fitted to an electron-probe microanalyser to cover the x-ray wavelength range of interest. Since the aim is to use crystals which give as high a reflection efficiency as possible yet provide good wavelength resolution (requirements which may be mutually exclusive in some cases), the number of crystals supplied should preferably be greater than this. With some designs of spectrometer as many as four crystals can be accommodated, all fitted to a mounting which permits crystal interchanging by remote control. In this way full wavelength coverage can be achieved with a single spectrometer. If fewer crystals can be accommodated, it is necessary to increase the number of spectrometers and, customarily, one might be fitted with crystals to cover the higher atomic number elements and the other with crystals selected for lighter element studies. This arrangement is generally preferable since it provides an opportunity for including x-ray counters in each spectrometer which have been optimised for detecting shorter and longer x-ray wavelengths respectively (section 3.3).

The earliest crystals used for x-ray spectrometry were cut from naturally occurring materials such as rocksalt, gypsum, quartz, mica and the chlorites (clinochlore, etc.). Nowadays these tend to be replaced by laboratory-made crystals such as PET (pentaerythritol, $C_5H_{12}O_4$), ADP (ammonium dihydrogen phosphate, NH_6PO_3), and KAP (potassium acid phthalate, $C_8H_5O_4K$) or its relatives containing rubidium (RAP) or thallium (TAP).

In the longer x-ray wavelength (light element) region, fatty acid crystals are used (Henke, 1963; Ong, 1966; Ehlert and Mattson, 1967). These can be made with effective *d*-spacings ~50Å, using a deposition technique (see for example Blodgett and Langmuir, 1937) which builds up the material, molecular layer by molecular layer, from an appropriate fatty acid solution. Incorporation of a suitably heavy metal ion such as lead into the fatty acid results in the formation of planes of heavily scattering atoms separated, by organic material, a distance of two molecular chain lengths. These planes in the pseudo-crystal then act as the diffraction planes in a true crystal. Usually several hundred molecular layers are deposited, commonly on a curved substrate of stainless steel or glass. Lead stearate decanoate (LSD with *d* ~50Å) is the most popular 'crystal' for light element analysis, while other useful fatty acid formulations include lead laurate (*d* ~ 35Å) and lead melissate (*d* ~ 80Å). Table 3.1 lists crystals in most common use and Fig. 3.7 gives an indication of their range of applicability. The subject of analysing crystals has been discussed by many workers (see, for example, the reviews by Blokhin (1965) and Beaman and Isasi (1972)).

Table 3.1 – Analysing crystals used in x-ray spectrometers.

Crystal material	Reflecting plane	2*d*Å	Reflectivity *a*	Resolution *b*
Lithium fluoride, LiF	200	4.02	high	medium
Calcite	104	6.07	high	medium
Quartz	$10\bar{1}1$	6.69	medium	high
Pentaerythritol, PET	002	8.74	high	medium
Ethylene diamine *d*-tartrate EDDT	020	8.81	high	
Ammonium dihydrogen phosphate, ADP	101	10.64	medium	medium
Gypsum	020	15.2	medium	
Mica	0002	19.84	medium	
Thallium acid phthalate, TAP	$10\bar{1}0$	25.9	high	medium
Potassium acid phthalate, KAP	$10\bar{1}0$	26.6	high	medium
Clinochlore	0001	28.39	low	medium
Lead laurate		70	low	low
Lead stearate decanoate		100	medium	low
Lead melissate		160	low	low

a These comments are intended merely as a guide since crystal performance varies with Bragg angle (element analysed).

b See also Fig. 6.5.

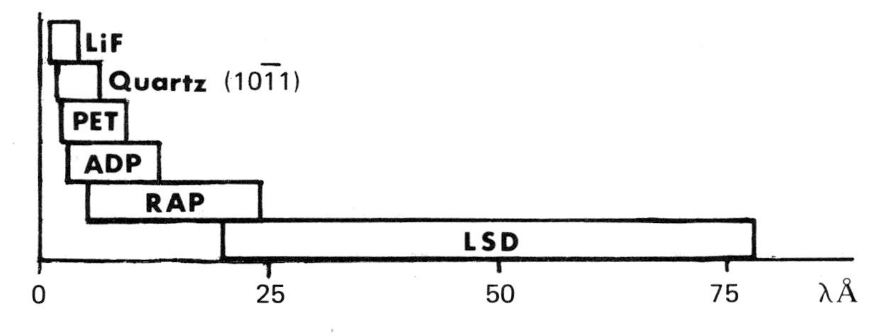

Fig. 3.7 – Wavelength ranges of crystals commonly used in spectrometers.

Most crystals will give higher order reflections, the relative intensities of the different orders varying from one crystal material to another, that is, x-ray wavelengths of $\lambda/2$, $\lambda/3$, etc. will also satisfy the Bragg equation. Mica is noted for producing quite strong high order reflections which may cause problems especially if it is used as a support for LSD layers. LiF, another commonly used crystal, gives a fairly strong second order reflection, some 10% of the first order, and so some care may be needed here in interpreting x-ray spectra. On the other hand, the acid phthalates give fairly weak high-order reflections, ~2% of the intensity of the first order. The higher order peaks may, fortunately, be removed using pulse-height analysis (section 3.5).

A good quality analysing crystal should have high reflectivity since this results in high intensity of the diffracted x-ray beam, and high spectral resolution since this improves its ability to discriminate between x-rays of closely similar energy and leads to a high peak-to-background ratio. An indication of the characteristics of various crystals is included in Table 3.1 (and later in Fig. 6.5) although, for practical microanalysis purposes, the essential criterion is the overall performance of the WD system, which includes for example design features of the spectrometer. Such data are presented as characteristic peak intensities and peak-to-background ratios for different target materials under specified analysis conditions. However, there would appear to be little point in giving extensive performance figures here owing to the fact that they are continually being revised and they differ significantly from instrument to instrument. As a guide, count rates of several hundred thousand per second would now be expected from elements such as iron (LiF crystal), titanium (PET crystal) and aluminium (RAP crystal) when irradiated with a beam current of 100 nA at 20 kV; peak-to-background values would be ~500:1. With lower atomic number elements, the figures would not be so good, typical count rates from sodium (RAP crystal) and carbon (LSD crystal) being $\sim 10^5\ s^{-1}\ \mu A^{-1}$ (both at 10 kV); although the RAP crystal still gives a good peak-to-background value, for sodium ~500:1, the LSD crystal manages only ~100:1 for carbon.

The generally poor performance of analysing crystals used for light element studies has generated interest in an alternative diffracting system, the ruled grating. Since diffraction gratings offer, potentially, superior wavelength resolution and reflected x-ray intensities compared with fatty acid crystals some comments should perhaps be included, even though they are not strictly in the context of this section. The idea is to use a conventional optical grating at small angles of incidence to foreshorten its effective spacing and to take advantage of the correspondingly stronger specular reflections. The improvement in diffraction efficiency produced by using blazed-angle gratings instead of those with the classical grooved form is described by Holliday (1966) and by Nicholson and Hasler (1966) who fitted a blazed grating with a curved profile to an electron-probe microanalyser. Work has also been carried out on 'phase' gratings (Sayce and Franks, 1964; Franks, 1972) where reflection takes place from parallel top and bottom surfaces of the ruled grooves. Although peak-to-background values of ~500:1 for carbon Kα x-rays have been obtained with such systems, they have not been very widely pursued as microanalyser attachments because they require a special spectrometer arrangement. Their high wavelength resolution makes them, however, particularly suitable for studying the effects of chemical bonding on shape of the x-ray emission band (section 2.2.3), and recent results (Kozlenkov *et al.*, 1981) would indicate that interest in blazed diffraction gratings for long wavelength x-ray analysis is still very much alive.

3.3 X-RAY DETECTORS

X-ray detectors suitable for fitting to WD systems include gas-filled counters operating in the ionisation, proportional or Geiger range, and electron multipliers. The latter are little used in connection with electron-probe microanalysers, except in some designs of grating spectrometer (Franks and Lindsey, 1966; Kozlenkov *et al.*, 1981), most spectrometers having a gas-filled proportional counter (Parrish and Kohler, 1956).

The gas-filled counter consists usually of a cylindrical metal tube through which passes an axial wire mounted on insulating seals at each end of the cylinder (Fig. 3.8). The wire is maintained at a positive potential while the tube serves

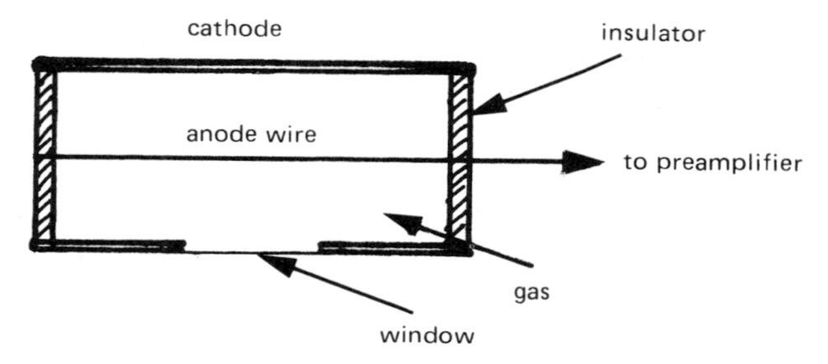

Fig. 3.8 – Gas proportional counter with coaxial geometry.

as the cathode. Commonly, the anode consists of 50 μm diameter wire of smooth finish and which has been thoroughly cleaned to minimise distortion of the electric field. The tube is filled with a suitable gas and a window in its side allows x-rays to enter the chamber.

The incident x-ray photons are absorbed in the counter gas by virtue of the photoelectric effect (section 2.2.1), an inner electron being ejected from the gas atom followed by the emission of photoelectrons, Auger electrons and secondary (fluorescence) x-rays; these latter are themselves absorbed to produce more electrons. On average, one ion pair (electron plus positive ion) is created for each 25–30 eV (for argon gas) of incident x-ray energy and hence the number of ion pairs is proportional to the energy of the incident x-ray photon. Provided that the voltage applied to the counter is sufficient to ensure collection of all the ion pairs, say >100 V, the amplitude of the pulse collected at the anode is proportional to the energy of the incident x-ray photon and the counter can, therefore, be used to help discriminate between x-rays of different energy. This assumes that the events associated with the absorption of a single x-ray occur virtually simultaneously and that the pulse can be processed before the next x-ray quantum enters the chamber. At high count rates this is unlikely, an aspect referred to in section 3.6 when discussing 'dead time' of the system. This counter operating range is termed the 'ionisation chamber region'.

The signal produced by a single x-ray using the above counter operating conditions is, however, small and difficult to detect but it can be increased by using a higher anode potential, 1000 V to 2000 V. The effect of the higher voltage is to accelerate the free electrons created in the initial x-ray absorption process such that they can cause further ionisations and hence more electrons (the Townsend 'avalanche'). The increased output pulse is still proportional to the incident x-ray photon energy, since each electron produces on average the same number of secondary electrons. Such amplification by internal multiplication is referred to as the 'gas gain' and it lies in the range 10^2 to 10^4. This is the counter operating condition normally adopted in x-ray crystal spectrometry and it is termed the 'proportional region'.

The gas gain increases rapidly with anode voltage and when the avalanche spreads to the whole wire, the signal becomes constant irrespective of x-ray photon energy. Called the 'Geiger region', it is rarely employed in x-ray spectrometry since not only does the counter then lack energy resolution but its 'dead time' is very long.

The choice of counter filling gas depends not only upon its providing sufficiently high gas gain but also upon its ability to absorb efficiently a range of x-ray energies. An inert gas is used because the chances are reduced of ion pairs being recombined before they are collected on the electrodes. (In electronegative gases such as oxygen, the negative ions become attached to gas atoms, reducing their mobility and increasing the probability of recombination with positive ions.) For detecting high energy x-rays ($\lambda < 3$ Å) xenon with its higher absorption is generally preferred, but counters containing argon may be used quite effectively over the entire range of x-ray energies of interest in electron-probe microanalysis. With argon a polyatomic gas must be added, its purpose being to absorb the ultraviolet radiation which would otherwise produce electrons from the counter walls and thereby contribute to the avalanche and advance the Geiger region. Typical polyatomic gas additions are 2.5% CO_2 or 10% methane. When dealing with very low energy x-rays, it is necessary to reduce the degree of absorption of the gas or else most of the primary ionisation occurs close to the counter window where the field is too weak to prevent ion-pair recombination. This can be achieved by reducing the pressure of the gas or by changing its composition (Henke, 1963; Manzione and Fornwalt, 1965; Wardell and Cosslett, 1966); for example, McFarlane (1972) recommends argon – 75% methane at a pressure of 125 torr. Finally, the presence of impurities in the gas, particularly oxygen and water vapour, must be avoided. With flowing gas counters it is therefore essential to use pure cylinder gases, while with sealed gas counters it may be necessary to replace them routinely owing to the occurrence of some slight but inevitable window leakage.

The window of the gas counter has to prevent gas leaking into the surrounding vacuum chamber yet transmit the x-rays of interest; consequently they are made very thin and with low atomic number materials. Beryllium of thickness ~125μm is commonly used for sealed counters because it can withstand atmospheric pressure over quite a large area without rupture. Such a window becomes increasingly opaque with decrease in x-ray energy and, since thinner beryllium windows tend to be very fragile, plastic windows are preferred below ~4 keV ($\lambda > 3$ Å). 'Mylar' ($C_{10}H_8O_4$) of 6μm or 3μm thickness is a widely used material but for longer x-ray wavelengths (>5Å) window thicknesses must be reduced much further, either by stretching commercially available sheet of, for example, polypropylene (CH_2) or polycarbonate ($C_{16}H_{14}O_3$), or by casting from solution thin layers of collodion ($C_{12}H_{11}O_{22}N_6$) or cellulose acetate ($C_{10}H_{21}O_{15}$). These thin layers (~0.1μm thick) usually require supporting on a grid. The transmission properties of several counter windows plotted as a function of x-ray energy are indicated in Fig. 3.9. It should be noted that in the long x-ray wavelength region the transmission properties of plastic windows show marked effects associated with absorption edges of the constituent elements. Hence, when analysing nitrogen x-ray emissions for example, it would be preferable to use a plastic film with a high nitrogen content such as collodion rather than say polypropylene, assuming of course that it is strong as well as thin since strength may be a decisive factor in choosing a window material. Usually plastic windows are coated with a conducting layer to avoid charge collection and the resulting distortion of the electric field in the counter.

From the foregoing it is clear that a counter which has been designed for detecting lower energy x-rays will not produce optimum performance for higher energy radiations. Some manufacturers have, therefore, achieved full coverage of

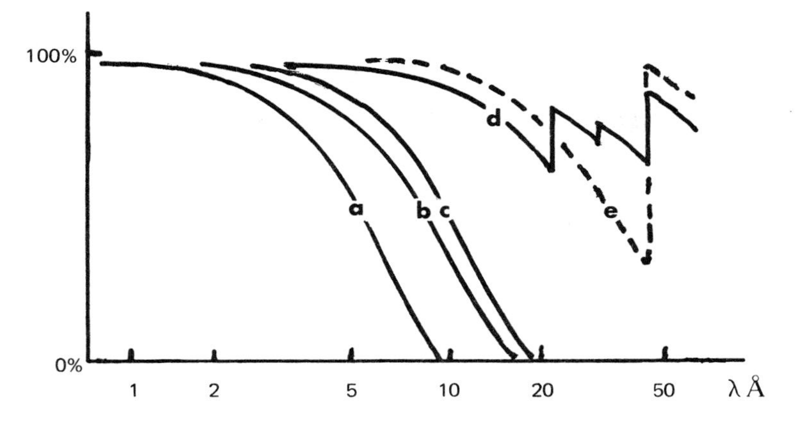

Fig. 3.9 – Transmission characteristics of different counter windows: (a) beryllium 120 μm; (b) mylar 6 μm; (c) mylar 3 μm; (d) collodion 0.1 μm; (e) polypropylene 0.1 μm.

the energy range in a single spectrometer by fitting two counters in tandem, a thin window flow counter, with a second thin window at the rear, being positioned in front of a beryllium window sealed counter. More usual practice, however, it to have a separate 'light element' spectrometer with additional x-ray analysis system(s) (WD and/or ED) for the heavier elements.

3.4 PERFORMANCE CHARACTERISTICS OF A GAS PROPORTIONAL COUNTER

There are several effects, related to the ionisation events occurring within the counter chamber, which have to be considered in analytical work. These involve the production of an escape peak, depression of the height of the pulse, and non-linearity between the number of incoming x-ray photons and the number of output pulses ('dead time').

The escape peak is an additional output signal from the counter having an energy which does not equate directly with any of the characteristic lines of the emitting element(s). Fig. 3.10 shows an escape peak obtained with an argon-filled counter when recording Cu Kα x-radiation. The energy of the escape peak is 2.96 keV lower than the characteristic line and the intensity is very much less, ~5% of the main peak; the difference in energy is a function of the gas species. The effect is related to processes involved when the incoming x-ray photon ionises the gas atom. As mentioned in section 2.2.1, either Auger electrons or characteristic x-rays may be produced as the excited atom lowers its energy and, while the Auger electron may be readily absorbed by other gas atoms and deposit its energy in the counter, the more penetrating x-ray may escape from being

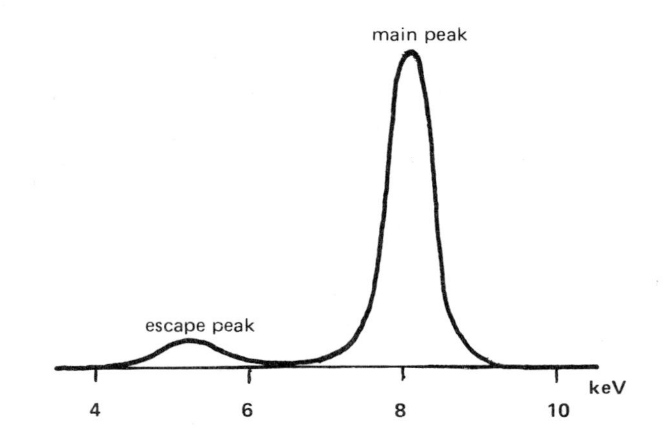

Fig. 3.10 – Argon escape peak for Cu Kα x-rays.

absorbed in the gas. When this happens the energy of the output pulse is reduced by the energy of the escaping x-ray, by 2.96 keV in the case of argon K x-rays. Clearly, lower characteristic x-ray energies would not give an associated escape peak. Approximately 10% of the initial gas atom ionisations result in argon K x-rays and approximately one half of these stand a chance of escaping, as indicated in Fig. 3.10. The escape peak produced in a xenon counter is of negligible intensity since, firstly, the K absorption edge for xenon is of very much higher energy than the range of energies involved in microanalysis and, secondly, xenon L x-rays are more heavily absorbed in xenon gas. In WD analysis, where the x-ray wavelength is measured before the counter, the counter's main function is to record the number of incident x-rays irrespective of their energy and the escape peak effect is, therefore, of little consequence. However, escape peaks are important in ED analysis (section 4.6.1).

Depression of the height of the pulse may be caused by a reduction in the gas gain arising from changes in the electric field in the counter. Such a change may result from screening of the anode wire by positive ions created in the avalanche. It takes the ions a few milliseconds to diffuse to the cathode and an x-ray arriving within this time interval will experience a reduced electric field. Hence gas amplification is less and the pulse height will be correspondingly reduced. Pulse-height depression becomes more serious at higher count rates when the time between successive avalanches becomes shorter, and at high gas gains when each incident photon produces a greater number of ion pairs. In practice counters should, therefore, be operated at a sufficiently low voltage to restrict the gas gain (say to no more than ~1000), relying upon electronic amplification to provide the required pulse height for processing (section 3.5). It should be mentioned that the geometry of the counter may accentuate problems associated with distortion of the electric field, such as the use of small diameter wires or finely pointed electrodes as the anode. Contamination of the wire or of the counter gas may also create a pulse-height depression effect.

Even if the counter were to be operated under ideal conditions, the pulses produced by a beam of monoenergetic x-rays would not have identical height because the number of ion pairs produced in the initial ionisation and the subsequent avalanche are subject to statistical fluctuations. On average, the number of gas atom ionisations ($\bar{N}$) is given by E/ϵ, where E is the energy of the x-ray photon and ϵ is the average energy required to create an ion pair. If a random series of ionisation events was to be assumed then the standard deviation would be $\sqrt{E/\epsilon}$ but, since ionisation is not entirely random (Fano, 1947), the standard deviation is better expressed as $\sqrt{FE/\epsilon}$, where F is the Fano factor. It is customary to express the resolution of a counter in terms of energy and then the energy spread (ΔE) corresponding to the standard deviation is given by $\epsilon\Delta N$ which equals $\sqrt{FE\epsilon}$. The additional statistical spread due to fluctuations in the size of the avalanche may be taken into account by adopting a modified Fano factor. With an argon-filled counter, $\epsilon = 26$ eV and the effective Fano factor is ≈ 0.8 which gives $\Delta E \approx 4.56\sqrt{E}$. For iron x-rays ($E = 6400$ eV), $\Delta E \approx 365$ eV but it is more usual to quote the energy spread at one half the maximum value of the distribution. The full width at half maximum (FWHM) is given by $2.355\Delta E$, that is 861 eV; the resolution may also be expressed as the percentage ($861/6400 \times 100\%$) $= 13.5\%$. However, such values are never realised in practice owing to variations in gas amplification produced by non-uniformity of the electric field in the counter and contamination effects, and 18% is probably closer to experimentally realised values for a good coaxial counter.

3.5 COUNTING CIRCUITRY

The essential components which form the counting electronics are shown schematically in Fig. 3.11. Their purpose is to amplify the pulses from the counter and to convert them into various signals which can be used for measurement and display.

The charge collected by the anode wire of the counter is first stored in a capacitor which then delivers a voltage pulse. This pulse requires amplification before it can be further processed. Most of the increase (up to $\sim\times 1000$) takes place in the main amplifier but an intermediate stage is introduced, a preamplifier, which is positioned close to the counter in order to maintain a low noise level in the system. A typical preamplifier produces $\sim 0.1\,\mu$V per ion pair and an output signal of a few millivolts for each pulse registered in the counter. As well as increasing the voltage signal, the amplifier has pulse-shaping circuitry in order to shorten the pulse and reduce noise levels; in turn, the signal-to-noise ratio is increased and the degree of pulse overlap occurring at high count rates is reduced. The time constant of the circuit is usually $\sim 1\,\mu$s which produces output pulses a few microseconds wide, sufficiently short to allow count rates of 10^5

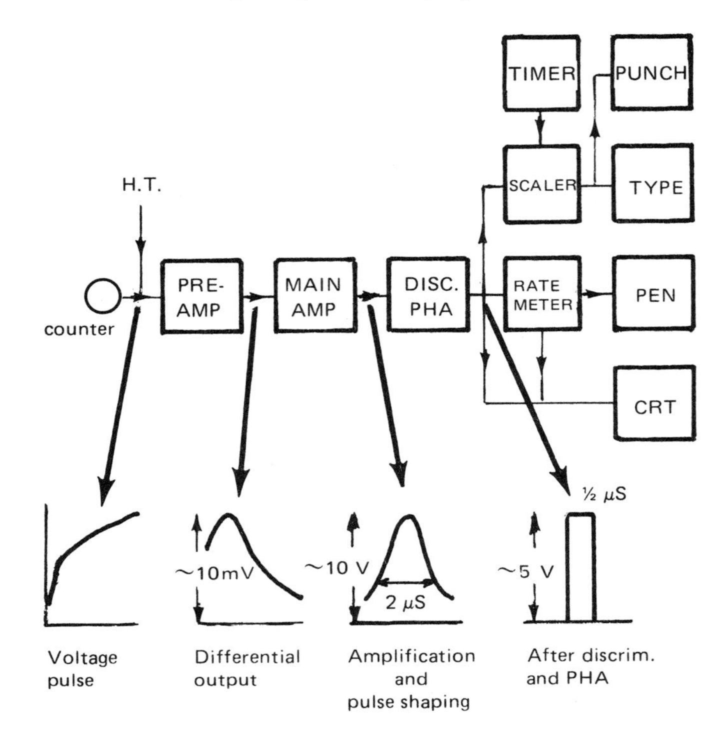

Fig. 3.11 – Electronics used with a gas proportional counter.

per second to be recorded without serious dead-time effects being created.

The pulse-height analyser has two functions. Firstly, it converts the input pulses into a suitable form for activating the counting circuits (ratemeter and scaler), and, secondly, it rejects unwanted pulses such as may be produced by stray electrons, x-rays and cosmic rays entering the counter chamber. The discriminator setting of the pulse-height analyser provides a low-energy threshold, below which pulses are rejected. Every time the threshold is exceeded a rectangular logic pulse is triggered and recorded by the ratemeter/scaler. The width of the logic pulse contributes to the dead time of the electronic system. The pulse-height analyser contains also an upper level discriminator so that a 'window' of adjustable width may be used. Now, only those input pulses whose height exceeds the lower level but not the upper level will generate an output pulse. The window setting is useful for suppressing any higher order reflections from the analysing crystal. In general, a narrow window will tend to give improved peak-to-background ratios although some of the peak intensity may be lost. If servo-control of the spectrometer is being used it may be preferable to adjust automatically the settings so that analysing conditions are optimised for each x-ray peak (Weber and Marschal, 1964). 'X-ray pictures' showing the distribution in the specimen of an element selected by the spectrometer may be formed using the output directly from the pulse-height analyser.

The ratemeter monitors continuously the rate at which pulses leave the pulse-height analyser. Each pulse adds a fixed increment of charge to a capacitor. The charge is leaked continuously through a resistor and the voltage on the capacitor is determined by the relative rates of charging and leaking and hence is proportional to the rate of arrival of pulses. The voltage may be monitored by a pen recorder or by observing the displacement of the spot on a CRT. There is a time constant associated with the ratemeter, and conditions must be set such that background noise and response times are optimised.

The scaler counts the arrival of the logic pulses in a given time interval. The time interval is measured by a separate scaler which counts pulses generated by an electronic clock (a high frequency oscillator). Usually the time is preset and the number of pulses recorded on a digital display, although it is possible to use a 'preset count' mode.

3.6 DEAD TIME OF THE COUNTING SYSTEM

During the short period of time that the detection and counting system is processing a pulse it is unable to deal with any which immediately follow. The interval between the time of arrival of a pulse and the time that the system is ready to receive the next one is referred to as the dead time (τ). As a result of this effect pulses are likely to be missed and the longer the system dead time and the higher the rate of arrival of pulses, the greater the probability of counting losses. The dead time of the gas proportional counter is affected by the operating conditions but, since it is much smaller than the dead time of the electronics, it can be ignored.

If the arrival of a pulse during the 'unreceptive' period does not cause an increase in that period, then the dead time is termed 'non-extendable'. Assuming n' is the recorded count rate, the system will be dead for $n'\tau$ seconds. Thus the 'live time' is $1-n'\tau$ and the true count rate (n) is given by (Ruark and Brammer, 1937),

$$n = n'/(1-n'\tau) \ . \tag{3.1}$$

A typical value for the dead time of a system would be 1 μs. Hence a high recorded count rate of 10^5 per second would be 10% less than the true count rate but below count rates of 10^4 per second the error may be considered negligible.

If, however, the arrival of a pulse during the 'unreceptive' period affects the dead time by introducing its own period of paralysis, then the dead time is 'extendable'. In this case a different relationship applies. Since the intervals between random pulses obey a Poisson distribution, the fraction of intervals greater than τ is $\exp(-n\tau)$, and

$$n = n'/\exp(-n\tau)$$

For $n\tau \ll 1$, $\exp(-n\tau) \approx (1 - n\tau)$, and the two expressions for dead time are almost equivalent. For the previously cited system with its dead time of 1 μs, the true count rate corresponding to a recorded count rate of 10^5 per second would then be within 0.1% of the value calculated using Ruark's expression.

Both equations may be regarded as simplifications, particularly at high count rates where coincidence losses will be large. It may, therefore, be preferable to calibrate the system by, for example, measuring the count rate as a function of the electron probe current (i). The true count rate will be proportional to the current, that is, n = const. i, and equation (3.1) may be written n'/i = const. $(1 - n'\tau)$. Hence a plot of n'/i versus n' will give a straight line of slope − const. τ. The value of the constant may be found from the intercept on the n'/i axis and thus τ can be calculated. Alternatively, the graph may be used directly to obtain the true count rate, an approach which may offer advantages for a system with an extendable dead time. As we shall see in section 4.9, in ED systems a correction for dead time is accomplished in the design of the electronic circuitry.

4

Energy-dispersive x-ray spectrometers

S. J. B. REED

4.1 THE LITHIUM-DRIFTED SILICON, Si(Li), DETECTOR

Lithium-drifted germanium and silicon detectors originated in nuclear physics laboratories as γ-ray detectors. The first application to electron-probe micro-analysis was by Fitzgerald *et al.* (1968) but at that stage the resolution was inadequate to separate adjacent Kα peaks and the sensitivity was poor at low energies. However, by about 1970 resolution and sensitivity had improved to the extent that the Si(Li) detector became viable for x-ray spectrometry down to ~1 keV. Germanium detectors are the preferred choice for energies above about 25 keV, but are less suitable for lower energies, especially below 5 keV (Barbi and Lister, 1981; Fink, 1981).

The mechanism by which x-rays are detected in these devices is similar to that of the gas-filled proportional counter (section 3.3), but with certain important differences arising from the fact that the detection medium is solid. Incident x-rays are absorbed by silicon atoms through the photoelectric effect, and the x-ray photon energy is transferred to photo- and Auger electrons, which in turn lose energy by processes which include raising valence band electrons into the conduction band (Fig. 4.1). These electrons can move freely through the lattice, thereby causing electrical conductivity; also the 'holes' left in the valence band act as free positive charges. In an ideal semiconductor, conductivity arises solely from thermally excited electrons and holes, and is low because the mean thermal energy at normal temperatures is much less than the energy gap between valence and conduction bands (1.1 eV in silicon). Thus, when a bias voltage is placed across a Si(Li) detector, only a very small current normally flows. An incident x-ray photon produces a brief burst of current, which is amplified to produce a pulse suitable for counting.

Even the purest silicon always contains some residual impurities, which have the effect of introducing additional energy levels between the valence

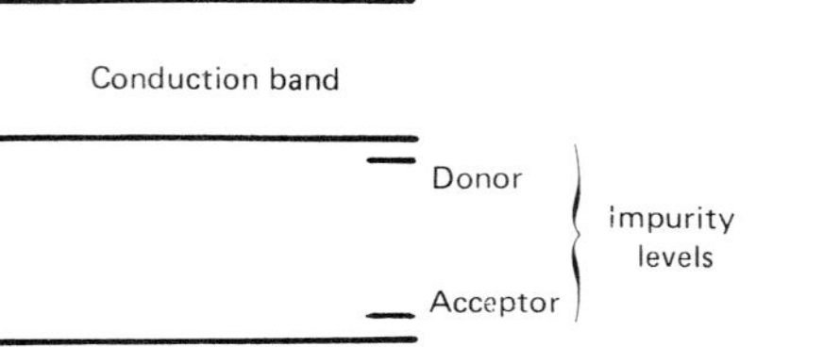

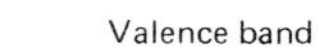

Fig. 4.1 – Electron energy bands in a semiconductor.

and conduction bands. For example, boron gives rise to impurity levels to which electrons are relatively easily raised from the valence band by thermal excitation. In this case the silicon is said to have 'p-type' properties, its conductivity being caused predominantly by positive holes. This conductivity is unwelcome since the resulting current tends to obscure the signal produced by x-rays. In order to compensate for p-type impurities, lithium is introduced by the process of 'drifting', that is, diffusion at about 100°C with an applied electric field controlling the distribution of lithium ions (Pell, 1960). The element lithium is chosen for its low ionisation energy and small radius, which confers high mobility. The lithium atoms donate electrons to impurity levels and nullify their effect, resulting in material which approximates to 'intrinsic' (ideally pure) silicon.

The mean energy used in producing one electron-hole pair in silicon is 3.8 eV (some energy being dissipated in other processes), which is about one-seventh of the equivalent figure for a gas proportional counter (section 3.3). The number of charge carriers generated for a given x-ray energy is correspondingly greater; however, the Si(Li) detector lacks the internal gain of the gas counter and therefore produces much smaller output pulses. For ~5 keV x-rays the number of electron-hole pairs generated is ~1000, and the pulse is equivalent to a current of only 1 nA flowing for 0.1 μs.

The number of charge carriers (electrons and holes), and hence the size of the output pulse, is proportional to the energy of the incident x-ray photon. Pulse-height analysis thus enables the x-ray spectrum to be resolved with respect to energy. Compared with the gas-filled counter, the number of charge carriers produced by a photon of a given energy is greater; hence the statistical variation is less and the energy resolution better. The Si(Li) detector is thus preferable for use in the 'energy-dispersive' (ED) mode, where the spectrum is resolved solely by means of pulse-height analysis (thereby obviating the need for a diffracting crystal). However, it is of interest to note that ED spectrometry originated in the use of gas proportional counters for the detection of light elements such as carbon, nitrogen and oxygen (Dolby, 1959).

4.2 CONSTRUCTION OF THE Si(Li) DETECTOR

In order to make an Si(Li) detector, a wafer of high-purity single-crystal silicon is first subjected to the lithium-drifting process, which produces a lithium distribution of the form shown in Fig. 4.2. In the quasi-intrinsic region the lithium concentration is constant and compensates exactly for the p-type impurities, as described above. The adjacent lithium-deficient region still has p-type properties, whereas excess lithium confers n-type properties owing to conduction band electrons donated by lithium atoms. Only the 'intrinsic' silicon is suitable for x-ray detection and in order to expose this, surplus p-type material is removed. A layer of gold approximately 20 nm thick is then evaporated onto the surface to form a contact, by means of which the negative bias voltage is applied to the front surface of the detector. The output signal is obtained via a connection to the n-type layer at the back.

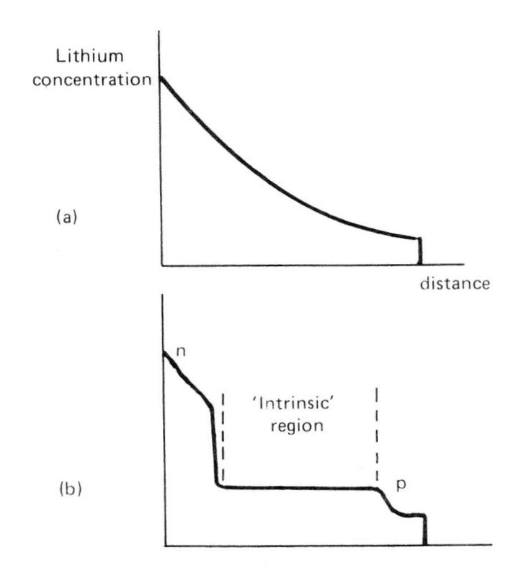

Fig. 4.2 – Lithium drifting: distribution of Li (a) after diffusion and (b) after drifting.

Fig. 4.3 shows schematically a typical detector configuration. The active area is most commonly about 10 mm^2. Detectors of larger area are made, but the increased capacitance results in some sacrifice in resolution and in most applications the larger area is not a significant advantage (exceptions are where the x-ray intensity is particularly low). A typical thickness is 3 mm, this being enough to absorb completely x-rays of energy up to at least 15 keV.

In order to stabilise the lithium and to minimise noise, both the detector and the field-effect transistor (FET) used for the first stage of amplification are cooled by means of a copper rod leading to a liquid nitrogen reservoir. Although limited periods at room temperature with the bias voltage off can be tolerated, it is preferable in the interests of maintaining optimum performance that the detector should be kept cold permanently. The cryostat containing the detector assembly is permanently evacuated in order to protect the sensitive surfaces of the detector. It is usual to incorporate zeolite or other adsorbent in order to maintain a good vacuum in the cryostat during its working life. Loss of vacuum is indicated by deterioration in the performance of the detector and an increase in liquid nitrogen consumption.

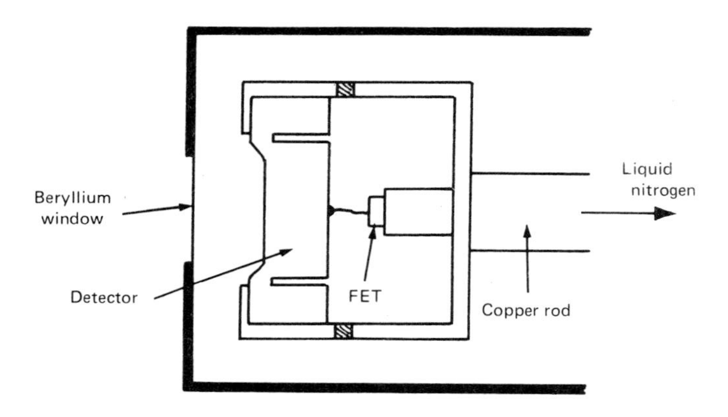

Fig. 4.3 – Si(Li) detector assembly (schematic).

Close attention to the mechanical design of the detector-FET assembly is necessary in order to minimise microphonics (sensitivity to vibration), which can significantly affect the resolution. Ice crystals in the liquid nitrogen Dewar flask serve as nuclei for boiling and are a potential source of microphonic effects. Also vibration derived from the instrument to which the detector is attached should be avoided. Acoustic noise and electrical interference are further hazards which may degrade the performance of the detector.

A 'window' in the cryostat is required to allow x-rays to enter. This must be vacuum-tight and strong enough to withstand atmospheric pressure on the outside. The window is usually made of beryllium foil with a nominal thickness of 8 μm. However, this represents a lower limit, the true thickness generally being in the range 8 to 12 μm. The x-ray transmission of such windows is close to 100% for energies down to 2 keV, below which it falls quite rapidly, to about 40% at 1 keV.

If the beryllium window is removed, the detector can record x-rays of energy well below 1 keV, such as the K lines of light elements ($Z < 11$) (Jaklevic and

Goulding, 1971; Russ and Sandborg, 1981; Musket, 1981). However, the detector is then exposed to the specimen chamber vacuum, which may be relatively poor, and there is a risk of surface contamination leading to a degradation of performance. It is therefore desirable to use a thin film of an organic material such as formvar in place of the beryllium. This should be coated with aluminium in order to exclude light, which otherwise could produce a spurious signal in the detector. Furthermore, a magnetic trap is essential to prevent scattered electrons from reaching the detector. ED analysis of light elements is further discussed in section 5.7.

The detector is attached to a port on the specimen chamber of the instrument (electron microprobe, etc.), suitable cryostat designs to fit all commercial models being available from the detector manufacturers. The minimum practicable diameter of the end of the detector housing is about 15 mm, which sets a limit to how close to the specimen the detector can be placed, especially since it is desirable to have a collimator between detector and specimen to limit the angle of acceptance and thereby exclude as far as possible x-rays produced by scattered electrons striking objects in the specimen chamber. The solid angle (Ω) subtended by the detector is given by: $\Omega = A/d^2$ (steradians), where A = detector area and d = detector–source distance. In a typical case, A = 10 mm^2, d = 70 mm, and hence $\Omega = 0.002$ steradian.

For conventional electron-probe microanalysis little is gained by bringing the detector closer to the specimen, except perhaps when dealing with specimens which can tolerate only a very low electron beam current without suffering damage. It may even be advantageous to move it further away in order to carry out energy-dispersive (ED) and wavelength-dispersive (WD) analysis simultaneously, as the latter demands a much higher current (for example, 100 nA compared to a few nA for ED analysis). For this reason some detectors are fitted with a racking movement which allows the distance to be altered. In the case of thin (TEM) specimens the x-ray intensity is generally much lower and high x-ray collection efficiency becomes important. The closest practicable distance is about 20 mm, the corresponding solid angle being 0.02 steradians.

4.3 THE FET PREAMPLIFIER

The purpose of the preamplifier is to amplify the pulses produced by x-ray photons in the detector with the least possible added noise, and to provide output pulses suitable for further amplification prior to pulse-height analysis. The first stage of the preamplifier consists of a field-effect transistor (FET). These are manufactured in quantity for other applications and individually selected for use in Si(Li) systems. The FET is mounted as close as possible to the detector in order to minimise stray capacitance and noise pick-up, and both are maintained at a temperature of about 100 K (just above the boiling point of liquid nitrogen).

D.c. coupling between detector and preamplifier gives less noise than a.c. coupling where the pulses pass through a capacitor. However, the current produced by x-ray photons flows only in one direction, and hence the output of a d.c.-coupled preamplifier consists of a 'staircase' waveform, which eventually reaches the working limit of the preamplifier unless corrective measures are applied. The simplest remedy is to use a feedback resistor (Fig. 4.4(a)) with a time constant equal to the product of R_f and C_f which restores the input voltage. However, this is somewhat unsatisfactory owing to the noise contributed by the feedback resistor. Hence, most systems now use 'pulsed opto-electronic feedback' (Landis *et al.*, 1971) in which the output voltage, on reaching a certain level, triggers a pulse which activates a light-emitting diode (LED) mounted close to the FET (Fig. 4.4(b)). The light pulse resets the FET output to zero, during which process the main amplifier output is clamped to prevent overloading.

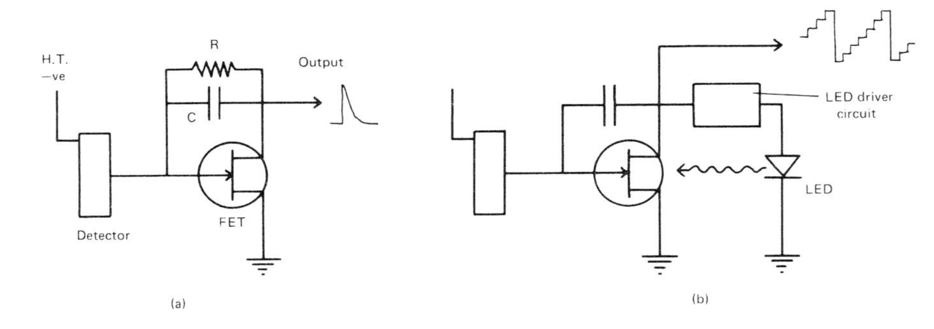

Fig. 4.4 – Detector feedback: (a) resistive and (b) opto-electronic.

The output of the preamplifier consists of a series of voltage steps, proportional in size to the x-ray photon energy, which are further amplified and shaped in the main amplifier. Noise is also present, the main sources being the FET and leakage current in the detector. The resulting fluctuation in the baseline upon which the signal pulses are superimposed has the effect of broadening the pulse-height distribution obtained for x-rays of a given energy – thus noise plays an important part in determining the resolution of the system, as discussed in section 4.5.

4.4 THE MAIN AMPLIFIER

The preamplifier produces pulses of a few millivolts, which need to be amplified to a size suitable for pulse-height analysis (a few volts). In addition, the main amplifier has the function of shaping the pulses in such a way as to minimise the effect of noise. (See, for example, Fairstein (1975) for a discussion of this topic.) Pulse shaping can be understood most easily in terms of the frequency response of the amplifier. The frequency spectrum of the noise has two main components, attributable to: (a) the detector leakage current, which produces a noise spectrum showing an inverse dependence on frequency, and (b) the FET, with the opposite trend. The overall noise spectrum thus has a minimum (see Fig. 4.5). A pulse waveform can be represented as the sum of sine waves of different frequency (Fourier's theorem), and the object of pulse shaping is to optimise the signal-to-noise ratio by matching the frequency spectrum of the pulse to the minimum in the noise spectrum. It is desirable that the sides of the pulse should have the lowest possible slope in order to minimise the high-frequency content. The low end of the frequency spectrum is governed by the pulse width, for which a typical optimum value is about 100 μs. The ideal shape of the pulse waveform is a cusp but other, more easily realised, shapes are used in practice as described below.

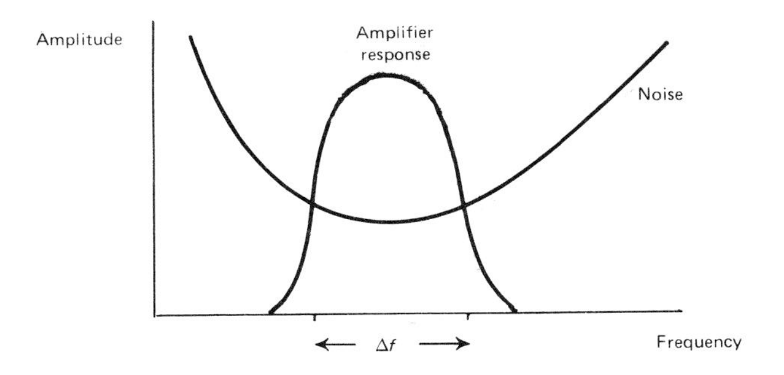

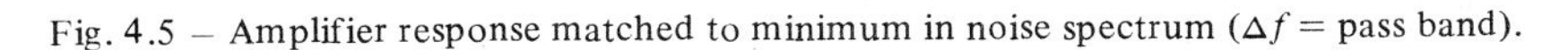
Fig. 4.5 – Amplifier response matched to minimum in noise spectrum (Δf = pass band).

Pulse shaping can be carried out by means of differentiating and integrating circuits employing resistors and capacitors (Fig. 4.6). These circuits attenuate low and high frequencies respectively and hence the required bandpass characteristics may be obtained by combining them. It can be shown that one differentiation and an infinite number of integrations produces a pulse shape corresponding to the Gaussian function, which is a reasonable approximation to the ideal cusp. With a finite number of integrations (for example, four) there is some asymmetry, but this has only a marginal effect. The resulting pulse shape is described as 'semi-Gaussian' (Fig. 4.7), a shape commonly used in Si(Li) detector systems.

The time-constant (τ_0) of an RC circuit is given by the product $R \times C$, and generally the same value is used for both differentiation and integration. With a semi-Gaussian pulse shape the total width of the pulse is about $5\tau_0$ and the rise-time (time from zero to maximum voltage, also known as the 'peaking time') is

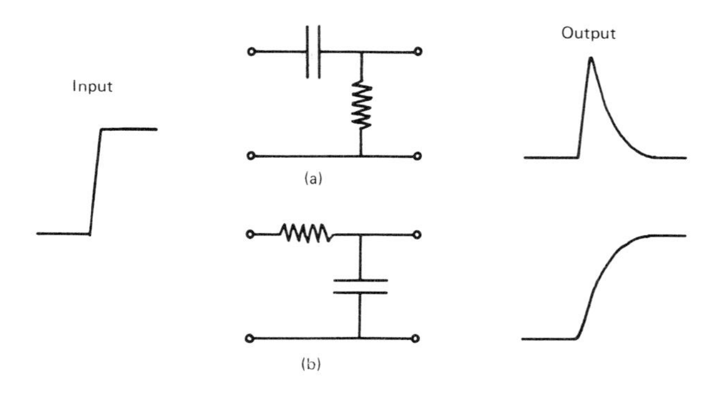

Fig. 4.6 – Pulse shaping circuits: (a) RC differentiation, (b) RC integration.

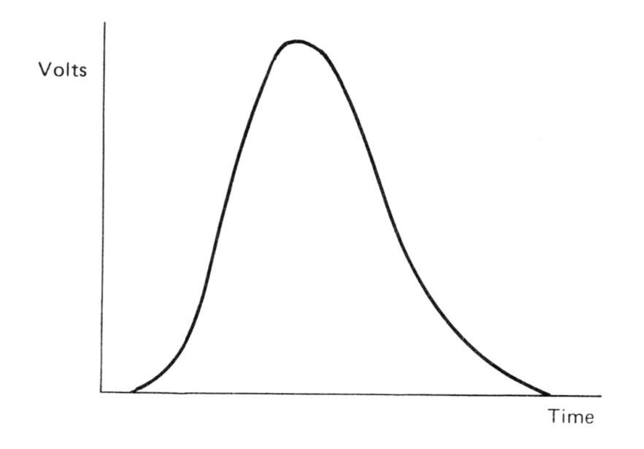

Fig. 4.7 – Semi-Gaussian pulse shape.

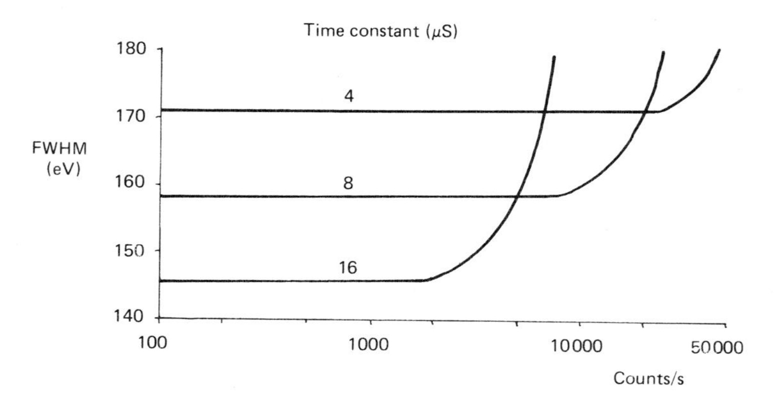

Fig. 4.8 – Energy resolution as a function of count rate for different amplifier time constants.

approximately $2\tau_0$. The best energy resolution is obtained with a pulse width of, say, 100 μs (that is, $\tau_0 = 20\,\mu$s), but only at low count rates (for example, less than 1000 counts s^{-1}), since at higher rates the probability of pulses overlapping is significant. It is therefore usual to compromise between count-rate capability and resolution by using time-constants somewhat shorter than the optimum (Fig. 4.8).

4.4.1 Pole-zero cancellation

The output pulses from an RC differentiation circuit may exhibit an undesirable 'undershoot', which is caused by the existence of additional differentiation time constants in preceding stages, due to coupling capacitors for example (Fig. 4.9(a)). This can be corrected by means of 'pole-zero cancellation' (Nowlin and Blankenship, 1965), whereby a fraction of the input voltage is fed to the output with the appropriate time-constant to cancel out the undershoot (Fig. 4.9(b)). Incorrect adjustment of the preset resistor in the pole-zero circuit (Fig. 4.9(c)) is indicated by asymmetrical peaks in the recorded spectrum.

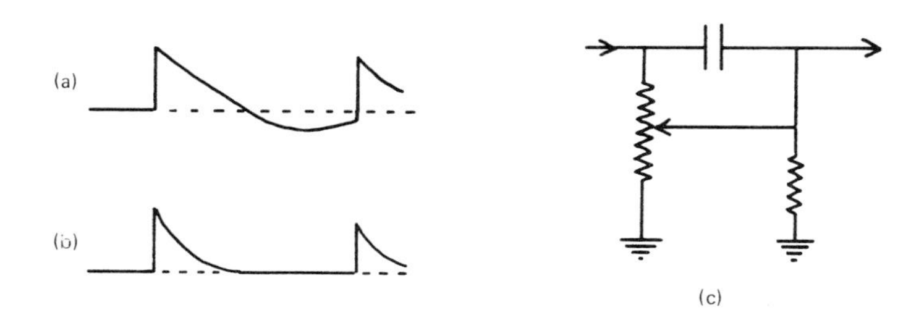

Fig. 4.9 – Pole-zero cancellation: (a) 'undershoot' produced by differentiation; (b) undershoot corrected by pole-zero cancellation circuit (c).

4.4.2 Baseline restoration

Pole-zero cancellation alone is not a complete solution: long secondary time constants in the system cause baseline shift and displacement of peak positions at high count rates. This is undesirable, especially in quantitative analysis and various methods of 'baseline restoration' have been proposed (for example, see Karlovac, 1975). An 'active' baseline restorer senses negative excursions of the output voltage and applies a corrective current to the coupling capacitor. However, this involves sampling the baseline voltage for a certain integrating time and, if a rapid response is required, this time must be short which makes noise fluctuations significant. Hence, the action of the baseline restorer degrades the energy resolution somewhat. As in the case of the choice of amplifier time constants, resolution is therefore balanced against high count-rate capability.

4.5 ENERGY RESOLUTION

The resolution of an ED system is measured in terms of the width (in electron volts) of the pulse-height distribution obtained when monoenergetic x-rays are incident on the detector. In the case of the gas proportional counter, the resolution is governed by the statistical fluctuation in the number of ion pairs produced (section 3.4). The same expression can be adapted to give the standard deviation (σ) of the number of electron-hole pairs generated in an Si(Li) detector by x-rays of energy E. Thus, $\sigma = \sqrt{FE/\epsilon}$, where ϵ is the mean energy used to produce one electron-hole pair and F is the 'Fano factor', and the FWHM (full width at half maximum) due to ionisation statistics is given by

$$\Delta E_s = 2.355\sqrt{\epsilon FE} \;. \tag{4.1}$$

The origin of the Fano factor can be explained qualitatively as follows: if the production of electrons and holes were infrequent compared with other energy-loss processes, they would be generated completely randomly and the fluctuations in their numbers would be given by the square root of the number (that is, $F = 1$). In the absence of such alternative processes, the number of electrons and holes produced by x-rays of a given energy would be constant ($F = 0$). The actual situation lies between these extremes, about 30% of the energy deposited by incident x-rays involving the production of electrons and holes; a typical value of F for an Si(Li) detector is about 0.12.

Random fluctuations of the baseline due to noise originating in the detector and FET also affect energy resolution. This contribution to the pulse-height distribution can be expressed as ΔE_n (in eV), which is added to ΔE_s in quadrature, so that the total FWHM (ΔE) is given by:

$$\Delta E^2 = \Delta E_n^2 + \Delta E_s^2 \;. \tag{4.2}$$

Substituting $F = 0.12$ and $\epsilon = 3.8$ eV in equation (4.1) we have, from equation (4.2):

$$\Delta E^2 = \Delta E_n^2 + 2.53E \;. \tag{4.3}$$

Fig. 4.10 shows the relationship between ΔE and E calculated for a typical system ($\Delta E_n = 80$ eV). Resolution is usually specified for $E = 5.9$ keV, this being the energy of the Mn Kα peak, which is conveniently available from radioactive ^{55}Fe. (The ^{55}Fe atom decays by 'K capture', that is, a K electron is captured by the nucleus, decreasing the atomic number by one and transforming the atom to manganese, at the same time leaving a K shell vacancy, which gives rise to the emission of K x-rays in the usual way.) Differences in resolution are mainly attributable to variations in ΔE_n, which have most effect at low energies as can be seen by inspection of equation (4.3). There is little apparent prospect of reduction in the ionisation statistics term, unless a superior detection medium to silicon is found. On the other hand improved FET performance is certainly conceivable and will mostly benefit the resolution at low energies.

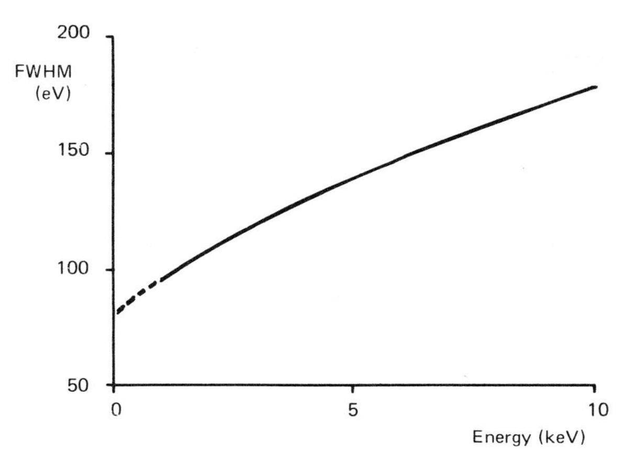

Fig. 4.10 – Energy resolution as a function of x-ray energy for a typical ED system.

4.6 DETECTOR RESPONSE FUNCTION

As discussed above, an Si(Li) detector upon which monoenergetic x-rays are incident will ideally produce a pulse-height distribution conforming to the Gaussian function:

$$y = a \exp\left[-(E-E_1)^2/2\sigma^2\right] ,$$

where y represents the number of counts, E the energy, E_1 the position of the peak on the energy axis and σ the standard deviation (see previous section). A correctly functioning system does indeed produce peaks which conform closely to this shape (Fig. 4.11). A useful test is to measure the width at one-tenth maximum height (FWTM), which theoretically should be 1.823 times the FWHM (anything less than 1.85 is acceptable). Closer inspection reveals significant non-Gaussian features in the detector response function, which can be

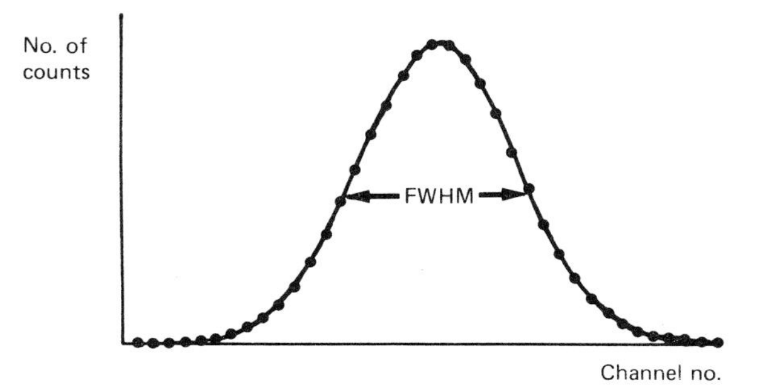

Fig. 4.11 – Gaussian pulse-height distribution for monoenergetic x-rays.

observed most readily with a radioactive ^{55}Fe source, since the manganese x-ray spectrum emitted is free of background. It is advantageous to use the logarithmic mode of display, which has the effect of expanding the vertical scale for low intensities. This reveals features on the low-energy side of the main peak caused by the reduction in pulse height which occurs when not all of the electrons and holes reach the electrodes ('incomplete charge collection'), as shown in Fig. 4.12. The 'shelf' immediately below the main peak is attributable to x-rays absorbed close to the front surface of the detector where there is a greater probability of charge carriers being trapped. The size and shape of this shelf depend on x-ray energy; it is especially prominent for x-rays just above the silicon absorption edge (1.84 keV). Below the shelf there is a 'continuum' arising from incomplete charge collection elsewhere in the detector. The shape and magnitude of these non-Gaussian features in the detector response function differ significantly from one detector to another; for example, the size of the shelf is related to the thickness of the silicon dead-layer. As an indication of detector quality, the ratio of the Mn Kα peak count to the background at 1 keV should be not less than 1500.

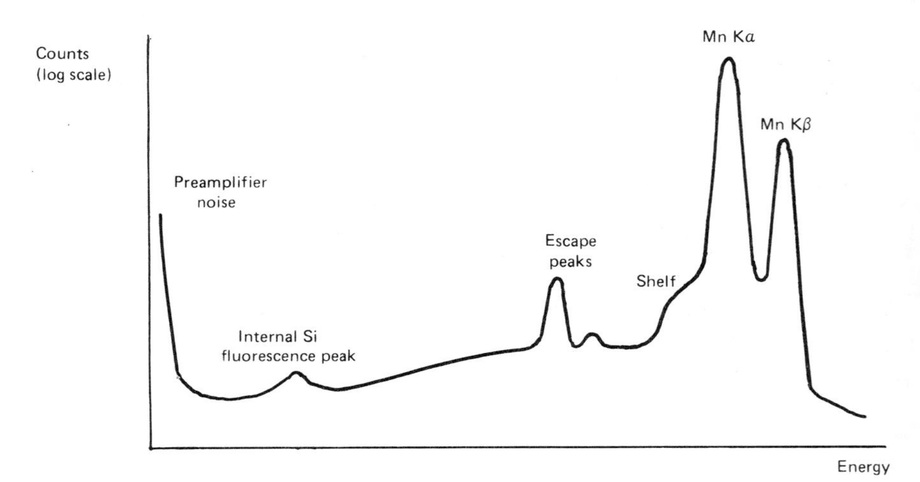

Fig. 4.12 – Detector response for Mn *K* x-rays from ^{55}Fe source.

4.6.1 Escape peaks

Another feature of the detector-response function is the escape peak, which appears at an energy of 1.74 keV below the parent peak (see Fig. 4.12). This occurs because there is a finite probability that an Si Kα photon produced in the detector following absorption of an incident x-ray photon will escape from the detector. In this event the energy of the escaped photon (1.74 keV) is not deposited in the detector and the height of the pulse is reduced by the corresponding amount. (An Si Kβ escape peak also exists, but for practical purposes is insignificant.) The size of the escape peak is a function of the incident x-ray energy (Fig. 4.13) and does not vary appreciably between detectors.

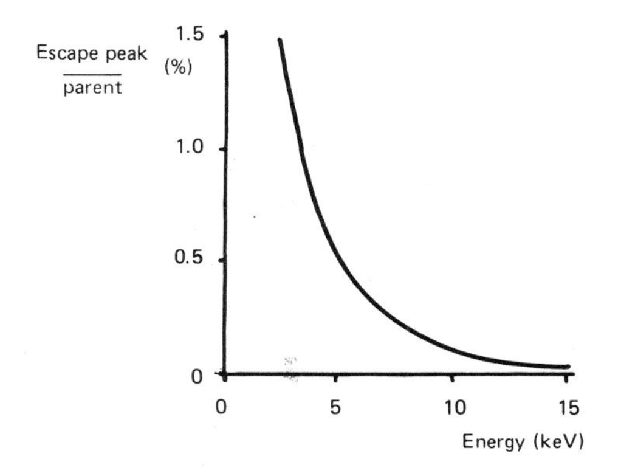

Fig. 4.13 – Relative size of escape peak as a function of x-ray energy.

4.6.2 Internal silicon fluorescence

Another detector artefact which can often be observed is a small spurious Si Kα peak (see Fig. 4.12) produced by the inverse of the escape peak mechanism, whereby the photo-electron is lost and the Si Kα photon is absorbed in the detector (Reed and Ware, 1972). The 'dead-layer' of silicon adjacent to the front surface of the detector plays an important part in the production of this 'internal fluorescence' peak, since within this layer electrons have a high probability of being trapped. The size of the silicon fluorescence peak is therefore related to the proportion of the incident x-ray photons absorbed in this region, and this is greatest for energies just above the absorption edge at 1.84 keV. It is also dependent on the thickness of the dead-layer, which varies significantly from one detector to another.

4.7 DETECTOR EFFICIENCY

Knowledge of detector efficiency as a function of energy is required, for example, in methods of background subtraction which depend upon calculating the shape of the x-ray continuum (section 5.3), and in 'no-standards' analysis where calculated factors are used to convert peak intensities into concentrations (sections 11.5 and 13.2.6). Above 3 keV the efficiency is close to 100%, as discussed below, but at lower energies becomes critically dependent on absorption in the beryllium window, etc.

One factor affecting detection efficiency is incomplete charge collection (section 4.6), insofar as a pulse appearing in the 'wrong' place in the spectrum is effectively lost. The loss can be estimated by integrating the low-energy 'background' in the ^{55}Fe spectrum and comparing it with the Mn Kα peak intensity: for a typical detector the loss is of the order of a few per cent. The dependence of this loss on energy is somewhat difficult to establish owing to the lack of suitable radioactive sources of other energies (electron-excited spectra having an excessive background for this purpose). Incomplete charge collection is most significant near the edges of the detector and is reduced by limiting the working area with an aperture. Keith and Loomis (1976), using x-rays of widely varying energy from a crystal spectrometer, measured detection efficiency with a small aperture in front of the detector and found the count loss due to incomplete charge collection to be well under 1% for energies from 2.6 keV upwards.

Counts are also lost from peaks by virtue of the escape peak mechanism (section 4.6.1) but, since the maximum loss is only about 1% (Reed and Ware, 1972) it is of little importance in the present context. At high energies (above 20 keV), other factors come into play: for example, incomplete absorption in the active volume of the detector and also the Compton effect (section 2.4). Furthermore, where there is an aperture in front of the detector, this may be partially transparent to x-rays of high energy, thereby changing the effective area of the detector.

Below 3 keV the detection efficiency falls owing to absorption in the beryllium window, evaporated gold layer and silicon dead-layer, and it is here that variation in detection efficiency is of the greatest interest. Calibrated radioactive x-ray sources have been used successfully for detection efficiency measurements at higher energies, but few suitable sources are available in the low-energy region and they suffer from self-absorption effects, correction for which is difficult.

An alternative approach is to measure the intensities of two or more suitable lines in an x-ray spectrum excited either by electron bombardment or x-ray fluorescence, and to deduce the window thickness from the observed intensity ratio. One possibility is to measure the Lα/Kα ratio of an element for which the Lα peak occurs at around 1 keV (for example, Co, Ni, Cu or Zn) where window absorption is significant, using an electron microprobe or SEM. This ratio can be utilised, for example, to monitor changes in window transmission due to contamination (Smith, 1981). The relative generation efficiency of Lα and Kα x-rays is dependent on accelerating voltage which, together with x-ray take-off angle, also affects the differential absorption of the two peaks in the sample itself. Hence, comparisons between detectors are valid only for constant accelerating voltage and specimen/detector geometry. Also, it is difficult to determine the absolute window thickness from Lα/Kα ratio measurements.

Some success has been achieved with fluorescence sources in which suitable elements are excited by ^{55}Fe. For example, Maur and Rosner (1978) measured the Kα peaks of chlorine and potassium in KCl; the generated Cl/K ratio was calculated from first principles and the relative window-absorption factor derived by comparison with the observed ratio. However, the accuracy of such a procedure is limited and it is not possible to obtain independent values for the beryllium, gold and silicon thicknesses. In principle, measurements at energies on each side of the silicon absorption edge might enable the thickness of the silicon dead-layer to be estimated, but this is difficult in practice.

Walter *et al.* (1981) have proposed a standard test procedure using an annular ^{55}Fe source to excite a glass containing various elements with peaks from 1.0 to 6.4 keV. From the observed intensity of each peak in the ED spectrum relative to that of backscattered Mn Kα, a 'window attenuation index' is derived. This provides a standardised basis for comparing detectors, where knowledge of absolute window thicknesses is not important.

4.8 PULSE PILE-UP

If a second pulse reaches the main amplifier during the rise time of the preceding one, the two effectively combine to produce a single pulse of enhanced amplitude (Fig. 4.14) and neither appears in its proper place in the spectrum. The probability of this occurrence is $1 - \exp(-n\tau_1)$, or approximately $n\tau_1$, where n is the count rate and τ_1 the rise time. For a semi-Gaussian pulse, $\tau_1 \sim 2\tau_0$, where τ_0 is the shaping time-constant (section 4.4). Thus, for example, if $\tau_1 = 10\,\mu s$ ($\tau_0 = 5\,\mu s$), the probability of pile-up is 5% at a count-rate of 5000 counts s^{-1}.

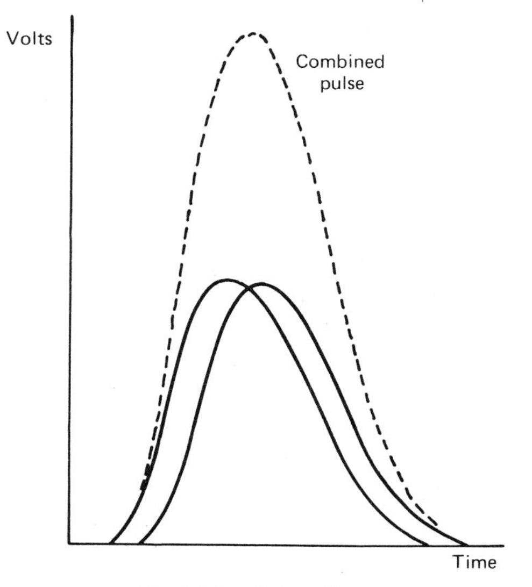

Fig. 4.14 – Pulse pile-up.

One effect of pile-up is the appearance of spurious peaks in the spectrum corresponding to the sum of the energies of the major peaks present (Fig. 4.15). These 'sum peaks' result from pulses coinciding almost exactly, so that the combined pulse height is equal to the sum of the heights of the superimposed pulses. When there is a significant displacement between the pulses on the time axis, their combined height is reduced by an amount which depends on the time interval between them. This gives rise to a 'shelf' on the low-energy side of each sum peak (Fig. 4.15). When both of the pulses involved belong to the x-ray continuum, the result is a pulse added to the background at a higher energy.

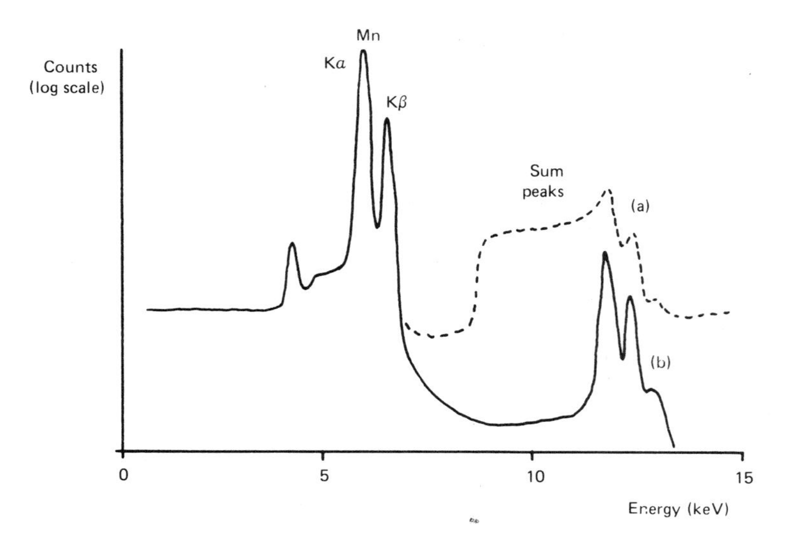

Fig. 4.15 – Sum peaks (a) without and (b) with pile-up rejection; ^{55}Fe spectrum.

Pile-up effects can be eliminated to a large extent by electronic pile-up rejection (Williams, 1968). For this purpose a second pulse amplifier is used; this has short time constants and hence a much faster response than the main amplifier, and incorporates a discriminator, the threshold of which is set just above the noise level. A 'pile-up recognition' circuit detects when one pulse follows another within a predetermined time interval, in which case a 'reject' signal is generated and the system is made inactive so that such pulses are not recorded. A pulse-pair resolution of 100 ns can be obtained, and in the example cited above, the probability of pile-up is reduced from 5% to 0.05%. Small residual sum peaks remaining in the spectrum (Fig. 4.15) result from pulses which are almost exactly coincident and thus escape rejection.

As explained in section 4.4, there is a relationship between pulse-shaping time constants and signal-to-noise ratio: thus, time constants shorter than those used in the main amplifier will result in an increased noise amplitude relative to the height of the pulses. The noise level determines the discriminator threshold setting in the pile-up rejection system and, if very short time constants are used in order to maximise the pulse-pair resolution, pulses below a certain energy (for example, 2.5 keV) will not be 'seen'. Thus, longer time constants are necessary if pile-up protection is required for low-energy x-rays (Reed, 1972). Consequently the best resolution obtainable at 1 keV is about 1 μs, and this entails a considerable sacrifice in pile-up rejection efficiency for high-energy x-rays. The best overall performance is obtained by combining two rejection channels with different time constants (Kandiah, 1975). For a more detailed discussion, see Statham (1977a).

4.9 DEAD TIME

In an ED system, dead time is associated with the main amplifier and is determined by the relatively large time constants required in order to optimise the signal-to-noise ratio (section 4.4). In this case the dead time is 'extendable' (see section 3.6) and the true count rate (n) is related to the recorded count rate (n') by

$$n = n'/\exp(-n\tau) \tag{4.4}$$

where τ is associated with the width of the main amplifier pulses (for semi-Gaussian pulses, $\tau \sim 5\tau_0$, where τ_0 is the pulse-shaping time-constant). Thus, for example, if $\tau = 50 \mu$s and $n = 5000$ counts s^{-1}, then $n' = 3900$ (that is, 22% of the counts are lost). It follows from equation (4.4) that n' passes through a maximum with increasing n (Fig. 4.16), n'_{max} being equal to $1/e\tau$ (or 7400

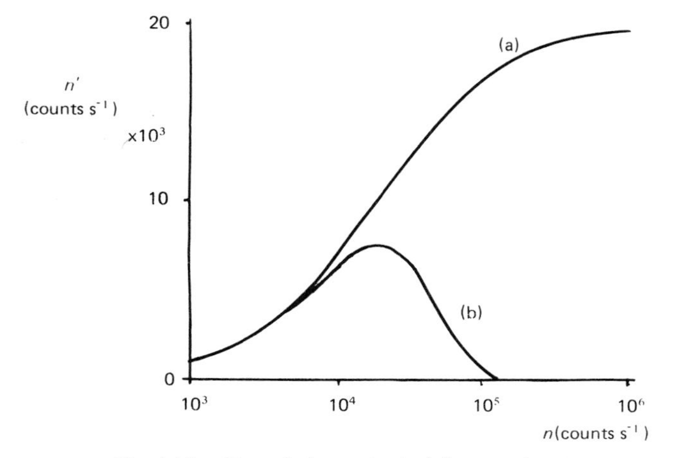

Fig. 4.16 – Recorded count rate (n') versus input count rate (n) for (a) non-extendable and (b) extendable dead time.

counts s^{-1} for $\tau = 50\mu s$). Note that n in equation (4.4) is the *total* spectrum count rate and the same dead-time correction factor applies equally to all the peaks in the spectrum (by contrast with dead time in WD analysis). In quantitative analysis, dead time affects the absolute concentrations obtained when comparing specimen and standard spectra if the total count rates differ (though relative concentrations are unaffected).

The usual method of correcting for dead time is to extend the counting time to compensate for the loss of counts (Covell *et al.*, 1960). The counting period is controlled by a 'clock' consisting of a pulse generator of precisely defined frequency, together with a scaler. When the scaler has accumulated the required number of clock pulses, spectrum acquisition is terminated. Correction for dead time is obtained by gating the clock pulses so that they are counted only when the system is 'live', that is, ready to receive a pulse. Thus, in order to obtain a 'live time' of 100s, the actual elapsed time might be, say, 125s, and will vary with the input count rate.

The 'live-time clock' as described above ensures that the total number of counts in the spectrum is corrected for dead-time losses. However, this is not the complete answer, since it takes no account of pulse pile-up which has the effect of transferring counts from the peaks to other parts of the spectrum, as described in section 4.8. Thus, in connection with peak intensity measurements, pile-up gives rise to additional dead time which is governed by the rise time of the pulses. It is sometimes incorrectly stated that pile-up rejection increases dead time; however, pulses suffering from pile-up are lost in any case as far as the peaks are concerned – they are merely prevented from appearing elsewhere in the spectrum when pile-up rejection is used. The additional loss of pulses due to pile-up effectively adds to the dead time of the system and can be corrected by modifying the live-time clock so that when a pile-up event is recognised, it remains switched off until the next pulse has been processed (Bartosek *et al.*, 1972). A useful test of the accuracy of the correction is to measure the intensity of a selected peak as a function of probe current: any departure from direct proportionality indicates failure of the correction procedure.

4.9.1 Beam switching

As noted above, the spectrum accumulation rate with a conventional ED system is subject to an upper limit imposed by the extendable dead time. However, higher recording rates are possible if the x-rays are switched off as soon as the arrival of a pulse is recognised. The dead time then becomes non-extendable and the maximum accumulation rate is theoretically $1/\tau$, that is, higher by a factor of 2.7. This technique was applied first to x-ray fluorescence analysis (Jaklevic *et al.*, 1972) but may also be used in the electron-probe microanalyser, where the electron beam can be switched off by a signal originating from a fast 'pulse recognition' amplifier (Statham *et al.*, 1974). Switching is achieved most easily by means of deflection plates in the column. An additional benefit is that pile-up is eliminated; indeed, beam switching is somewhat more efficient than conventional pile-up rejection, since it is only necessary to recognise single pulses rather than pairs (Statham, 1977a).

4.10 THE HARWELL PULSE PROCESSOR

So far, only 'conventional' pulse shaping has been considered, in which the time constants are fixed. There are, however, significant advantages in using time-variant time constants; for example, if the integrating time constant is switched to a small value after the pulse waveform has reached its maximum, the system is restored more rapidly to a quiescent state ready for the next pulse. In the Harwell pulse processor, a central control unit generates switching signals for this and other functions at appropriate times during the processing of each pulse (Kandiah *et al.*, 1975). The preamplifier output is divided between a 'recognition' and a 'measurement' channel. The former has time constants of about 1 μs and, when the arrival of a pulse is detected, a signal is sent to the control unit in order to initiate the processing cycle. The first stage of pulse shaping is an RC integrator, in which the resistance value is reduced by a large factor at the end of the pulse processing time (T) as described above. This is followed by a differentiating circuit which is switched to a long time constant on the arrival of a pulse, and back to a very short time constant at the end of the processing time. Finally, there is a gated integrator, which is switched on at the beginning and off at the end of the processing time. The resulting analogue pulse passes to the multichannel analyser via an output gate. At this point a 'restore' pulse is generated, and this resets the preamplifier output to zero by means of opto-electronic feedback (see section 4.3, but note that in this case restoration occurs after each pulse). A further 'protection time' is allowed before the next pulse is accepted, to permit transients to decay. The total 'busy time' per pulse is about $2.5T$. The pulse processing time is approximately equivalent to the rise time of conventionally shaped pulses, and is related to the energy resolution in a similar manner to a conventional system (section 4.4). However, the Harwell system can operate at higher count rates without significant loss of resolution. In situations where high-energy x-rays reach the detector (for example, in a TEM), a conventional system is subject to increased dead time and peak broadening owing to overloading of the amplifier, whereas with switched time constants such effects are minimised.

The Harwell processor incorporates a dual pile-up rejection system, providing optimum pile-up protection for all energies (see section 4.8). On recognising a pile-up event, processing is terminated immediately and a restore pulse is generated, followed by the usual protection time. This somewhat reduces the effective dead time compared with a conventional system.

A unique feature of the Harwell processor is the 'zero strobe', which produces a 'noise peak' at zero energy by carrying out the pulse-height analysis operation periodically when no x-ray pulse is present. The FWHM of this peak corresponds to the noise term (ΔE_n) in the energy resolution expression (equation (4.2)). Thus, performance can be monitored directly by observation of the noise peak. Also the position of the noise peak indicates the location of the true zero of the energy scale and can be used to generate an automatic correction for zero drift.

4.11 THE MULTICHANNEL PULSE-HEIGHT ANALYSER

In order to produce an ED spectrum, the output pulses from the main amplifier are processed by an analogue-to-digital converter (ADC). In the Wilkinson type of ADC, the input voltage charges a capacitor through a diode; when the voltage waveform of the pulse passes its peak, the diode stops conducting and the capacitor holds the peak voltage. It is then discharged from a constant current source and during this process clock pulses are counted by an 'address scaler'. The discharge time is proportional to the height of the pulse; hence the scaler gives the digital address of the appropriate channel. It is desirable to use an ADC in which the 'conversion time' is less then the duration of the main amplifier pulses and therefore does not add to the dead time of the system.

A complete multichannel analyser includes a memory for storing the spectrum and some means of display and read-out. However, since it is now usual to have a small computer interfaced with the ADC, it is convenient to accumulate the spectrum in the memory of the computer. Computer-based systems offer greater flexibility than earlier 'hard-wired' multichannel analysers; furthermore, the computer can be used for spectrum stripping and calculating ZAF corrections.

A typical multichannel analyser with 1024 channels and a calibration of 10 eV/channel, giving an energy coverage of 0–10 keV, is suitable for most purposes. There is no significant advantage in narrower channels in view of the finite width of the peaks. Exact calibration is obtained by fine adjustment of the main amplifier gain and zero controls using two peaks of known and preferably widely spaced energy.

The spectrum is displayed on a cathode ray tube as a histogram of the number of counts per channel versus channel number (energy) along the horizontal axis. The vertical and horizontal scales can be expanded in order to facilitate inspection of fine detail. Usually there is also an optional logarithmic display mode, which is advantageous for displaying large and small peaks at the same time. Information such as energy calibration, count rate, elapsed live time, etc., may be displayed on the screen along with the spectrum. A colour display has some advantages, for example in identifying different regions of interest in the spectrum.

Computer-based systems can provide useful facilities for peak identification (section 5.1); for example, if a marker is aligned with an unknown peak, the computer can determine its possible identity by reference to a stored list of peak energies. Often there is more than one possible candidate, in which case it is helpful to be able to superimpose, on the spectrum, markers corresponding to the other peaks of the element of interest. It is also useful to be able to compare two spectra by displaying both simultaneously with a suitable vertical offset. 'Unknown' spectra can then be compared with stored spectra from known samples for the purpose of identification.

5

Processing energy-dispersive spectra

S. J. B. REED

Although the energy-dispersive (ED) spectrometer performs essentially the same task as the wavelength-dispersive (WD) type, differences in their characteristics, mode of use and form of output, are such as to require quite different techniques of operation and data handling, especially with regard to quantitative analysis.

The most obvious difference is that the peaks in the ED spectrum are much wider, making background corrections both more important on account of the lower peak-to-background ratio (due to the broader band of continuum recorded) and at the same time more difficult because of the need to extrapolate the

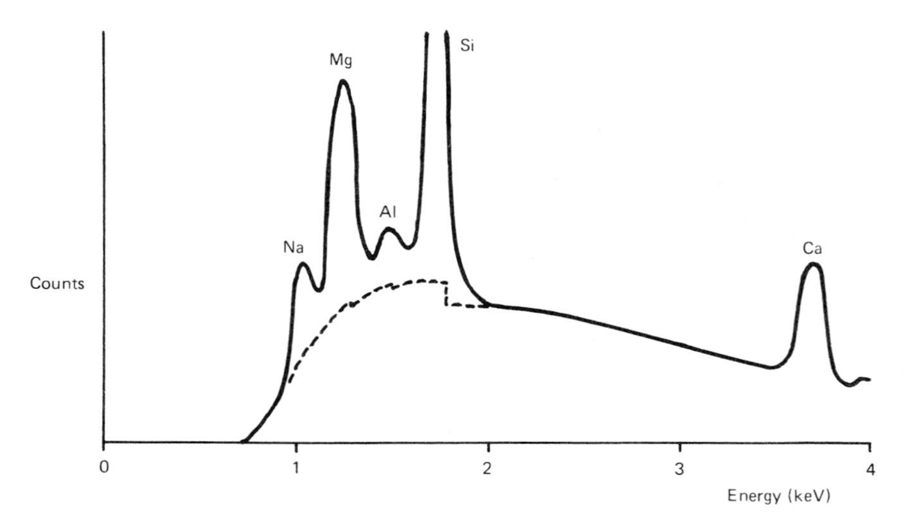

Fig. 5.1 – Typical ED spectrum showing partially overlapping peaks superimposed on continuum background.

background intensity over larger energy intervals (see Fig. 5.1). Furthermore, the probability of significant interference between peaks due to overlap is much greater. Consequently, most of the effort in developing methods of quantitative ED analysis has been devoted to finding satisfactory solutions to the problems of background and overlap corrections.

Having extracted the true peak intensities from the ED spectra of specimen and standard, concentrations can be derived using ZAF corrections as in 'classical' quantitative analysis. However, since ED spectrometer efficiency as a function of x-ray energy is both more predictable and more reproducible than is the case with WD spectrometers, standard data can be stored and need be updated only at such intervals as long-term changes in detector behaviour may require. In this case operating parameters such as accelerating voltage and specimen tilt angle must be kept rigorously constant. By avoiding the necessity for recording standard spectra with each batch of analyses, a considerable saving of time can be effected. This approach is taken a stage further in 'no-standards' analysis, in which standard intensities are predicted rather than actually measured (sections 11.5 and 13.2.6).

5.1 PEAK IDENTIFICATION

The energy of a peak may be determined by identifying the channel containing the maximum number of counts and converting channel number to energy. The peak can then be identified by reference to tables which, in the case of a computer-based system, may be stored in the computer memory for convenience. Aids are usually provided, such as a movable cursor on the display which can be aligned with an unknown peak, whereupon the computer will identify the peak by reference to stored tables of energies. Since there is an uncertainty of at least one channel (for example, 10 eV) in the peak position, such an identification may not be unambiguous, in which case the computer will suggest several possible alternative lines. The K spectrum is simple and poses few difficulties with regard to identification, apart from possible ambiguity between the Kα line of one element and the Kβ line of another (for example, V Kα – 4.949 keV, Ti Kβ – 4.931 keV). Greater complications arise when L or M lines are involved. In order to decide whether a given peak represents, for example, the Kα line of one element or an L line of another, the computer may be employed to display the complete K or L spectra in question with the correct relative peak heights: it is then easy to see if the peaks in the recorded spectrum coincide with one or other of the computer-generated spectra.

A useful facility for identifying phases is the ability to compare the spectrum just recorded with one recalled from the computer memory, where a library of relevant spectra may be stored. If two such spectra can be displayed simultaneously with a small vertical offset, similarities and differences can be appraised very easily.

5.2 SPECTRAL ARTEFACTS

As discussed in section 4.6, the Si(Li) detector exhibits a 'non-ideal' response to monoenergetic x-rays in certain respects. Thus the Gaussian peak shape is modified by incomplete charge collection effects and procedures for quantitative ED analysis must take this into account. Other mechanisms result in discrete spurious peaks which can give rise to significant errors and which will be considered now.

5.2.1 Escape peaks

'True' peaks in the spectrum are accompanied by 'escape peaks'. The subsidiary peak occurs 1.74 keV below the 'true' peak. Its relative size is dependent only on the energy of the latter and the ratio is the same for all Si(Li) detectors (see section 4.6.1). Although escape peaks are small (up to about 1% of the parent peak height), it is desirable that they are removed from the spectrum for quantitative analysis. This can be accomplished by subtracting an appropriate fraction of the content of each channel from the channel 1.74 keV lower in energy. The escape peak intensity ratio as a function of energy was calculated for Ge(Li) detectors by Fioratti and Piermattei (1971) and the resulting expression is also applicable to the Si(Li) detector with suitably modified values for certain physical parameters (Reed and Ware, 1972). However, since mass absorption coefficients are involved, calculation for every channel would be laborious. This led Reed and Ware (1973) to devise a simple polynomial expression for the purpose of 'stripping' escape peaks. A somewhat more widely applicable expression for the ratio (K) of escape peak to parent peak has been proposed by Statham (1976a):

$$K = 0.0202/\{1 + (mE + b)E^2\}$$

where $m = 0.01517 \cos\theta - 0.000803$ and $b = 0.0455 \cos\theta + 0.01238$; in this equation θ is the angle between the normal to the detector surface and the incident x-ray beam (thereby allowing for the non-normal incidence which is sometimes encountered, especially in scanning and transmission electron microscopes).

5.2.2 Internal silicon fluorescence peak

The mechanism by which a small spurious Si K peak appears in the spectrum is described in section 4.6.2. Given a knowledge of the thickness of the detector dead layer, it is possible in principle to calculate the size of the silicon fluorescence peak, but this entails an integration with respect to energy over the whole spectrum above 1.84 keV (Statham, 1976a) and is not very practicable as a routine procedure. In the absence of any correction, small spurious concentrations of silicon (usually less than 0.5%) may be recorded in silicon-free specimens. The size of the peak is dependent in a rather complex manner on the content of the spectrum concerned, and furthermore varies significantly from one detector to another.

5.2.3 Sum peaks

Pulse pile-up (section 4.8), which has the effect of transferring pulses to higher energies in the spectrum, can largely be eliminated by means of electronic rejection systems. However, owing to the limited pulse-pair resolution of such systems, small residual 'sum peaks' remain. Statham (1977a) has discussed in some detail the possibility of calculating the size of sum peaks for the purpose of removing them by a 'stripping' procedure. He concluded that, although possible in principle (subject to certain difficulties with regard to defining the pulse-pair resolution, especially for pulses of different size), it is scarcely a realistic proposition for an on-line mini-computer. The practical answer is therefore to limit the spectrum accumulation rate to below the point where sum peaks cause significant errors (typically 5000 counts s^{-1}). In this regard it is advisable to carry out practical tests for a particular system.

5.3 BACKGROUND SUBTRACTION METHODS

In WD analysis the usual way of correcting for background is to offset the crystal spectrometer by a small angle on each side of the peak. The mean of the intensities obtained is generally a good enough approximation to the true background under the peak. This approach is applicable to ED analysis only in the case of isolated peaks and is totally inadequate in the low-energy region, where the background is markedly non-linear and peaks are close together (Fig. 5.1).

Two types of background correction have been evolved specifically for ED analysis. The first involves setting up a model for the shape of the background, taking into account the generated continuum intensity, absorption of the continuum in the specimen, and the spectrometer efficiency. Various uncertainties exist in such models but fortunately for practical purposes the continuum intensity need be calculated to an accuracy of only about ±5%, which corresponds to an absolute concentration error of less than 0.05%. Obviously any contribution to background from sources other than the continuum will adversely affect the results but normally the non-continuum component is small.

The alternative is to use a mathematical filtering technique to separate peaks from background (section 5.3.4). In this case knowledge of the shape of the background is not required and therefore the method is more widely applicable (for example, to irregular specimens for which continuum modelling is not feasible, and to thin foils where there may be a significant non-continuum background).

5.3.1 Continuum modelling

If the dependence of the background intensity on energy is known, the background under the peaks can be determined by extrapolation from measurements made in peak-free regions (Ware and Reed, 1973). To a fair approximation, the continuum may be represented by the Kramers expression (section 2.2.4):

$$N(E) = bZ\frac{E_0 - E}{E} \tag{5.1}$$

in which only the *shape* factor, $(E_0 - E)/E$, is relevant here. A further term is required to allow for absorption of continuum x-rays in the spectrum and a spectrometer efficiency function must also be included. Taking all these considerations into account, Ware and Reed proposed the following expression for background intensity:

$$I(E) = b\left\{\frac{E_0 - E}{E} f(\chi)\exp\left[-\sum \frac{\mu}{\rho_i}\rho_i x_i\right] + F(E)\right\}\mathrm{d}E\ . \tag{5.2}$$

The absorption factor $f(\chi)$ is analogous to that used in ZAF corrections as discussed in Chapter 8, although for the continuum the method of calculation requires modification (see section 5.3.3). The exponential term refers to absorption in the detector window, evaporated gold layer and silicon dead layer. Typical thicknesses (x_i) are: Be – 8 μm, Au – 20 nm, Si – 0.1 μm, but since there are significant variations between detectors it is desirable to use empirical values, which can be obtained by finding the best fit to observed background spectra in the low-energy region where absorption is significant. A perfect fit cannot be expected owing to the fact that the beryllium window and gold surface layer are not uniform in thickness (Statham, 1981) and the empirical term $F(E)$ was added by Ware and Reed to allow for this and any other residual errors. Above 3 keV it was found that a fixed value F_0 could be substituted, though this might not be true for different operating conditions; below 3 keV empirical values of $F(E)$ for energies of interest corresponding to characteristic peaks were obtained from suitable standards. For each analysis the scaling factor b in equation (5.2) and F_0 are determined from background measurements made in peak-free regions, for example usually around 3 and 10 keV (the positions of the Kα peaks of argon and germanium, which are assumed to be absent). Fig. 5.2 is an ED spectrum of silicate mineral (hornblende) with the continuum calculated according to equation (5.2) (Myklebust *et al.*, 1979).

Kramers based his continuum expression on a simplified theoretical approach and it was verified by comparison with spectra from x-ray tubes in which window absorption was significant. It is therefore not unexpected to find that it is somewhat deficient in the energy range of interest in ED analysis. In the Ware and Reed procedure described above, the empirical term $F(E)$ takes care of any divergence from the Kramers expression with regard to shape. An alternative approach is to modify the shape factor $(E_0 - E)/E$, which on both theoretical and experimental grounds is now known to underestimate the continuum at

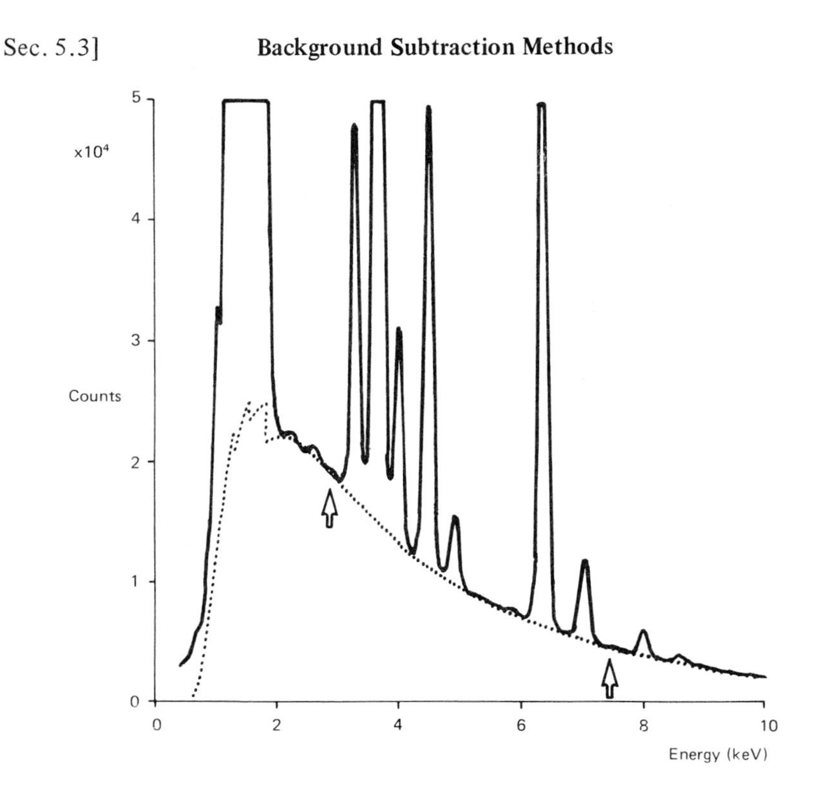

Fig. 5.2 – ED spectrum of silicate mineral (hornblende) with continuum calculated according to equation (5.2) (after Myklebust *et al.*, 1979). Arrows indicate fitting points.

low energies. The modified form $(E_0 - E)/E^{1+a}$ was proposed by Reed (1975a) on the basis of calculations using theoretical continuum production cross-sections, with a suggested value of 0.21 for 'a' (for $Z = 13$ to 29). Similar, more extensive calculations by Statham (1976a) gave 'a' values varying from 0.26 for $Z = 13$ to 0.15 for $Z = 79$, indicating some degree of dependence of the continuum shape on atomic number (by contrast with the original Kramers expression).

The required shape can be obtained by means of alternative expressions, for example that due to Lifshin (see Fiori *et al.*, 1976):

$$N(E) = b_1 \frac{(E_0 - E)}{E} + b_2 \frac{(E_0 - E)^2}{E} \tag{5.3}$$

which is used for background subtraction in the computer program 'FRAME C' developed at the US National Bureau of Standards (Myklebust *et al.*, 1979). Here the 'constants' b_1 and b_2 are determined empirically by finding the 'best fit' for each spectrum, and these may vary with atomic number.

5.3.2 Use of a reference background spectrum

There may be some advantage in using a *recorded* background spectrum as the basis for the correction, thereby automatically taking account of the detector efficiency function and obviating the need for explicit calculation of window absorption, etc. (Smith *et al.*, 1975). A suitable choice of 'background standard' is carbon (for example, diamond), which produces no visible peaks (carbon K radiation being totally absorbed in the beryllium window). The standard background spectrum requires scaling and correction for the difference in composition between 'unknown' and standard, before subtraction from the 'unknown' spectrum. This method has been applied in a computer program called 'EDATA 2' (Smith and Gold, 1979), in the latest version of which the following expression is used for the continuum intensity:

$$N(E) = b\bar{Z}^{n}\left(\frac{E_0 - E}{E}\right)^{x} \tag{5.4}$$

where $n = E(0.0739 - 0.0051\ln\bar{Z}) + p$, $x = a - 0.00145(\bar{Z}/E)$, and $\bar{Z}$ is the mean atomic number. Both p and a depend on E_0, and for $E_0 = 15$ keV the best fit is obtained by assuming $p = 1.6561 - 0.1150\ln\bar{Z}$ and $a = 1.76$.

As discussed previously, a correction is required for self-absorption of the continuum in the specimen. In EDATA 2 (unlike other procedures) an explicit correction is also applied for electron backscattering, which causes a loss of continuum intensity that varies as a function of the x-ray energy and therefore affects the shape. After applying these corrections to the carbon spectrum, it is divided by that part of equation (5.4) which affects the continuum *shape* only, giving a background shape normalised to $\bar{Z} = 1$. In use, the stored normalised background is multiplied by the same shape function calculated for the appropriate $\bar{Z}$ and corrected for absorption and backscattering, in order to derive the continuum shape for the 'unknown'. A scaling factor is then obtained by minimising the difference between experimental and calculated backgrounds in peak-free regions.

It may be of interest to note that a similar approach can also be applied to WD analysis and is useful when the number of peaks is large and it is difficult to find places to measure background (Smith and Reed, 1981).

5.3.3 The problem of continuum absorption

In equation (5.2), $f(\chi)$ is the factor by which the continuum intensity is reduced in emerging from the specimen. However, expressions used for characteristic radiation in ZAF correction procedures, such as that of Philibert (equation (8.9)), are not strictly applicable to the continuum without modification, since continuum production is concentrated more towards the ends of the electron trajectories and therefore occurs at a greater mean depth (Fig. 5.3). This can be taken into account by modifying σ in equation (8.9). (Note that in calculating σ

by methods given in section 8.2.2, the continuum energy E should be substituted for E_c.) Reed (1975a) estimated $f(\chi)$ for the continuum by means of Monte Carlo calculations and proposed that the difference between continuum and characteristic radiation should be taken into account by changing the constant in the Heinrich equation for σ from 4.5×10^5 to 4.0×10^5. The effect on $f(\chi)$, however, is not very great; for example, for $f(\chi) = 0.7$, the change is only 5%, which is roughly equivalent to 0.05% in terms of absolute elemental concentration. (The use of empirical fitting procedures tends to mask the effect in any case.)

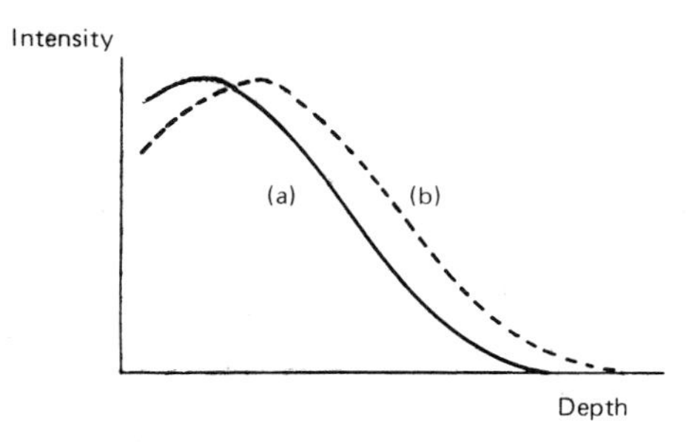

Fig. 5.3 – Depth distribution of (a) characteristic and (b) continuum x-ray production (after Reed, 1975b).

Since the absorption factor is composition-dependent, the background correction has to be calculated iteratively when the continuum modelling method is used, and can be included in the ZAF iteration. The rate of convergence is not significantly affected since the background correction has only a small effect on the major peaks.

5.3.4 Mathematical filtering

An alternative to continuum modelling for background corrections in ED analysis is to use mathematical filtering techniques which discriminate between the peaks and background by virtue of their different frequency content, where the spectrum is considered as a waveform with the 'time' axis equivalent to energy. In terms of Fourier analysis, the background contains predominantly low frequencies, by contrast with the relatively high frequencies represented by the peaks in which the waveform exhibits a rapid rate of change. Therefore background can be suppressed by means of a frequency filter with the appropriate pass band (Russ, 1972). This procedure is reasonably effective, though peaks and background are not perfectly separated in the frequency domain. However, Fourier analysis in the strict sense is not the best approach in practice and the digital filtering procedure described below is preferable.

Digital filtering was developed originally for detecting peaks in γ-ray spectra.

Various methods have been proposed but, as pointed out by Robertson *et al.*, (1972), the rectangular, symmetrical, 'top hat' filter function has the advantage of speed of calculation when using a mini-computer. The 'top hat' consists of a central group of channels assigned a positive weighting value, with a group of negatively weighted channels on each side, the sum of the weighting factors being zero (see Fig. 5.4). The weighted sum of the contents of the channels lying within the width of the filter function is calculated with the 'top hat' centred on each channel of the recorded spectrum in turn. This sum is zero (apart from statistical fluctuations) when the spectrum consists of a straight line, even if the line is sloping, whereas a peak produces a characteristic response consisting of a positive central lobe with smaller negative lobes on each side (Fig. 5.4). Curvature of the background somewhat complicates the issue and is a potential source of error in the 'filter-fit' procedure described in section 5.5.1. The filtering operation approximates to obtaining the second derivative (with the sign reversed), combined with smoothing (as a result of averaging the counts over a number of channels). For optimum results the width of the positive centre lobe should be about the same as the FWHM of the unfiltered Gaussian peak and each side lobe should be about half this width (Statham, 1977b). In the context of ED analysis the main purpose of filtering is to eliminate the background, as shown in Fig, 5.5, prior to peak fitting (see section 5.5.1).

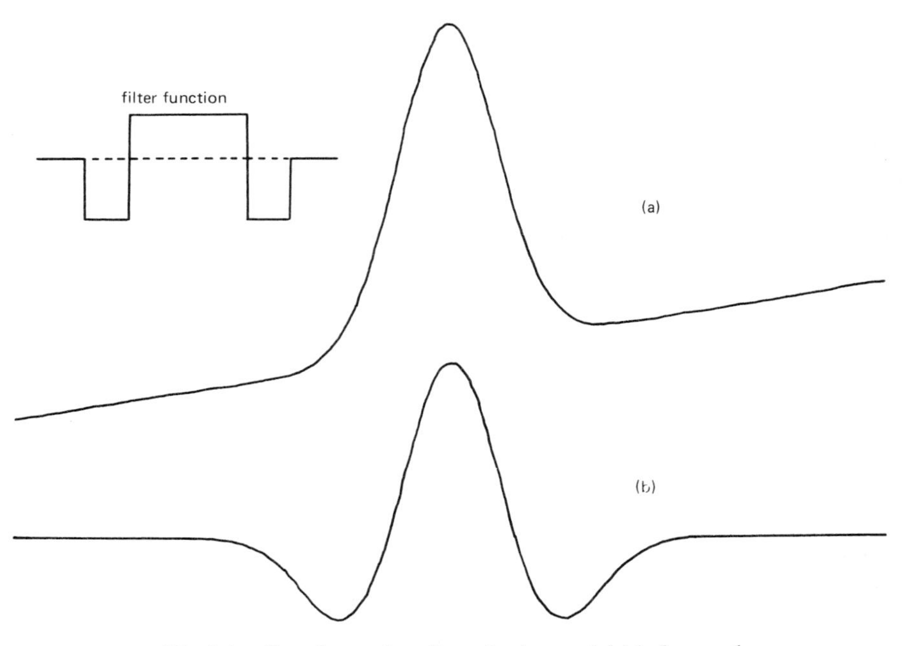

Fig. 5.4 – Gaussian peak on linear background (a) before, and (b) after digital filtering (after Schamber, 1977).

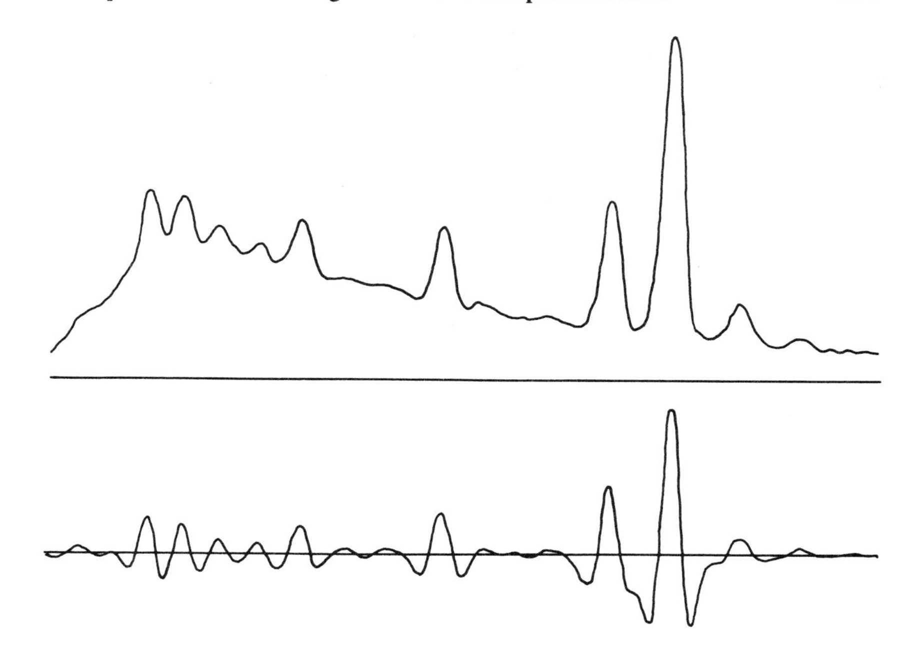

Fig. 5.5 – Complex spectrum (a) before, and (b) after digital filtering.

5.4 PEAK INTEGRATION AND OVERLAP CORRECTIONS

The most obvious measure of peak intensity is the number of counts in the centre channel but this gives poor statistical precision since the rest of the counts in the peak are wasted. It is thus much better to obtain the *integrated* peak intensity by summing the contents of a group of channels spanning the peak. For a Gaussian-shaped peak, 99% of the counts are contained within a band 2.2 times the FWHM in width. However, if such wide integration limits are used, there is a relatively high probability of overlap between neighbouring peaks; also the peak-to-background ratio is low owing to the large amount of continuum included. A reasonable compromise is to integrate over a region approximating to the FWHM, which still includes 74% of the total intensity of the peak.

This method of peak integration is particularly appropriate for use with continuum modelling procedures for background correction (section 5.3.1), it being necessary to calculate the continuum intensity only for the energies of the peaks. The correction required for interference between adjacent peaks can also be implemented fairly easily by means of 'overlap coefficients', which represent the fraction of the integrated peak intensity of one element which appears in the integrated region of its neighbour (see Fig. 5.6). The true peak intensities can be derived from the uncorrected intensities (after background subtraction) by solving a set of linear simultaneous equations, although in practice it is more convenient to use an iterative procedure. Overlap corrections are thus calculated initially using the uncorrected intensities; the process is then repeated using corrected intensities in order to obtain a better approximation for the corrections, and so on until convergence is achieved.

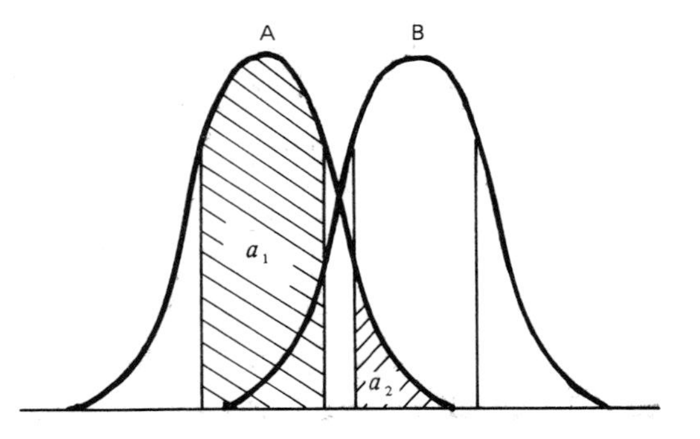

Fig. 5.6 – Peak overlap: overlap coefficient = a_2/a_1.

Overlap coefficients can be obtained experimentally with the aid of suitable standards (Reed and Ware, 1973). Alternatively, the need for such standards can be obviated by calculating the coefficients from an assumed peak profile function, as in the FRAME C program (Myklebust *et al.*,1979) in which a Gaussian function is used with a modification to allow for incomplete charge collection (section 4.6). Overlap coefficients are quite sensitive to the FWHM of the peaks (section 5.4.1) and a given set of coefficients is therefore specific to a particular system. Also, the count rate should be kept reasonably constant in order to avoid variations in FWHM, though the calculation method has the advantage that detector resolution parameters derived for each spectrum from the width of suitable major, interference-free peaks can be used to generate new overlap coefficients for each analysis.

The experimentally determined overlap coefficients in Table 5.1 show several interesting features. Overlap between adjacent Kα peaks is small, even at low atomic numbers where the peaks are closest (for example, the Mg → Al overlap coefficient is only 0.2%). However, the non-Gaussian low-energy tail has a noticeable effect, especially in the case of peaks just above the Si K absorption edge (see section 4.6); for example, the P → Si overlap coefficient is 4.1% and, furthermore, the influence of the tail extends over several elements.

The relatively severe overlap caused by Kβ peaks in some cases (for example, 13.7% for Ti → V) requires slightly different treatment from that described above, since the intensity of the interfering Kβ peak is not measured as part of the analysis routine. However, the Kβ intensity can be estimated from that

Table 5.1 – Empirical overlap coefficients % (after Smith and Gold, 1976).

Overlapping element (Z)	Overlapped element (Z)																		
	11	12	13	14	15	16	17	19	20	21	22	23	24	25	26	27	28	29	30
11	100	0.7																	
12	1.6	100	0.2																
13	0.5	1.3	100	0.1															
14	0.2	0.3	0.9	100	0.2														
15	1.1	1.6	2.2	4.1	100	0.7													
16	0.7	0.8	1.0	1.4	2.8	100	1.2												
17	0.5	0.4	0.5	0.6	0.9	2.0	100	0.1											
19	0.4	0.3	0.2	0.1	0.1	0.2	0.3	100	7.2										
20	0.4	0.3	0.1	0.1	0.1	0.2	0.3	0.6	100	10.0									
21	0.3	0.2	0.1	–	–	0.1	0.1	0.2	0.5	100	12.4								
22	0.3	0.2	–	–	–	–	0.1	0.1	0.2	0.4	100	13.7							
23	0.2	0.1	–	–	–	–	–	–	0.1	0.2	0.3	100	12.7						
24	0.2	0.1	–	–	–	–	–	–	–	0.1	0.2	0.3	100	11.1					
25	0.2	0.1	–	–	–	–	–	–	–	–	0.1	0.1	0.2	100	8.2				
26	1.5	0.2	–	–	–	–	–	–	–	–	–	0.1	0.1	0.2	100	5.0			
27	1.1	0.3	0.1	–	–	–	–	–	–	–	–	–	0.1	0.1	0.2	100	2.5		
28	1.2	0.9	0.4	0.1	–	–	–	–	–	–	–	–	–	0.1	0.1	0.1	100	1.3	
29	24.5	0.6	0.1	–	–	–	–	–	–	–	–	–	–	–	–	0.1	0.1	100	0.7
30	282.1	2.1	1.6	–	–	–	–	–	–	–	–	–	–	–	–	–	–	0.1	100

of the Kα peak of the same element, given a knowledge of the Kβ/Kα intensity ratio (see Table 5.2). In this case differential absorption must be taken into account.

Table 5.1 indicates a very strong Zn → Na overlap due to the Zn Lα peak. It is necessary to know the Zn Lα/Kα ratio in order to calculate the required correction and (unlike the Kβ/Kα ratio) this is a function of the incident electron energy due to the difference in the excitation energy of the K and L shells. A correction for differential absorption is again necessary.

Table 5.2 – The ratio of the intensities of Kβ and Kα lines (after Khan and Karimi, 1980).

Z	Kβ/Kα	Z	Kβ/Kα
19	0.120	28	0.136
20	0.127	29	0.137
21	0.131	30	0.139
22	0.131	31	0.146
23	0.132	32	0.152
24	0.133	33	0.156
25	0.134	34	0.161
26	0.134	35	0.166
27	0.135		

5.4.1 Errors due to peak shift and broadening

A necessary condition for the use of integrated peak measurements and overlap corrections in quantitative ED analysis is that the widths and positions of the peaks should remain constant, and it is of interest to estimate the errors that may arise if they do not. Broadening of a peak will obviously reduce the area lying between fixed integration limits. Assuming a Gaussian peak of 150 eV FWHM and integration limits set at the half-maximum points, the loss will amount to 1% for a broadening of 2.5 eV and is an approximately linear function of broadening. (For other peak widths, broadening by the same fraction of the FWHM has the same effect.) Care is evidently required to ensure that count-rate dependent broadening, due for example to a difference in total count rate between specimen and standard, is within appropriate limits.

Peak shift (due either to drift or to count-rate-dependent electronic effects) also has the effect of reducing the area between the integration limits: thus, for a 150 eV Gaussian peak, a shift of 12 eV causes a 1% loss, rising to 5% for a 25 eV shift (the variation being non-linear). Potentially there is also an effect on overlap corrections but the spacing of neighbouring Kα peaks is large, even in the low energy region, compared with typical peak widths (100–180 eV) and therefore the effect of drift is unlikely to be significant. A possible exception here is sodium, though for a different reason, namely that the background in the vicinity of the Na Kα peak is steeply sloping and the background correction is prone to error due to drift. A rough estimate for a typical slope suggests an error of 0.1% Na for a shift of 25 eV.

Much more significant errors due to peak shift can arise in cases of Kβ peaks overlapping Kα peaks. Although the overlap correction is at a maximum when the peaks are most nearly coincident (for example, V Kα and Ti Kβ which are only 18 eV apart), the greatest sensitivity to drift occurs when the peaks are separated by about half the FWHM. Thus in the case of Cr Kβ and Mn Kα (54 eV apart), a shift of 10 eV changes the overlap by an amount corresponding to a manganese concentration of 1%. Hence, if the correction does not take account of drift, large relative errors in small manganese concentrations can easily occur. (Cases involving L spectra can be even worse owing to the higher relative intensity of the Lβ lines.) Long-term drift in the system should be corrected by readjustment of the gain and zero controls at suitable intervals; alternatively peak positions and overlap factors can be redetermined by the computer on the basis of a standard spectrum containing suitable lines. Count-rate-dependent effects are more difficult to deal with and should be minimised by suitable choice of parameters such as amplifier time constants (section 4.4) and by limiting the maximum count rate.

5.5 LEAST SQUARES FITTING

For a Gaussian peak profile, the number of counts n_i in channel number i is given by:

$$n_i = A \exp[-(E_0 - E_i)^2/2\sigma^2]$$

where E_0 is the energy of the centre of the peak, E_i the energy of channel i, σ the standard deviation of the Gaussian, and A the peak amplitude.

A complex spectrum consisting of several peaks may be represented by the sum of a corresponding number of such expressions;

$$n_i = \sum_j A_j \exp[-(E_{0j} - E_i)^2/2\sigma_j^2] \tag{5.5}$$

An important consideration is that only the amplitudes A_j are *linearly* related to n_i (the number of counts in channel i): it follows that fitting procedures involving linear equations are limited to determining A_j for predetermined values of the peak positions (E_{0j}) and widths (σ_j). In principle it is desirable to fit E_{0j} and σ_j as well as A_j in order to allow for variations in peak position and width which might otherwise cause significant errors. However, this entails the use of non-linear methods (see section 5.5.2) which present some technical difficulties and require more computing time.

In quantitative ED analysis we require to extract from the recorded spectrum the intensities (areas) of the peaks of interest where these peaks are broadened by the detector response function, are subject to overlap and are superimposed on continuum background. Provided the Gaussian function adequately represents the true peak shape, a procedure that determines optimum values of A_j by minimising the difference between the real spectrum and that calculated from equation (5.5) will automatically take care of peak width and overlap effects. For this purpose the familiar concept of 'least squares fitting' can be applied. A measure of the degree of 'misfit' between real and calculated spectra is provided by the parameter χ^2 (chi-squared) given by:

$$\chi^2 = \sum_i \frac{(n_i - \sum_j A_j P_{ij})^2}{\sigma_i^2} \qquad (5.6)$$

where σ_i is the standard deviation of n_i, and the peak shape function (previously assumed to be Gaussian) is now written in the more general form P_{ij}. The set of equations represented by (5.6) can be differentiated with respect to A_j and solved to find A_j values corresponding to minimum χ^2. The value of χ^2 at its minimum is an indication of 'goodness of fit' and may reveal, for example, the presence of unsuspected peaks or changes in peak widths and positions. The question of estimating the errors in A_j is discussed by Schamber (1977).

The least squares procedure can be used for fitting peak functions, derived either from a theoretical expression (for example, modified Gaussian) or from recorded pure element spectra, to the observed spectrum. However, background presents something of a problem, since a channel by channel simulation using one of the continuum models described in section 5.3.1 would be unduly time-consuming. Gehrke and Davis (1975) fitted complete spectra (including the continuum) from the oxides of pure elements to the spectra of compounds consisting essentially of mixtures of oxides (for example silicate minerals). However, this procedure is not rigorous with regard to the continuum and it is preferable to use digital filtering prior to fitting in order to eliminate background, as described in the following section.

5.5.1 The filter-fit method

Since the filtered spectrum represents the sum of the constituent (filtered) peak functions, linear least-squares fitting procedures are still applicable. The filtered peak shape is, of course, no longer Gaussian, but in any case it is better to use pure element reference spectra since the actual detector response is then automatically taken into account. Also the complete K, L or M spectrum can be treated as one, instead of requiring several separate peak functions. A long counting time should be used for recording reference spectra in order to minimise statistical 'noise', thereby obtaining a more faithful representation of the peak shape. Computer memory can be economised by storing only the region of each reference spectrum that contains the required peaks. This 'filter-fit' procedure (Schamber, 1973, 1977; Statham, 1977b) is now widely used for quantitative ED analysis.

For perfect elimination of background by filtering, the background must be linear within the width of the filter function and, in principle, curvature may cause errors in apparent peak heights obtained subsequently by fitting. In practice such effects are small, although according to Statham (1977b) significant errors (up to 0.3% in terms of concentration) can arise in the region of maximum curvature at the low-energy end of the spectrum.

Steps in the continuum caused by absorption edges are another potential source of error. These are never directly visible, since they always lie within the region occupied by the associated peak (Fig. 5.7). Continuum models that include an absorption term (section 5.3.3) automatically take edges into account and failure to do so is a possible drawback of the filtering technique. However, the errors are generally small in practice, partly because the step height, being related to the concentration of the element concerned, is approximately proportional to the peak height. Furthermore, the presence of edges in pure element reference spectra tends to cancel out the effect. The worst case is that of a small peak close to a large one with an associated absorption edge, when the error in the small peak could be up to one-third of the height of the edge (Statham, 1977b); however, this would seldom amount to more than 0.1% in terms of concentration.

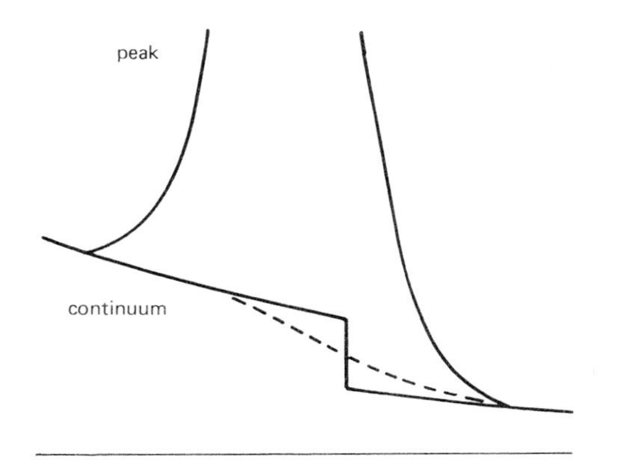

Fig. 5.7 – Step in continuum underlying peak due to absorption edge; dashed line shows edge 'smeared' by detector response function.

Differential absorption effects are likely to be significant when the absorption edge of another element in the specimen lies between the peaks of the element of interest (for example, a K edge between Lα and Lβ lines). Schamber (1981) showed that in the case of a simulated 10% Pt–Zn alloy, a 4% relative error in the platinum concentration could occur owing to the Zn K edge between the Pt Lα and Lβ lines. A possible strategy for avoiding such errors is to divide the spectrum for fitting purposes into segments between absorption edges, so that in the above example the Pt Lα and Lβ peaks would be treated separately.

It is essential that the reference spectra used in the filter-fit procedure should represent accurately the peak functions in the actual spectrum. This requires the detector resolution and calibration to be constant to within limits which it is relevant to attempt to define. Schamber (1981) calculated that for an isolated peak of 160 eV FWHM, a shift of only 6 eV would cause an error of 1% (a somewhat greater sensitivity than when simple integration is used, see section 5.4.1). The effect is most serious in the case of a small peak adjacent to a large one, when the former is very susceptible to shift in the latter. An example considered by Schamber is Ba Lα and Ti Kα, which are separated by 44 eV; here a 10 eV shift causes an apparent barium concentration of 0.7% to be recorded in a pure titanium specimen.

Another possible hazard is the presence of unsuspected peaks which, depending on their size and location, can cause significant errors. Prior inspection of the spectrum is a sensible precaution but does not always reveal 'hidden' peaks. Most fitting programs give some indication of 'goodness of fit' and this should be monitored by the user. It is very useful to be able to display the channel-by-channel fitting error, since neglected peaks will show up very clearly. The fitting procedure can then be repeated, taking account of the missing element.

5.5.2 Non-linear fitting

Small amounts of peak shift and broadening can cause significant errors when using the methods described above and it is therefore of interest to explore procedures that are most tolerant in this respect. In linear fitting methods only the peak heights are fitted, the positions and widths of the peaks (whether calculated or experimental) being predetermined. Equation (5.5), in which the number of counts per channel (n_i) is obtained as the sum of the contributions from Gaussian expressions representing the individual peaks, is non-linear with regard to the parameters representing peak position and width, and there is no analytical solution from which 'best fit' values can be determined other than in very simple cases. However, non-linear fitting equations can be reduced to approximate 'linearised' forms (for example, by using the Taylor expansion and neglecting high-order terms) which are valid for small differences (for example, see Schamber, 1981). This allows an iterative approach whereby solutions are obtained for the difference between the initial estimate of each parameter and its 'best fit' value. By using the results of the previous calculation as the initial estimates for the next, the 'true' values can be arrived at fairly rapidly. However, problems may arise due to non-convergence or convergence on a local false minimum of the chi-squared function, especially if the initial estimate is not reasonably good (Statham, 1978).

It is undesirable, and unnecessary, to allow the positions and widths of the peaks to be independently variable. By making use of the fact that the ED system can at least be assumed to be *linear*, the number of peak position parameters can be reduced to two (corresponding to zero error and gain factor), and likewise peak widths can be derived from an expression of the form of equation (4.3), thereby again reducing the number of variables to two (related to the Fano factor and noise term). Nullens *et al.* (1979) showed that non-linear fitting constrained in this manner is superior to linear fitting procedures, especially with respect to sensitivity to peak shift and and broadening effects, although the computing time required is considerably greater.

5.5.3 Sequential simplex method

'Trial and error' methods offer an alternative to the analytical approach described in the preceding section. For example, chi-squared can simply be determined for a 'grid' of values of the parameters to be fitted, from which the 'best fit' values may be obtained directly. Although reliable, this method is very uneconomical of computing time especially if the number of parameters is large.

A more elegant procedure particularly well-suited to the capacity of the small computer is the 'sequential simplex' method (Spendley *et al.*, 1962). The 'simplex' is a polyhedron with $n + 1$ vertices in n-dimensional space, where n is the number of parameters to be fitted. Each vertex point represents a set of values of these variables, for which chi-squared can be calculated. At each step in the calculation the vertex for which chi-squared is maximum (that is, worst fit) is identified and the values of the variables are changed in such a way as to move the point in the direction of decreasing chi-squared. This can be visualised more easily in two dimensions (Fig. 5.8) in which case there are two variables (x and y) and the simplex consists of a triangle (ABC). If the highest chi-squared value is obtained for point C, this point is moved to a new position, C′, representing the 'reflection' of C in the line AB. Chi-squared is then re-calculated for the new values of x and y and the process repeated (in general, one of the other vertices will be moved next). The rate of convergence can be controlled by varying the ratio C′D/CD: thus in order to avoid oscillatory behaviour this ratio can be reduced as the 'best fit' point is approached.

When fitting real spectra, n is in general much larger than 2, but exactly the same principle applies. As noted in the preceding section, the number of variables can, however, be reduced by using suitable expressions for peak width and position, thereby effecting a considerable saving in computer time and reducing the likelihood of obtaining incorrect results.

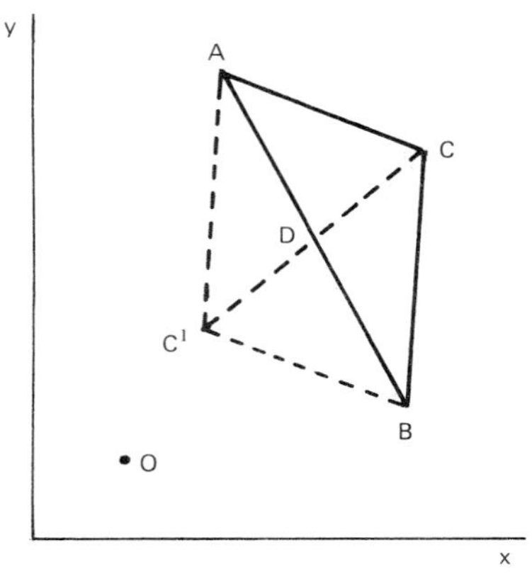

Fig. 5.8 – Sequential simplex method: ABC represents initial simplex in two dimensional case and C′ is the new position of vertex C; C′ is nearer to O, the coordinates of which represent the 'best fit' values of x and y.

In the interests of reducing the probability of converging on a false minimum, it is desirable that the initial estimates of the variables should be as close as possible to the 'true' values. Peak position is by far the most sensitive parameter but fortunately is also generally the best known (Fiori *et al.*, 1981). The vertices of the initial simplex are obtained by perturbing the estimated initial values of the variables by small arbitrary amounts (for example, 10% for amplitudes, 1% for peak widths, 0.1% for peak positions). Convergence is assumed to have been attained when the change from one calculation to the next is negligibly small.

5.6 ACCURACY AND DETECTION LIMITS

Quantitative ED analysis is subject to errors in ZAF corrections to the same degree as 'classical' WD analysis and, depending on the extent of the compositional difference between standard and unknown, this limits the accuracy to ±2% (relative) or thereabouts. Additional errors might, however, be expected in ED analysis owing to the larger background corrections and overlap effects. For concentrations of several per cent or more, background errors should not be significant when one of the methods developed specially for ED analysis (section 5.3) is used but at low concentrations errors are liable to be greater than for WD analysis at a similar concentration. Combining filtering and continuum modelling methods might lead to some improvement in the accuracy of background corrections: thus an initial background subtraction based on a simplified continuum model could be followed by digital filtering, thereby removing most of the background curvature which can give rise to errors in the filter-fit method as normally applied.

Overlap between Kα peaks can be corrected quite easily by the methods already described and need not be a significant source of error. However, the same cannot be said of Kβ overlap, which can be serious for neighbouring elements of atomic number around 23 (and in analogous situations involving Lβ lines). In such circumstances the accuracy of the result is inevitably affected and this should be kept in mind when interpreting ED analyses.

Assuming an adequate background correction and excluding cases of severe overlap, the limit of detection is governed by statistical considerations. Typically the total number of counts recorded in the whole spectrum is around 500 000, of which about half are contained in the characteristic peaks. For a pure element the Kα peak will thus contain 250 000 counts, of which approximately 180 000 (75%) can be considered as 'useful' (that is, contained within integration limits equal to the FWHM). With a typical peak-to-background ratio of 100, it follows that the integrated background count is 18 000. A just-detectable peak, corresponding to three times the standard deviation of the background count, thus contains 400 counts which is equivalent to a concentration of 0.25%. This is a fairly representative figure but the detection limit varies somewhat with the element concerned and the mean atomic number of the matrix (which governs the continuum intensity).

The total spectrum content of 500 000 counts assumed above corresponds to a count rate of 5000 counts per second and a live time of 100 seconds. As described in Chapter 4, the count rate is limited by electronic factors, although there is some prospect of improvement especially through advances in FET performance and the use of beam switching. Extending the counting time is not very profitable in view of the square-root relationship between number of counts and statistical precision. Also, improvements in precision are not beneficial beyond the point where the accuracies of background and overlap corrections become limiting factors.

5.7 LIGHT-ELEMENT ANALYSIS

As has been mentioned in section 4.2, the ED detector can record x-rays of energy less than 1 keV provided that the beryllium window is removed. X-ray attenuation will still occur of course in the thin gold electrode on the Si(Li) detector and in the inactive or 'dead' surface layer of silicon, but the spectral response of the windowless device is now sufficiently extended at the low energy end (Fig. 5.9) for useful light-element analysis. Some commercially available ED detectors incorporate a mechanical system whereby the beryllium window can be rotated out of the x-ray beam and replaced either by an aperture or by a thin window made of plastic which is relatively transparent to soft x-rays (see Fig. 3.9). The plastic option is extremely useful because with this in position the vacuum integrity of the detector is maintained and any volatile contaminants in the vacuum system of the microprobe are prevented from condensing directly upon the detector surface.

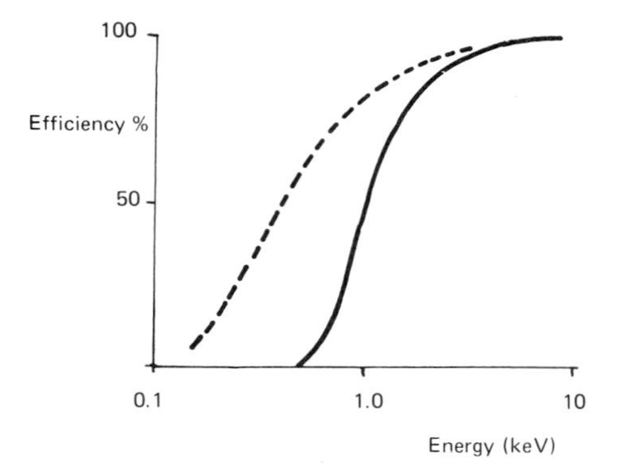

Fig. 5.9 – Efficiency of Si (Li) detector as a function of x-ray energy; solid line, with beryllium window; dashed line, without window.

Fig. 5.10 shows a spectrum from alumina which was taken with a formvar window positioned in front of the detector. The peak-to-background ratio for oxygen is about one-third that of aluminium which indicates that the sensitivity for light-element analysis should not be markedly worse than for elements with $Z > 11$. In fact figures of 0.21 wt% and 0.11 wt% are indicated here, in accord with the information given by Russ and Sandborg (1981). They also suggested that minimum detection levels of 0.5 wt% for nitrogen and carbon were achievable, although, as discussed later, there are a number of reasons why such figures might be difficult to achieve in more general applications.

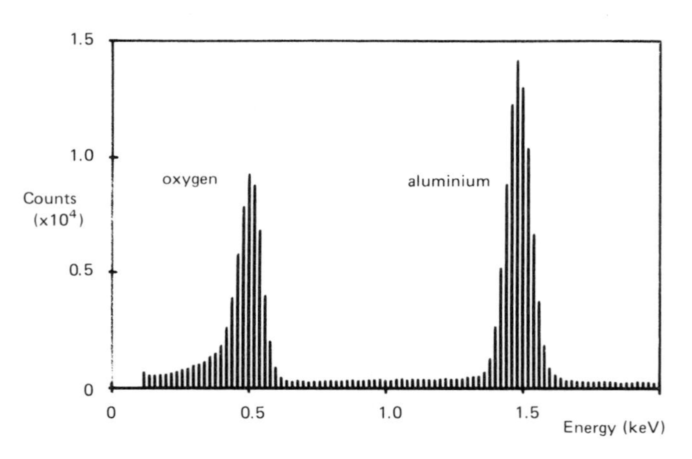

Fig. 5.10 – ED spectrum from alumina, 7 kV, formvar window.

Electronic noise of the ED system determines the lightest element (boron) which can be detected. The Gaussian noise peak is centred at zero and noise levels are at present ~100 eV FWHM. Consequently the tail of the peak extends to the positions of the boron and carbon lines thereby limiting the sensitivity with which these elements can be detected. In practice the baseline is set between 60 eV and 120 eV to prevent the system being swamped with noise.

Quite clearly qualitative analysis of the light elements is feasible but the question which requires an answer is whether an ED system is capable of providing quantitative results. Generally the problems involved here are not new. They arise in conventional ED analysis but appear in a more acute form when processing light-element spectra.

Pulse pile-up correction has been discussed in section 4.8 but in the soft x-ray region pulse pile-up cannot be dealt with in a simple way because the x-ray energies fall below the threshold of the fast amplifier. It is possible to minimise pile-up by working at low count rates (not more than 1000 counts per second, say) but it is difficult to eliminate. An additional problem involving the fast amplifier arises if it is used to determine the live time of the system (Russ *et al.*, 1973), an incorrect live time being indicated whenever the rate of soft x-rays entering the detector becomes appreciable (Bloomfield *et al.*, 1981).

The poor energy resolution of the ED system (Fig. 6.6) creates formidable problems when dealing with the proliferation of soft x-ray lines from heavier elements (L, M and N spectra) which are likely to interfere with the light-element x-ray peaks. There are many overlap possibilities and Sutfin and Ogilvie (1971), for example, have pointed out that 87 x-ray lines lie within 100 eV of the carbon K line. Similar situations exist for the other light elements and to quote just two instances the chromium Lα peak (573 eV) lies close to the oxygen peak (525 eV) and Ti Lα (452 eV) is difficult to separate from the nitrogen peak (392 eV). Coates (1980) has referred to ambiguities which can occur when M line overlaps are involved and points out the need for reliable data on the relative intensities of the spectral components. He quotes, in particular, the case of the M_ξ line which he considers is often more intense than published data would suggest.

Another peak identification problem may arise from silicon escape peaks (sections 4.6.2 and 5.2.1); for example, the escape peak associated with the phosphorus Kα line occurs at 274 eV, close to the carbon peak (277 eV), and the silicon escape peak from sulphur Kα (at 568 eV) interferes with oxygen line. Data provided by Reed (1975a) appear to indicate that the intensity ratios of escape peak to parent peak are higher in this region of the spectrum (for example, 1.41% for P Kα compared with 0.28% for Fe Kα.)

It is evident that the complex nature of the light-element spectrum will mean that some of the methods of measuring true peak intensities described in this chapter will not be practicable. For example, the technique of overlap coefficients (section 5.4) would not appear promising because of the large number of overlaps and the lack of data on relative intensities for M and N lines. Compounding these difficulties is the fact that the peak shapes occurring below 1 keV are markedly non-Gaussian because of the phenomenon of incomplete charge collection. In the soft x-ray region some background subtraction methods may also be prone to error. Continuum modelling, which is based on the Kramers expression or a derivative of it (section 5.3.1), has at least two serious limitations. Firstly, prediction of the background shape is less accurate in the soft x-ray region because of the much larger continuum absorption correction required (see section 5.3.3). Secondly, even if the true continuum emission could be modelled precisely it would still not correspond to the observed background because the latter is distorted by the presence of degraded events (Statham, 1980). The method of Smith *et al.* (1975), mentioned in section 5.3.2, suffers similarly from these limitations. Although mathematical filtering techniques of background subtraction (section 5.3.4) are not affected directly by continuum absorption or by degraded events, the marked curvature of the background which occurs at low energies may be expected to produce significant errors here also.

Summarising, quantitative ED analysis of the light elements with an accuracy comparable to that obtainable for elements of atomic number above 10 is likely to be an elusive goal.

6

Experimental determination of x-ray intensities

M. G. C. COX

6.1 SPECIMEN

Metals, ceramics, polymers, minerals and bio-materials number among the wide range of substances examined by electron-probe microanalysis, each class of material presenting its own special preparation problems. Whilst it is outside the scope of the present book to discuss in detail individual requirements, some of the more general criteria which need to be satisfied in quantitative analysis are mentioned here.

6.1.1 Mounting and polishing

For quantitative analysis the surface being examined should be flat on a micro-scale. Any irregularities in surface topography will alter the electron backscattering and x-ray absorption characteristics and hence introduce errors into the analysis. For example, the surface roughness of a piece of sulphur (Fig. 6.1(a)) has resulted in variation in the sulphur x-ray emission across the surface (Fig. 6.1(b)). Fig. 6.2 illustrates schematically the increase in path length caused by the proximity of a step and Salter (1970) has shown that errors in analysis of ~10% may be introduced by surface irregularities of 0.5 μm or less.

Often the first stage in preparing materials for electron-probe microanalysis is to cut a section through the specimen and then embed it in appropriate mounting media to facilitate mechanical polishing. Specimen embedding is generally performed in a press (~150 bar pressure) at elevated temperatures (~150°C) using either bakelite or diallyl phthalate as a mounting medium. With those types of specimens which are unstable under such conditions, embedding in a cold-setting epoxy resin or a liquid polyester may be adopted. After the mounting operation the sample is ground using successively finer grades of abrasive to achieve a macroscopically flat surface. The grinding process usually commences with silicon carbide papers, starting on 220 grade (70 μm) and finishing

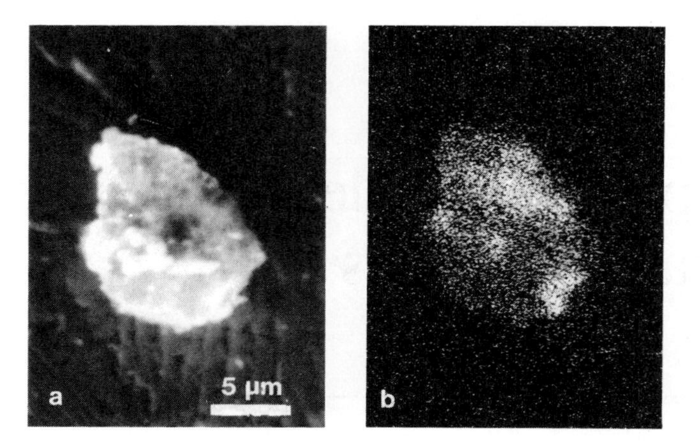

Fig. 6.1 – Sulphur particle showing variation in x-ray emission caused by surface roughness; (a) electron and (b) sulphur x-ray images.

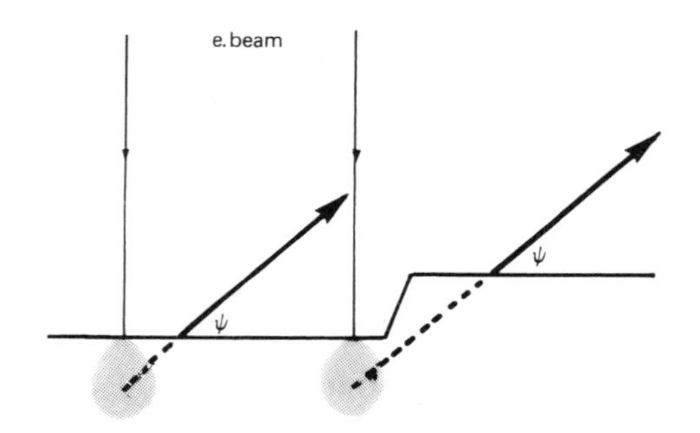

Fig. 6.2 – Changes in x-ray path length in specimen (dotted lines) close to surface step; ψ is x-ray take-off angle.

on 600 grade (20 μm). Use of copious amounts of lubricant during grinding is advised in order to wash away grinding debris and to prevent heating of the sample but lubrication should be reduced when the specimen contains inclusions so that the likelihood of particles being pulled out is avoided. The final stage of preparation usually involves polishing on cloth pads impregnated with fine abrasives such as diamond pastes varying in grade from 14 μm to 0.25 μm.

Ideally the specimen and mounting material should be of similar hardness but since the specimen is generally the harder of the two some rounding of its edge is almost inevitable. The effect should of course be minimised, especially when studying surface regions of a specimen by means of a transverse section. One way in which this can be achieved is to use harder mounting materials such as composite resins which contain glass fibre fillers or metallic particles. Another way of protecting the edge of a specimen is to plate with a metal layer, care being taken to avoid contamination from elements in the plating bath. A simpler method is to surround the specimen with a hard metallic tube but unless this is in close contact with the specimen satisfactory protection of the edge will not be achieved. Indeed, the gap may entrap polishing compounds and create even more difficulties. With soft specimen materials such as lead, where abrasive material may become embedded in or smeared across the surface, Heinrich (1981) has indicated that satisfactory results can be achieved by microtomy, although the relatively small (~2 mm) width of the diamond knife restricts the area which can be sectioned.

With multi-phased materials, harder constituents tend to polish more slowly than softer ones. This differential polishing can result in the formation of steps at phase boundaries which, as mentioned earlier, will affect x-ray intensity measurements (Fig. 6.2). Light pressure during specimen polishing is advised in these circumstances and also the adoption of a hard lap rather than a soft polishing cloth.

Electropolishing can be used for preparing surfaces on pure metals and homogeneous alloys and Kehl (1949) has described techniques for a number of such materials. However, with multi-phased specimens the rates of removal of the individual phases may be different, and even in homogeneous alloys the less noble metal may be dissolved more quickly. Hence non-metallic inclusions tend to be preferentially removed to leave a pit whilst other micro-constituents may be left proud of the surface. Consequently, because mechanical polishing methods are in general satisfactory, electropolishing is not widely used in preparing specimens for electron-probe microanalysis although the technique is extremely useful in channelling-pattern investigations where the surface layer of the specimen must remain undisturbed.

Throughout the whole of the specimen preparation process thought should be given to the nature of the x-ray measurements required. Wherever possible mounting media and polishing compounds should be chosen such that they do not contain any elements of interest. For example, when measuring carbon levels specimen polishing should be carried out using alumina rather than diamond paste.

After polishing, the sample must be thoroughly cleaned to remove all traces of polishing compound and detritus. Particular care is necessary with porous samples, since any debris forced into the pores during polishing can be difficult to extricate. Cleaning is often carried out using an ultrasonic bath containing a suitable solvent. However, it is almost impossible to remove residual traces of carbon by usual cleaning methods and Love *et al.* (1981) showed that the only way carbon contamination could be entirely eliminated from their iron specimens was to etch away the first 5 nm of the surface by ion bombardment; otherwise a residual level, equivalent to ~0.1 wt % carbon, was always recorded. For best results the operation should be carried out within the vacuum system of the microanalayser but fortunately this is not necessary unless very low levels (<0.1 wt %) of carbon are being measured. An alternative solution to the problem is to construct a calibration curve (Fig. 6.3) from measurements on low carbon steels of known composition (Fisher and Farningham, 1972); the carbon contamination on the surface, which is assumed to be the same on specimen and standards, produces a small offset in the curve and is thus readily taken into account.

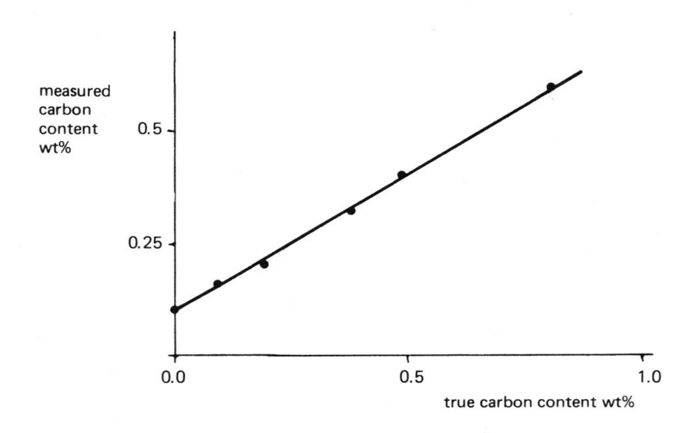

Fig. 6.3 – Calibration curve for analysis of carbon in steels; Fe_3C standard, 10kV.

When analysing particulate materials, it may be possible to prepare sections through them using the types of polishing method mentioned above. However, special care is required in ensuring that the mounting compound does not create charging-up problems, particularly since the position of the electron beam is never far from the edge of a particle. For this reason it is unwise to mount particles in bakelite even if the final polished surface is given a metallic coating. A better method is to mix the particles with a small quantity of conducting silver epoxy resin. A small hole is then drilled in a block of mounting compound and filled with the resinous mixture. After drying, the specimen is polished in the usual manner. The technique is probably most suitable for particles >10 μm diameter where fluorescence from the silver epoxy is relatively insignificant, but Small *et al.* (1978) have suggested that it is useful for particles down to 1 μm diameter.

Although sectioning of the particle is the only way of ensuring a quantitative analysis, a number of other methods, described by Goldstein *et al.* (1981) essentially for scanning electron microscopy, may enable semi-quantitative data to be obtained. These techniques involve the transfer of whole particles to a specimen stub for examination. Stubs should be either graphite or beryllium since these have low backscattering coefficients and do not produce characteristic fluorescence of elements in the sample. The particle may be transferred to the stub in a wet or dry state. In the former case the particles are first suspended in a volatile liquid by ultrasonic agitation (note; prolonged ultrasonic treatment especially at low frequencies may cause break up of agglomerate particles) and then the solution is pipetted onto the substrate. The liquid quickly evaporates and particle adhesion usually proves satisfactory, although if necessary the particles may be fixed more firmly in position by subsequent coating with a thin conducting layer (section 6.1.2). Dry transfer of particles is to be preferred, however, especially where analysis of an individual particle is the objective. Placement on the support stub may then be accomplished using a vacuum pick-up, stainless steel tweezers or a tungsten needle, the choice depending upon the size of particle. A coating of carbon may suffice to hold the particles in position. Alternatively a thin layer of graphite or collodion could be first spread on the stub, allowing the adhesive to dry partially before the particles are placed in position so that liquid does not spread over their upper surfaces.

Preparation of biological tissues and fluids is often very complex and specialised and is outside the scope of this book. For further information on this topic the reader is referred to Chapter 12 in Goldstein *et al.* (1981) and to Erasmus (1982).

The specimen must be electrically earthed by connecting it to the sample holder. This is usually accomplished using colloidal graphite or silver paint, the minimum amount being applied to avoid excessive outgassing into the vacuum system. As an alternative, and to reduce the possibility of contaminating the specimen surface, the electrical contact may be made by drilling a hole in the mount and making connection to the rear of the specimen, for example by means of Wood's metal.

6.1.2 Coating

In electron-probe microanalysis the x-ray source is typically $\sim 1\,\mu m$ diameter and extends a similar distance below the surface (section 2.8). Within this small volume ($\sim 10^{-18} m^3$) most of the energy of the incident electrons is dissipated, which means the energy density may be quite high; for example with an incident electron energy of 30 keV and a beam current equal to $0.1\,\mu A$ the energy density at the surface is $\sim 3 \times 10^9\,W\,m^{-2}$. Although such a high energy flux is unlikely to affect materials which are good thermal and electrical conductors, it can lead to severe problems with other materials. For example when using the above conditions the temperature rise at the point of impact is $\sim 200°C$ for a typical ceramic (thermal conductivity $K \sim 0.1\,W\,cm^{-2}\,K^{-1}$) compared with $\sim 20°C$ for a metal ($K \sim 1\,W\,cm^{-2}\,K^{-1}$).

Moreover the build-up of charge at the surface of an insulating specimen as the electrical potential at the point of impact rises produces deflection of the beam and causes specimen current instability.

A conducting coating on the specimen may eliminate problems due to heating and charging but not necessarily those due to ionic migration of constituent elements. Ion movement occurs as a result of the potential gradient which exists between the conducting coating (at earth potential) and the interior of the sample. Alkali metal glasses show some ionic mobility at temperatures a little above ambient and as a result the alkali metal x-ray intensity decreases during electron bombardment (Lineweaver, 1963). Similarly, metals sometimes 'plate' out (often as whiskers) on the surface of metal chalcogenides.

The application of a coating on specimens with poor electrical and thermal conductivity can alter the x-ray emission from the specimen by changing the backscatter and absorption properties of the target but the effects are small if the coatings are thin, say ~100 Å or less. In practice, metallic coatings of this thickness are usually sufficient to eliminate both charging and thermal degradation of the target although thicker coatings may be necessary with materials which are especially beam-sensitive such as polymers. It will be necessary also to apply a coating to metallic specimens which are contained within non-conducting mounting blocks where regions close to the metal/block interface are to be examined.

Carbon, aluminium, copper, silver and gold are common coating materials. (It should be noted that carbon is a much poorer conductor then the metals and hence a thicker coating must be used, but since carbon scatters electrons weakly and is almost transparent to most x-rays this is of little consequence).

Reed (1964) has expressed the fractional drop in x-ray intensity due to electron energy losses within the coating as

$$\frac{\Delta I}{I} = \frac{8.3 \times 10^5}{E_0^2 - E_c^2} \rho \Delta z$$

where Δz is the thickness (cm) of coating, ρ its density (g cm^{-3}) and E_0 and E_c are the incident electron energy and the critical excitation energy (in keV) respectively. Hence for an aluminium coating of thickness 10 nm, $\Delta I/I \sim 0.6\%$ at $E_0 \sim 20$ keV, but at 5 keV the loss amounts to almost 10%.

In addition some x-rays will be lost by absorption in the coating. Absorption effects will be most marked when measuring soft x-rays; for example, the mass absorption coefficient of oxygen Kα x-rays in carbon is ~12 000 cm^2 g^{-1} and the fraction lost in a 200 Å thick layer is given by

$$\frac{\Delta I}{I} = 1 - \exp\left(-\frac{\mu}{\rho}\rho z \operatorname{cosec} \psi\right)$$

where ψ is the x-ray take-off angle. Assuming $\rho \sim 2$ and $\psi = 30°$

$$\frac{\Delta I}{I} \sim 0.072, \quad \text{i.e. the loss is } 7.2\% \; .$$

In contrast the absorption of Cu Kα radiation for the conditions cited above is only 0.2%.

Because of the effect of coating thickness on intensity of x-ray emission from the target the amount deposited on the specimen and standard must be maintained the same. Reasonable reproducibility may be obtained by introducing a film thickness monitor into the coating equipment. These instruments work by recording the change in natural frequency of a quartz crystal which is coated simultaneously with the specimen. Hence coatings of consistent thickness (variation $<5\%$) may be evaporated and the system calibrated for different coating materials, thereby eliminating the need to coat specimens and standards simultaneously.

The choice of coating material is determined largely by the need to avoid x-rays from the coating interfering with x-ray lines from the sample. In wavelength-dispersive analysis with the high spectral resolution of the spectrometer it is necessary only to ensure that the sample does not contain the coating element. In energy-dispersive analysis however, greater care in selection is required and it is best to ensure that x-ray lines from the coating are separated by several hundred electron volts from any x-ray peak being measured in the sample. This criterion is readily satisfied by carbon coating since most ED systems do not detect elements with $Z \leqslant 9$.

Specimen coating is conventionally carried out either by thermal evaporation or by sputtering.

For evaporation a coating unit is required which is capable of achieving a vacuum of $\sim 10^{-8}$ bar (10^{-5} torr). The specimen is placed inside the workchamber and the evaporant metal (gold, silver, copper or aluminium), usually in the form of wire, is inserted into a molybdenum boat or tungsten filament. The chamber is evacuated and an electric current passed through the boat or filament raises its temperature by resistance heating and causes evaporation of the contents. Carbon coating is carried out differently, usually by passing a high current through a pair of carbon rods arranged with pointed ends in contact. Localised resistance heating gives rise to carbon evaporation from the tips of the rods. The process is difficult to control and as a result it is not easy to obtain reproducible coating thicknesses. A satisfactory alternative, which has recently become available, is to replace the carbon rods with braided strands of carbon fibre stretched across the terminals (Vesely and Woodise, 1982).

Sputtering involves producing a glow discharge between a disc of the coating material which forms the cathode and the specimen holder which acts as the anode. The unit is first evacuated to $\sim 2 \times 10^{-5}$ bar and then argon gas is admitted until the pressure becomes stabilised at $\sim 8 \times 10^{-5}$ bar. The potential between anode and cathode is raised above 1 kV to produce a glow discharge. Under these conditions argon atoms are ionised and strike the cathode causing the ejection of atoms of coating material. Film thickness is controlled by the magnitude of the applied voltage ($\sim$1.5 kV), the discharge current ($\sim$10 mA) and the duration of the process (a few minues). The sputtering of gold, silver and copper is straightforward but the sputtering yield of carbon is extremely low and it is virtually zero for aluminium. Hence the technique is somewhat less versatile in this respect than thermal evaporation but it has the useful feature that atoms of the coating element are scattered by residual gas on their passage from cathode to anode thereby coating specimen surfaces not in direct line of sight of the cathode. Early models of sputtering equipment were prone to damage heat-sensitive materials during the coating process but later designs appear to have overcome the problem with the use of Peltier-cooled specimen holders and magnets to deflect electrons away from the specimen (Panayi *et al.*, 1977).

6.1.3 Locating features of interest

At this stage a preliminary examination by optical microscopy is frequently useful to locate areas of interest with respect to major specimen landmarks (for example, surface boundaries, grain boundaries, cracks, etc.), thus facilitating rapid identification of features once the sample is in the microanalyser.

Most microanalysers are equipped with an optical microscope coaxial with the electron beam and location of areas of interest is usually straightforward. In addition, modern instruments usually have high resolution scanning electron optics which can be used to form a variety of electron images depending on the detected signal (for example, secondary electrons, reflected or backscattered electrons, specimen current, etc.). These images may be enhanced using signal processing (for example, gamma control, differentiation, sum and difference processing from two detectors, etc.) and, consequently, a range of contrast mechanisms can be invoked to aid identification of the areas for analysis. For example, Fig. 6.4(a), taken from a polished section through the nitrided surface

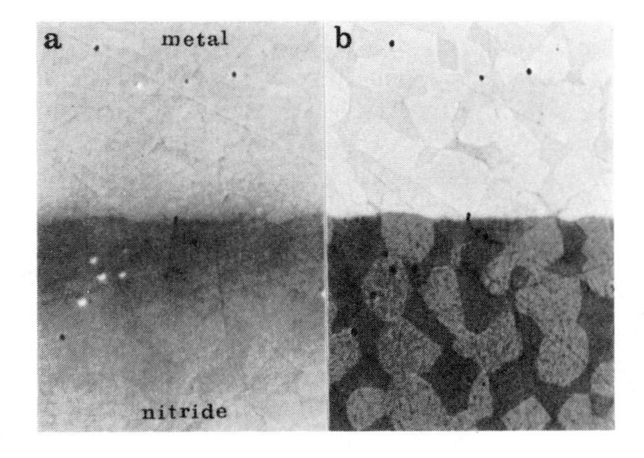

Fig. 6.4 – Metallographic section (un-etched) through nitrided semi-austenitic steel; (a) secondary electron picture and (b) backscattered electron picture (× 600).

of steel using the conventional secondary electron imaging mode, is relatively featureless. However, by using split electron detectors and adding the respective signals a picture is produced which enhances atomic number contrast and reduces effects due to topography (Fig. 6.4(b)). As well as delineating the nitrided zone more clearly, the austenite and ferrite grains in the metal are easily distinguished making it feasible to study on a single sample the nitriding response of these two different phases. It is worth noting that the difference in mean atomic number of the austenitic alloy phase and the ferritic is <0.3.

If such methods fail to identify features of interest it is sometimes necessary to resort to etching the sample in order to enhance contrast. In general, however, chemical treatment is not recommended since it may lead to selective removal or deposition of elements onto the specimen (Hallerman and Picklesimer, 1969). One way of avoiding such problems is to etch the specimen first to reveal the desired feature and then to mark its position accurately using microhardness indentations. The specimen can be repolished to remove the etched layer and the feature relocated in the microanalyser by reference to the residual indentation marks.

6.2 SELECTION OF STANDARDS

Standard specimens are required for conventional quantitative electron-probe microanalysis although some recent methods (section 11.4) obviate the need to retain standards for every element of interest. A standard specimen must be of known chemical composition, homogeneous and stable. It is also an advantage if it has similar physical and chemical properties to the material being analysed since this will minimise any systematic errors arising from inadequate ZAF corrections and from imperfect specimen preparation.

Metallurgical analysis is usually carried out using pure element standards, most of the metallic elements used in engineering alloys being readily obtainable in the pure state. Alloy standards are sometimes used, but should be carefully checked for homogeneity on the microstructural scale of the analysis. This is easily accomplished in the microanalyser by random point measurements on the prepared surface followed by application of standard statistical tests for homogeneity.

Some metals and almost all non-metals are too reactive or volatile to be used as standards and as a consequence more stable compounds are needed; for example, sodium chloride frequently serves as a standard for the elements sodium and chlorine. As well as ensuring that such compound standards are homogeneous, it is prudent to check they are stoichiometric. Compound standards are frequently employed in mineralogical studies where large single crystals can be analysed by classical chemical analysis and checked for homogeneity by optical methods. These may then be used for analysis of mineral specimens of similar composition. In light-element analysis difficulties arise because substantial changes occur in the distribution of x-rays within the emission band (see section 2.2.3). For example, in the analysis of oxygen in oxides the large changes arising from variations in the distribution of intensity in the oxygen K band mean that peak heights become inappropriate measures of the total intensity. Quantitative analysis is still possible however, provided that the intensity is integrated over the whole band for specimen and standard (Love *et al.*, 1974a).

A list of materials used as standards in microanalysis is published by the National Bureau of Standards, Washington, DC (see, for example, Heinrich *et al.*, 1971), who are also prepared to supply certain samples on request.

6.3 MICROANALYSER OPERATING CONDITIONS

6.3.1 Specimen geometry

In purpose-built microanalysers, the specimen holder and x-ray optics are usually designed such that the specimen surface is perpendicular to the incident electron beam. However, in scanning electron microscopes this geometry may not be available, the incident beam striking the specimen at some other angle. Indeed, for special purposes such as identification of surface films it may be desirable to increase the tilt angle in order to reduce the depth of penetration of the electrons and confine analysis to the more immediate surface regions.

In cases of tilted specimen geometry it is important that the detailed arrangement of the specimen chamber is known, that is, the specimen tilt angle (γ) and the azimuthal angle (α) subtended by the spectrometers, etc. The correct value of the x-ray take-off angle (ψ), which may be different for spectrometers at different positions around the column, can then be calculated; Statham and Ball (1980) give

$$\operatorname{cosec}\psi = \frac{1}{\sin\gamma\cos\alpha\cos\delta + \cos\gamma\sin\delta},$$

where δ is the elevation angle of the x-ray detector.

6.3.2 Electron accelerating voltage

The electron accelerating voltage strongly influences the intensity of x-ray emission and just above the critical excitation energy (E_c) very few x-rays are emitted. The intensity of emission rises rapidly with electron energy but, since absorption in the target also increases, it is usual to limit the incident electron energy to 2–3 times E_c. This may not be always possible for elements with low values of E_c due to a practical limit for the instrument, usually $\sim$5 keV; for example, with oxygen the critical excitation energy is $\sim$0.56 keV and an operating voltage even as low as 5 kV gives a high ($\sim$9) overvoltage ratio (E_0/E_c).

The electron accelerating voltage controls the penetration of the beam into the specimen and hence for analysis of thin films low operating voltages are advantageous. However, then the maximum current in the beam may be less and/or the spot size enlarged, and these factors must also be considered when selecting operating conditions. High accelerating voltages on the other hand, as well as leading to large ZAF factors, produce an increase in the excited volume of the target with consequent loss of spatial resolution (see section 2.8).

Most instruments have either preselected voltages or are equipped with a meter which monitors the value but in view of the strong dependence of ZAF correction factors on the magnitude of the accelerating voltage, it is unwise to rely too much on instrument readings. These should be checked periodically by maintaining a constant beam current and dropping the voltage until the emission of a known x-ray line ceases to occur (Sweatman and Long, 1969). This voltage is equal to the critical excitation voltage of the x-ray line. An alternative and more convenient method involves measuring the Duane–Hunt limit (the voltage at which continuum x-rays cease to be produced) using an energy-dispersive spectrometer (Beaman and Isasi, 1972). With care this can be determined to within a few electron volts and its value gives the electron accelerating voltage directly.

6.3.3 Beam current

The current in the electron probe is determined by the electron optics of the column, the nature of the electron source and the accelerating voltage. Large currents increase the emitted x-ray intensity but lead to larger spot sizes, less stable operation, and to greater specimen damage. In energy-dispersive analysis large currents may also cause greatly increased dead-time and pile-up corrections (or even total paralysis of the system). Beam currents usually adopted range from $\sim 10^{-9}$ A for energy-dispersive to $\sim 10^{-7}$ A for wavelength-dispersive analysis.

6.3.4 Minimising contamination

The vacuum system of most microanalysers consists of an oil diffusion pump backed by an oil-filled rotary pump. As a consequence a substantial fraction of the residual gases are hydrocarbons which can polymerise at or close to the point of impact on the target of the electron beam and cause an undesirable build up of carbonaceous material. The hydrocarbons arrive at the point of electron impact by a surface diffusion mechanism and contamination may form as a dark spot or ring depending on the beam current and spot size. If the beam is finely focused a sharp cone is produced lying on an area of more diffuse contamination. This effect is illustrated in Fig. 6.5(a), a transmission electron micrograph from a thin metal foil; the upper cone is associated with the incident electron beam and the lower cone with the emergent beam. If, however, the spot size is large then all the hydrocarbon molecules are deposited at the edge of the spot since they are immobilised before reaching the centre; this results in the formation of a ring (Fig. 6.5(b)).

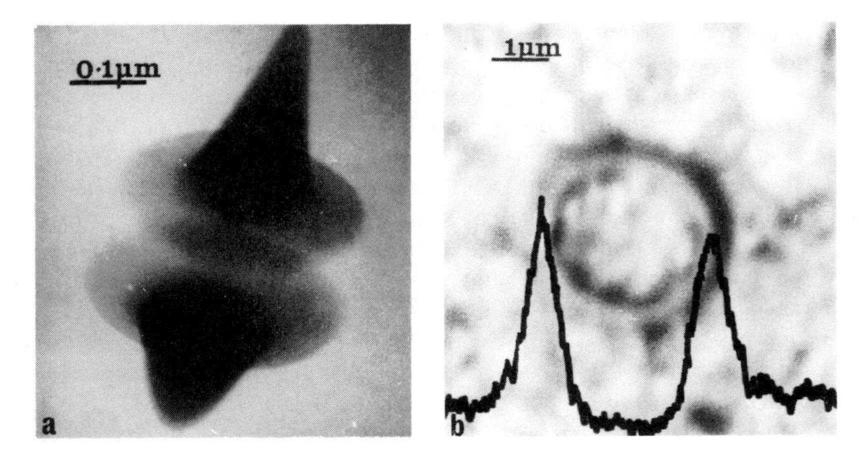

Fig. 6.5 – Contamination on specimens after electron irradiation (a) for several minutes with a focused beam ~10 nm diameter, thin section TEM image; (b) for 30 minutes with beam diameter of ~1 µm, solid sample SEM image, carbon profile superimposed. (Love *et al.*, 1981), courtesy Gerhard Witzstrock.

The effects of contamination on x-ray measurements can be very severe and the phenomenon has been studied experimentally by a number of workers (Ong, 1966; Ranzetta and Scott, 1966; Conru and Laberge, 1975). The observed decrease in x-ray intensity with time when the electron beam is stationary on a point is because some of the energy of the electrons is dissipated in the contamination leaving less available for x-ray generation. Further losses occur because the deposit will also absorb some of the generated x-rays. The effects of contamination are most severe in light-element microanalysis. This is particularly true when attempting to measure the carbon content of steels since contamination may set a limit to the minimum detectable concentration. Contamination is also a problem in nitrogen and oxygen analysis since their x-rays are strongly absorbed by carbon.

Most modern instruments incorporate water- or liquid nitrogen-cooled baffles immediately above the diffusion pump which help to minimise the back-streaming of hydrocarbon vapours and hence reduce specimen contamination. A further reduction in contamination rate is achieved by placing a large plate ('cold finger') cooled to liquid nitrogen temperature immediately above the specimen. This acts as a cryopump and reduces the partial pressure of hydrocarbons in the vicinity of the specimen. The efficiency of cold fingers varies greatly and depends critically on the solid angle subtended at the specimen and the separation of specimen and cooled surface. A less popular method of reducing contamination is to fit a gas jet in the specimen chamber which is aimed at the point of impact of the electron beam to reduce the partial pressure of hydrocarbons (Castaing and Descamps, 1954). The system is more difficult to engineer into the cramped space around the specimen and has the further disadvantage that the gas jet may scatter the electron beam to some extent. Gases such as oxygen and water vapour are reported actually to remove previously existing deposits of carbonaceous material from the surface (Moll and Bruno, 1967; Duerr and Ogilvie, 1972).

Other remedial measures to reduce contamination have been suggested. Some workers (Baker *et al.*, 1971) have recommended replacing all hydrocarbon oils and greases with fluorocarbons (Fomblin) which do not polymerise under electron bombardment. Fitting oil-free turbomolecular pumps which give a very clean vacuum is another possible solution. The replacement of polymeric materials containing plasticisers, as used for plastic tubing, wire insulation etc., is another suggestion (Love *et al.*, 1981) since these also provide sources of hydrocarbons; PTFE or polypropylene are better materials to use. The same workers point out the need to avoid introducing contamination into the system via the specimen itself.

6.3.5 Measuring x-ray intensities

In wavelength-dispersive spectrometry x-ray intensities are recorded only at the wavelength which satisfies the Bragg condition and in quantitative analysis it is important that the spectrometer is set exactly at the peak maximum when the measurement is being taken.This is ensured by rotating the spectrometer through the peak using the stepping motor, the counts being measured with a scaler at each angular setting. Once the maximum has been located the spectrometer can be repositioned on the peak and the x-rays collected over a sufficient length of time to achieve good counting statistics. The corresponding background measurement is obtained preferably by averaging the counts recorded an equal distance (say 2°) either side of the peak. The practice of taking readings directly from the trace on a chart recorder is not recommended if high precision is required.

In energy-dispersive spectrometry information is collected simultaneously for the whole spectrum and, consequently, simply to measure the intensity of the peak maximum is inefficient since this discards information in adjacent channels which are part of the same x-ray peak. It is usual, therefore, either to average the counts recorded in a number of channels either side of and including the peak maximum or to fit a response function to the total peak with suitable weighting of the individual channel counts to achieve optimum statistical accuracy. Furthermore, because dead time is more significant in ED systems the count is not recorded for a fixed real time but for a fixed system live time (section 4.9).

6.4 STATISTICS

6.4.1 Standard errors

Under carefully selected experimental conditions ZAF corrections can provide analyses accurate to $\sim$1% and hence it is usual in electron-probe microanalysis to try to achieve measured intensity data commensurate with this order of precision.

Now during electron bombardment x-rays are emitted from the target randomly with time. Consequently, the output pulses from the counting electronics also occur randomly in time and thus obey Poisson statistics. The properties of a Poisson distribution are such that if a number of pulses (N) are counted, then the variance (σ^2) of this count will also equal N and the standard deviation (σ) is given by $\sqrt{N}$. The standard error of the count, $\epsilon = \sigma/N$, is given by $1/\sqrt{N}$. It follows that if a standard error of 1% is to be achieved then

$$\frac{1}{\sqrt{N}} = 0.01 \ ,$$

that is, 10 000 counts must be collected.

If the count recorded with the spectrometer set at the peak maximum is N_{P} in the time of t_{P}, then the count-rate I_{P} is $N_{\mathrm{P}}/t_{\mathrm{P}}$ and the standard deviation of the count rate is given by $\sqrt{N_{\mathrm{P}}}/t_{\mathrm{P}}$. Similarly, the background count rate (I_{B}) is $N_{\mathrm{B}}/t_{\mathrm{B}}$ and the standard deviation of the background count rate is $\sqrt{N_{\mathrm{B}}}/t_{\mathrm{B}}$. Thus the nett peak count rate is given by $I_{\mathrm{P}} - I_{\mathrm{B}} = N_{\mathrm{P}}/t_{\mathrm{P}} - N_{\mathrm{B}}/t_{\mathrm{B}}$. The standard deviation of $I_{\mathrm{P-B}}$ can be calculated by noting that the variances of the two counts are additive, that is

$$\sigma^2_{\mathrm{P-B}} = \sigma^2_{\mathrm{P}} + \sigma^2_{\mathrm{B}}$$

$$\sigma_{\mathrm{P-B}} = \sqrt{\frac{N_{\mathrm{P}}}{t^2_{\mathrm{P}}} + \frac{N_{\mathrm{B}}}{t^2_{\mathrm{B}}}}$$

and thus the standard error of the nett peak count rate is

$$\epsilon_{\mathrm{P-B}} = \sqrt{\frac{N_{\mathrm{P}}}{t^2_{\mathrm{P}}} + \frac{N_{\mathrm{B}}}{t^2_{\mathrm{B}}}} \Bigg/ \left(\frac{N_{\mathrm{P}}}{t_{\mathrm{P}}} - \frac{N_{\mathrm{B}}}{t_{\mathrm{B}}}\right) \ . \qquad (6.1)$$

It should be noted that $\epsilon_{\mathrm{P-B}}$ increases rapidly as

$$\frac{N_{\mathrm{P}}}{t_{\mathrm{P}}} \rightarrow \frac{N_{\mathrm{B}}}{t_{\mathrm{B}}} \ , \qquad \text{i.e. as } I_{\mathrm{P}} \rightarrow I_{\mathrm{B}} \ .$$

The ratio of the intensity from the specimen to the intensity from a standard is

$$k = \frac{(I_{\mathrm{P-B}})_{\mathrm{SPEC}}}{(I_{\mathrm{P-B}})_{\mathrm{STND}}}$$

and if the standard error of the two intensities are ϵ_{SPEC} and ϵ_{STND} respectively, then the resultant standard error in the intensity ratio (ϵ_{k}) is given by

$$\epsilon^2_{\mathrm{k}} = \epsilon^2_{\mathrm{SPEC}} + \epsilon^2_{\mathrm{STND}} \ .$$

6.4.2 Counting strategy

If only a limited time (T) is available in which to measure the peak and background intensities, the error in the nett peak intensity ($I_P - I_B$) will depend on the proportion of the total time spent counting the peak and the background.

Since $N_P = I_P t_P$ and $N_B = I_B t_B$ it follows that

$$\epsilon_{P-B} = \sqrt{\frac{I_P}{t_P} + \frac{I_B}{t_B}} \Big/ (I_P - I_B)$$

or

$$\epsilon^2_{P-B} = \left(\frac{I_P}{t_P} + \frac{I_B}{t_B}\right) \Big/ (I_P - I_B)^2$$

Putting $t_P + t_B = T$ and $t_B/t_P = f$

$$\epsilon^2_{P-B} = \frac{\dfrac{I_P(1+f)}{T} + \dfrac{I_B(1+f)}{fT}}{(I_P - I_B)^2} \quad . \qquad (6.2)$$

Differentiating with respect to f gives

$$\frac{d(\epsilon^2_{P-B})}{df} = \frac{1}{T(I_P - I_B)^2}\left(I_P - \frac{I_B}{f^2}\right) \quad .$$

Equating this with zero to locate the minimum gives

$$I_P = \frac{I_B}{f^2} \qquad (6.3)$$

and hence $f = \sqrt{I_B/I_P} = t_B/t_P$, that is the minimum error occurs when the counting times on the peak and background are in proportion to the square root of the respective count rates.

Substituting this result in equation (6.2) gives

$$(\epsilon^2_{P-B})_{min} = \frac{\dfrac{I_P(1+\sqrt{I_B/I_P})}{T} + \dfrac{I_B(1+\sqrt{I_B/I_P})}{T\sqrt{I_B/I_P}}}{(I_P - I_B)^2}$$

$$= \frac{I_P + \sqrt{I_P I_B} + \sqrt{I_P I_B} + I_B}{T(I_P - I_B)^2}$$

$$= \frac{I_P + 2\sqrt{I_P I_B} + I_B}{T(I_P - I_B)^2} = \frac{(\sqrt{I_P} + \sqrt{I_B})^2}{T(I_P - I_B)^2}$$

$$(\epsilon_{P-B})_{min} = \frac{(\sqrt{I_P} + \sqrt{I_B})}{\sqrt{T}(I_P - I_B)} .$$

This equation enables the minimum error to be evaluated.

For example, suppose the x-ray intensity recorded at the peak maximum is 1000 counts per second and the intensity measured to one side of the peak is 50 counts per second. If both peak and background are measured for 50 seconds then the total counting time is 100 seconds. The number of counts recorded for the peak and background are 50 000 and 2500 respectively and the standard error for each is $1/\sqrt{50\,000} \equiv 0.447\%$ and $1/\sqrt{2500} \equiv 2.00\%$. Hence the standard error for the difference I_{P-B} is given by equation (6.1) as 0.482%. The *optimum* ratio of counting times for the peak and background is given by equation (6.3), that is

$$\frac{t_B}{t_P} = \sqrt{\frac{50}{1000}} = 0.224$$

and keeping $t_B + t_P = 100$ seconds it follows that $t_P = 81.7$ seconds and $t_B = 19.3$ seconds. The standard error of the difference I_{P-B} calculated from equation (6.1) is 0.407%. Thus by optimising the counting strategy the error has been reduced by 20% with no increase in total counting time. For machines operated manually this optimum counting strategy is hardly significant since the potential saving in time is minimal compared with operating time, but for automated instruments with computer control the benefits are significant.

6.4.3 Detection sensitivity

The ability to detect the presence of an element by electron-probe microanalysis is determined by whether the count in a peak is significantly greater than the fluctuations in the background.

If the background count rate is I_B and this is measured for a time t then the total background count is $I_B t$. The standard deviation of this count is $\sqrt{I_B t}$. It can be shown that to be 99% certain a peak is present the background level must be exceeded by at least three standard deviations. If the nett peak count rate for the pure element is I_{P-B} then the intensity ratio relative to the pure element for the minimum detectable concentration is given by $(3/I_{P-B})\sqrt{I_B/t}$. Now, to a first approximation, $k \approx c$ and

$$c_{MDL} = \frac{3}{I_{P-B}} \sqrt{\frac{I_B}{t}} . \tag{6.4}$$

where c_{MDL} is the minimum detectable limit. These values are often quoted as measurements of spectrometer performance and a time of 400 seconds is usually used. It can be seen that the minimum detectable concentration decreases as $1/\sqrt{t}$ and hence detection of trace elements requires long counting times.

Equation 6.4 may be arranged as

$$c_{MDL} = \frac{3}{\sqrt{I_{P-B}t}}\sqrt{\frac{I_B}{I_{P-B}}} \ .$$

Hence for $I_{P-B} \sim 10^5$ counts per second, $I_{P-B}/I_B \sim 1000$ and a counting time of 400 seconds

$$c_{MDL} = \frac{3}{\sqrt{(10^5 \times 4 \times 10^2 \times 1000)}} = \frac{3}{2 \times 10^5} = 1.5 \times 10^{-5}$$
$$= 15 \text{ p.p.m.}$$

In less favourable cases (for example, carbon analysis) $I_{P-B} \sim 10^4$ and $I_{P-B}/I_B \sim 100$; then for the same counting time

$$c_{MDL} = \frac{3}{\sqrt{(10^4 \times 4 \times 10^2 \times 10^2)}} = \frac{3}{2 \times 10^4} = 1.5 \times 10^{-4}$$
$$= 150 \text{ p.p.m.}$$

It should be noted that these examples are calculated for pure element standards and hence give an indication of the detection limit to be expected for the element in a matrix of similar atomic number. However, since the background intensity is a strong function of atomic number (see section 2.2.4) actual detection limits are likely to be higher when analysing a light element in a heavy matrix, and lower for a heavy element in a light matrix. Furthermore, if peak overlaps occur the effective background is dominated by the interfering peak which can give an optimistic value for the calculated c_{MDL}.

In addition to the uncertainty in the background associated with Poissonian counting statistics some uncertainty also arises due to instrumental instability which is not accounted for in the above treatment. Consequently, c_{MDL} values are best regarded as a guide to the minimum detectable concentration that can be measured in favourable circumstances.

If the analysed volume is assumed to be a cube with an edge length of 1 μm then its volume is 10^{-12} cm^3. For an element of atomic number 25 and density 8 g cm^{-3}, the number of atoms (M) contained in this volume is

$$\frac{10^{-12}\rho N}{2Z}$$

where N is Avogadro's number, that is,

$$M = \frac{10^{-12} \times 8 \times 6 \times 10^{23}}{50} \approx 10^{11} \text{ atoms} .$$

Assuming an electron microprobe can detect ~10 p.p.m. the number of atoms detectable is 10^6. When this number is compared with the number of atoms in a monolayer (~10^{15} atoms cm^{-2}) it is evident that one-tenth of a monolayer of material can be detected in favourable cases. These calculations are only estimates but nevertheless serve to illustrate the extreme sensitivity of the technique and the importance of surface cleanliness when analysing for trace elements.

6.5 COMPARISON OF WD AND ED METHODS

Wavelength-dispersive and energy-dispersive spectrometers (see Chapters 3 and 4 respectively) are available for fitting to most types of electron-optical column and the decision facing the microanalyst is often which system is best suited to his own particular research needs. Indeed, when the question of purchasing electron-probe microanalysis equipment is raised, the choice of which x-ray analysis system or combination of systems to buy is crucial, the more so when funding is limited.

In this section features of WDS and EDS are compared and their relative merits discussed, from which it will become evident that they should be regarded as complementary systems with their own special advantages.

6.5.1 Stability

Here we are concerned solely with the ability of the x-ray detection and analysis system to give the same result over a period of time whenever a measurement is repeated (within limitations imposed by counting statistics). Obviously, if there is any short-term (minutes or hours) instability in either system it would be unwise to proceed with quantitative analyses. However, variations over a longer period (weeks or months) will not necessarily affect the results when measurements are being taken sequentially on specimen and standards, although it may be dangerous to rely upon 'standardless' methods or reference to stored data on standards (section 11.4). It hardly needs stating that whether WDS or EDS is being employed it is sensible to check at regular intervals, using closely defined analysis conditions, that the system is working within tolerable limits.

There are a number of factors which can lead to long-term instability in WD spectrometers. Firstly, wear in mechanical components can produce backlash and small but significant deviations from exact Bragg focusing conditions. In addition, the analysing crystals may slowly deteriorate (especially fatty acid crystals which become less ordered with age) and this reduces their diffraction efficiency. Short-term instabilities may arise with changes in temperature since this affects the 'd'-spacing of analysing crystals and hence the reproducibility of analysis. (Organic crystals such as PET have particularly large coefficients of expansion.) Furthermore, the stability of the WD system is influenced by the characteristics of the gas proportional counter and, as well as effects caused by temperature and pressure variations in the counter gas, any impurities present (these usually increase as the gas cylinder becomes emptied) will cause a change in the efficiency of x-ray detection.

Although there are no problems of wear of mechanical components with an ED spectrometer, a reduction in its efficiency can occur in the long term owing to the deposition of contaminants on the detector window, an effect which is more serious when carrying out light element analysis (Love *et al.*, 1981). There may be a possibility too that the vacuum in the detection system will deteriorate, gradually causing an increase in the noise and a reduction in detector resolution. As regards problems of short-term instability these are likely to be associated with drift in the electronics; for example, a change in amplifier gain will displace the position of the peaks in the analyser and affect peak intensity measurements (section 5.4.1).

6.5.2 Efficiency of x-ray detection

The region around the specimen in an electron-probe microanalyser is very crowded due to the proximity of pole pieces, specimen stage, optical microscope, electron detectors, etc. In the case of a WD system this imposes severe restrictions on how closely the analysing crystal can approach the specimen and thereby limits the solid angle subtended by the spectrometer at the point of impact of the electron beam on the specimen. The situation cannot be improved by making the diffracting crystal larger since crystal perfection decreases and Bragg conditions can be maintained only over a limited area of its surface. The detector of an ED system subtends, however, a much larger solid angle since it is mounted on a thin tubular extension of the main cryostat and can therefore be positioned close to the specimen. Furthermore, for most of the x-ray energies (>2 keV) used in microanalysis, the detection efficiency of the Si(Li) crystal is approximately 100% compared with the diffraction efficiency of an analysing crystal (WDS) of say 10%. The factors combine to make EDS particularly suited to examining rough surfaces, to analysing thin films and small particles where x-ray intensities are low, and to studying beam-sensitive specimens such as polymeric and biological materials. These attributes also make EDS an attractive proposition for fitting to transmission electron microscopes (see section 13.2.2).

The speed with which ED data covering the whole spectrum can be acquired usually means that a single 100-seconds measurement will provide information on all major ($\sim$1 wt%) elements of atomic number above 11. Furthermore, because the detection efficiency is constant for a wide range of elements (Fig. 5.9) an indication of their relative concentrations can be obtained quickly by inspection.

In WDS it is necessary to search for each element individually and using manual operation this might take over an hour to cover the same range of elements; the actual time will of course depend upon the number of spectrometers fitted to the instrument. The situation is, however, changed by computer control when the search may be reduced to ten minutes or less. Nevertheless, WD detection efficiencies vary widely and discontinuously over the x-ray wavelength range and any estimate of concentration is difficult without recourse to standards or other additional information.

A disadvantage of ED geometry is that the system is more sensitive to extraneous sources of radiation than WDS since the field of view of the Si(Li) detector is limited only by collimation. (The focusing requirements of the crystal spectrometer on the other hand restrict the field of view to a small elliptical region at the point of impact of the electron beam.) As a consequence x-rays excited from the specimen chamber by backscattered electrons and x-ray fluorescence may be recorded along with x-rays from the target in ED spectra, for example, copper and zinc radiations from brass components, aluminium from specimen stubs and silver from conducting paint. These spurious x-ray signals may be largely eliminated by coating the offending components with colloidal graphite since this produces no measurable spectrum with most ED analysers.

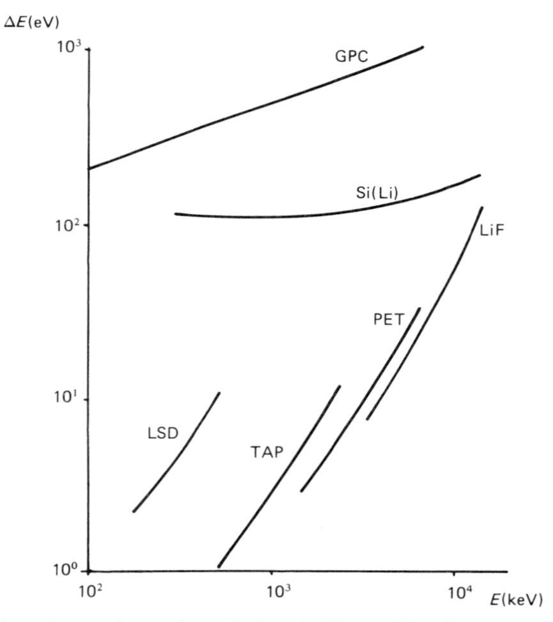

Fig. 6.6 – Comparison of resolution (ΔE) as a function of x-ray energy (E) of a gas proportional counter (GPC), Si(Li) detector and a number of common analysing crystals.

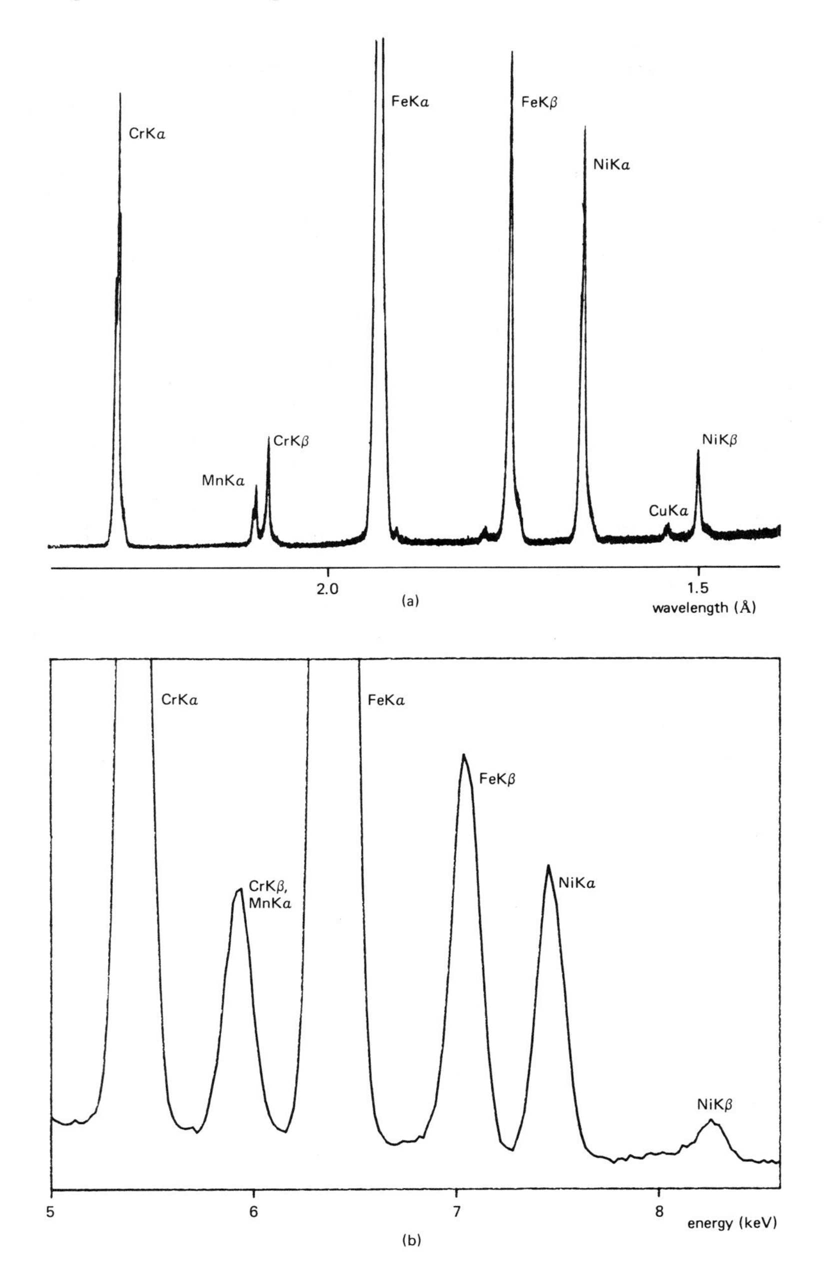

Fig. 6.7 – Comparison of spectra from a steel (1.7 wt % manganese) at 20 kV taken using (a) WDS and (b) EDS; note the manganese Kα peak is not resolved with EDS.

6.5.3 Spectral resolution

The energy resolution of the crystal spectrometer (WDS) is determined by a large number of factors including crystal perfection, Bragg angle, spectrometer focusing conditions, etc. Typically, it takes values of a few electron volts (Fig. 6.6), comparable with the natural linewidth of characteristic x-ray emissions. This resolution usually suffices to separate completely all the emitted lines and overlap problems are rarely encountered. If difficulties arise due to higher order reflections the use of pulse-height analysis (section 3.5) will usually remove them. A typical WD spectrum from an alloy steel is illustrated in Fig. 6.7(a); this shows that the Kα and Kβ peaks from the different elements are all clearly resolved. By comparison ED detectors have resolutions of ~140 eV (Fig. 6.6) and, as a consequence, the peaks are broadened considerably and their maxima correspondingly reduced. The effect is illustrated in Fig. 6.7(b), an ED spectrum taken from the same alloy steel in which a number of line overlaps are present. It may be seen that overlap of the weak Mn Kα line by Cr Kβ makes it difficult to measure this small concentration in the alloy. (The Mn Kβ line is too weak to be of much help in identification.) Such overlap problems may be particularly severe in light-element analysis as has been discussed in section 5.7. Compared with WDS, where peak-to-background ratios may be ~1000:1, corresponding ratios in EDS are 100:1. Hence, since detection sensitivity is a function of peak-to-background ratio (see equation (6.4)), minimum detectable limits for ED analysis are much inferior to those given in WD. As a guide one would expect the ED system to be an order of magnitude less sensitive depending upon the actual spectra being studied.

7

The atomic number correction

M. G. C. COX

When electrons enter a target they lose energy as a result of interactions with the constituent atoms. As well as being dependent upon electron energy, the rate of energy loss is a function of mean atomic number of the material. Hence electrons will behave differently in specimen and standard and a correction termed the stopping power factor (S) is required to take the effect into account. The assumption made in calculating S is that the incident electrons give up all their energy to the target but, in practice, a certain number will be backscattered with finite energy. This fraction is also dependent upon mean atomic number and, consequently a backscatter factor (R), is introduced into the calculation to take account of the corresponding losses in x-ray production which occur. Both backscatter and stopping power factors may, individually, differ greatly (~40%) in specimen and standard but because they tend to oppose one another, the atomic number correction is usually small and rarely exceeds 15%. For example, in targets of high atomic number the greater loss of x-ray production due to backscattering is largely compensated by the greater efficiency with which those electrons remaining in the target generate x-rays. The two factors may be evaluated separately as the following treatment shows.

Consider a multi-element target containing weight fraction c_A of element A. The number of A atoms per unit volume is $c_A(N\rho/A)$, where N is Avogadro's number and ρ the density of the target material. The number of ionisations in path length ds is

$$dn = c_A Q_A \frac{N\rho}{A} ds \qquad (7.1)$$

where Q_A is the ionisation cross-section for A atoms. Rewriting

$$dn = c_A Q_A \frac{N}{A} \frac{d\rho s}{dE} dE \ .$$

Thus the total number (n) of ionisations produced by an electron with an initial energy E_0 coming to rest in the target is obtained by integrating

$$n = c_A \frac{N}{A} \int_{E_0}^{E_c} \frac{Q_A}{dE/d\rho s} \, dE \; . \tag{7.2}$$

The integration limit is set at E_c since electrons below the critical excitation energy are incapable of exciting the characteristic x-ray of interest.

In order to account for those electrons which escape from the target with finite energy, the backscatter factor (R) is introduced and the total number (n_1) of ionisations per average incident electron is reduced accordingly, viz.

$$n_1 = c_A \frac{N}{A} R \int_{E_0}^{E_c} \frac{Q_A}{dE/d\rho s} \, dE \; .$$

Multiplying by the fluorescence yield (ω) gives the x-ray intensity.

$$I = c_A \omega \frac{N}{A} \frac{R}{S} \; ,$$

where S, the stopping power factor, is expressed by

$$\frac{1}{S} = \int_{E_0}^{E_c} \frac{Q_A}{dE/d\rho s} \, dE \; . \tag{7.3}$$

Since the number of electrons impinging on the target in unit time is the same for both specimen and standard and the factors N, A and ω are also common,

$$\frac{I(\text{spec})}{I(\text{stnd})} = \frac{c_A(\text{spec})}{c_A(\text{stnd})} \times \frac{[R/S](\text{spec})}{[R/S](\text{stnd})} = \frac{c_A(\text{spec})}{c_A(\text{stnd})} \left[\frac{R(\text{spec})}{R(\text{stnd})} \frac{S(\text{stnd})}{S(\text{spec})} \right] . \tag{7.4}$$

The bracketed term is known as the atomic number correction. We shall consider the stopping-power factor first.

7.1 STOPPING POWER FACTOR

Methods of evaluating the stopping power factor (S) are based upon equation (7.3) with appropriate expressions introduced for $dE/d\rho s$ and Q.

The formula for $dE/d\rho s$ most frequently used in microanalysis is the continuous-slowing-down approximation of Bethe and Ashkin (1953)

$$-\frac{dE}{d\rho s} = 78\,500 \frac{Z}{AE} \ln(1.166E/J) \tag{7.5}$$

where J, the mean excitation energy of the target atoms, represents the average electron energy loss at each inelastic collision. Bloch (1933) deduced that J was directly proportional to Z with a proportionality constant of 13.5 eV, but many workers prefer to use other expressions for J which incorporate a more complicated atomic number dependence.

As regards the ionisation cross-section (Q), equations based upon the Bethe (1930) formula are commonly used

$$QE_c^2 = \frac{a}{U} \ln (bU)$$

where U is the overvoltage ratio and a and b are constants. This may be simplified since x-ray emission resulting from ionisation of a particular atomic energy level ceases as soon as the energy of the exciting electron falls below the critical excitation energy, when Q must be zero. It follows that since $1/U$ can never be zero, $\ln(bU)$ must be zero when $U = 1$ and hence $b = 1$, that is

$$QE_c^2 = \frac{a}{U} \ln U \ .$$

Green and Cosslett (1961) showed that this was a good approximation to the measured values of Q for overvoltage ratios less than 20. It is convenient in electron-probe microanalysis to rewrite as

$$Q = \frac{c}{U} \ln U \ , \tag{7.6}$$

where c is now a constant for a particular shell of a specific element. Equation (7.6) progressively underestimates the ionisation cross-section at values of U in excess of 20 but, when calculating the atomic number correction using equation (7.3), the term Q appears in both the denominator and numerator and systematic errors at high overvoltage are therefore of little consequence. A more comprehensive discussion of Q values is given by Powell (1976a).

Differences between the various methods which have been proposed to deal with the stopping power factor involve the extent to which mathematical approximations are used in evaluating the integral and the choice of expression for J.

7.1.1 Duncumb–Reed method

Duncumb and Reed (1968) introduced some simplification into the expression for S and then showed that as a result the exact form of the ionisation cross-section was not required. In their method the ratio of $\mathrm{d}E/\mathrm{d}\rho s$ for specimen to standard is assumed to be approximately constant over the electron range, which is reasonable since the only variation is due to the slowly changing logarithmic term in equation (7.5). Thus writing the ratio of stopping power factors from specimen (AB) to standard (A), the $(\mathrm{d}E/\mathrm{d}\rho s)$ terms are taken outside the integral and the integrals then cancel, viz.

$$\frac{\displaystyle\int_{E_0}^{E_c} \frac{Q_A \mathrm{d}E}{(\mathrm{d}E/\mathrm{d}\rho s)_{AB}}}{\displaystyle\int_{E_0}^{E_c} \frac{Q_A \mathrm{d}E}{(\mathrm{d}E/\mathrm{d}\rho s)_A}} = \frac{\displaystyle\frac{1}{(\mathrm{d}E/\mathrm{d}\rho s)_{AB}} \int_{E_0}^{E_c} Q_A \mathrm{d}E}{\displaystyle\frac{1}{(\mathrm{d}E/\mathrm{d}\rho s)_A} \int_{E_0}^{E_c} Q_A \mathrm{d}E} = \frac{(\mathrm{d}E/\mathrm{d}\rho s)_A}{(\mathrm{d}E/\mathrm{d}\rho s)_{AB}} .$$

It is common practice (Thomas, 1963) to evaluate $\mathrm{d}E/\mathrm{d}\rho s$ at a mean energy $\bar{E}$ given by $(E_0 + E_c)/2$. Then

$$S = \frac{Z}{A} \ln \left[1.166 \, \frac{E_0 + E_c}{2J} \right] \tag{7.7}$$

and it is this form which is frequently used in electron-probe microanalysis. (Reed (1975b), however, has suggested $(2E_0 + E_c)/3$ might be a better choice since higher energy electrons contribute more to the total x-ray intensity).

In order to minimise errors, Duncumb and Reed derived a set of J values which gave a best fit to the analytical results from a series of binary alloys, these being carefully chosen to have small absorption and large atomic number corrections. Duncumb *et al.* (1969) later generalised the J values in the following empirical expression

$$J/Z = 14\{1 - \exp(-0.1Z)\} + 75.5/Z^{Z/7.5} - Z/(100 + Z) .$$

This equation is recommended for use with Duncumb and Reed's method since it should minimise any systematic errors resulting from the authors' approximations. However, other workers (for example, Yakowitz *et al.*, 1973) have used the formula of Berger and Seltzer (1964)

$$J/Z = 9.76 + 58.8Z^{-1.19} . \tag{7.8}$$

in conjunction with equation (7.7) without noticeable deterioration in accuracy.

In a multi-element specimen the stopping power is an additive function, that is

$$\frac{\mathrm{d}E}{\mathrm{d}\rho s} = \sum c_i \left(\frac{\mathrm{d}E}{\mathrm{d}\rho s} \right)_i$$

and it follows from Duncumb and Reed's treatment that

$$S = \sum c_i S_i ,$$

where S_i is the stopping power factor for the ith element.

7.1.2 Philibert–Tixier method

Philibert and Tixier (1968a) have demonstrated that it is possible to solve equation (7.3) exactly as follows.

For a multi-element specimen, from equation (7.5)

$$\frac{\mathrm{d}E}{\mathrm{d}\rho s} = \sum c_i \left(\frac{\mathrm{d}E}{\mathrm{d}\rho s}\right)_i = \frac{-78\,500}{E} \sum c_i \frac{Z_i}{A_i} \ln\left(1.166\,\frac{E}{J_i}\right).$$

$$= -\frac{78\,500}{E}\left[\sum c_i \frac{Z_i}{A_i} \ln(1.166E) - \sum c_i \frac{Z_i}{A_i} \ln(J_i)\right]$$

$$= -\frac{78\,500}{E} \sum c_i \frac{Z_i}{A_i}\left[\ln(1.166E) - \frac{\sum c_i \frac{Z_i}{A_i}(\ln J_i)}{\sum c_i \frac{Z_i}{A_i}}\right]$$

$$= -\frac{78\,500}{E}\left(\sum c_i \frac{Z_i}{A_i}\right) \ln\left(1.166\,\frac{E}{\bar{J}}\right), \tag{7.9}$$

where

$$\ln \bar{J} = \sum c_i \frac{Z_i}{A_i} \ln J_i \Big/ \sum c_i \frac{Z_i}{A_i}\ .$$

Since $U = E/E_c$,

$$\frac{\mathrm{d}U}{\mathrm{d}\rho s} = \frac{1}{E_c}\,\frac{\mathrm{d}E}{\mathrm{d}\rho s}$$

$$= -\frac{78\,500}{UE_c^2} \sum c_i \frac{Z_i}{A_i} \ln\left(1.166\,\frac{UE_c}{\bar{J}}\right).$$

Putting

$$m = \sum c_i \frac{Z_i}{A_i} \qquad \text{and} \qquad w = 1.166\,\frac{E_c}{\bar{J}}$$

we obtain

$$\frac{\mathrm{d}U}{\mathrm{d}\rho s} = -\frac{78\,500}{UE_c^2}\, m \ln wU\ .$$

Now

$$\frac{1}{S} = \int_{E_0}^{E_c} \frac{Q}{\mathrm{d}E/\mathrm{d}\rho s}\,\mathrm{d}E = \int_{U_0}^{1} \frac{Q}{\mathrm{d}U/\mathrm{d}\rho s}\,\mathrm{d}U$$

and, substituting for Q (equation (7.6)),

$$\frac{1}{S} = \int_{U_0}^{1} - \frac{\frac{c}{U}\ln U\,\mathrm{d}U}{\frac{78\,500}{UE_c^2}\,m\ln wU} = \frac{cE_c^2}{78\,500\,m}\int_{U_0}^{1} - \frac{\ln U}{\ln wU}\,\mathrm{d}U \ ,$$

The solution of this integral is given by

$$\int_{U_0}^{1} - \frac{\ln U\,\mathrm{d}U}{\ln wU} = U_0 - 1 - \frac{\ln w}{w}\left\{\mathrm{Li}\,(wU_0) - \mathrm{Li}\,(w)\right\} \ .$$

The logarithmic integral (Li) can be evaluated as an infinite series, that is

$$\mathrm{Li}\,(X) = \ln|\ln X| + \sum_{F=1}^{\infty} \frac{(\ln X)^5}{F\cdot F} + D \ ,$$

where D is Euler's constant. Omitting terms common to both specimen and standard gives the stopping power factor as

$$\frac{1}{S} = \frac{1}{m}\left[U_0 - 1 - \frac{\ln w}{w}\{\mathrm{Li}\,(wU_0) - \mathrm{Li}\,(w)\}\right] .$$

In deriving this equation no assumptions or approximations have been introduced other than those inherent in the expressions for the stopping power and ionisation cross-section. Philibert and Tixier used $J = 11.5Z$ for the mean excitation energy although Henoc *et al.* (1973) have adopted the Berger and Seltzer expression (equation (7.8)) in the program COR2.

7.1.3 Love–Cox–Scott method

Love *et al.* (1978a) pointed out that the Bethe expression for $\mathrm{d}E/\mathrm{d}\rho s$ is strictly valid only if $E \gg J$ since, as the electron energy approaches J, it predicts that $\mathrm{d}E/\mathrm{d}\rho s$ will increase to a maximum and drop rapidly to zero at $E = J/1.166$. Such behaviour is obviously unrealistic, the implication being that the electron will cease to lose energy and hence travel indefinitely through the target. The anomaly is a consequence of regarding J as constant whereas in reality it changes discontinuously to lower values whenever the energy of the electron drops below one of the atomic energy levels. Thus E never becomes comparable with J although both values converge as the electron energy approaches the Fermi level. Accordingly systematic errors will be introduced into the atomic number correction which are largest when the matrix has a high atomic number and the energies of incident electrons and characteristic x-rays are low, for example when measuring oxygen radiation (525 eV) in a matrix of lead ($J \sim 1$ keV). The method of Love avoids this problem by empirically modifying the Bethe expression to give better limiting behaviour as $E \to J$.

Equation (7.5) was rewritten in terms of the variable V, where $V = E/\bar{J}$ and $\bar{J}$ refers to a multi-element specimen, that is

$$\frac{\mathrm{d}E}{\mathrm{d}\rho s} = -\frac{1}{\bar{J}}\left(\sum \frac{c_i Z_i}{A_i}\right)\left(\frac{78\,500}{V} \ln 1.166\, V\right)$$

$$= -\frac{1}{\bar{J}}\left(\sum \frac{c_i Z_i}{A_i}\right)\frac{1}{f(V)} ,$$

where $f(V)$ was expressed as

$$f(V) = 1.18 \times 10^{-5}\, V^{1/2} + 1.47 \times 10^{-6}\, V .$$

Thus

$$\frac{\mathrm{d}E}{\mathrm{d}\rho s} = -\frac{\sum \frac{c_i Z_i}{A_i}}{\bar{J}[1.18 \times 10^{-5}(E/\bar{J})^{1/2} + 1.47 \times 10^{-6}(E/\bar{J})]} . \tag{7.10}$$

It can be seen (Fig. 7.1) that whilst equation (7.10) still accords closely with the Bethe expression for $E/\bar{J} > 9$, it has a more realistic limiting behaviour at lower $E/\bar{J}$. (It should be noted that Rao Sahib and Wittry (1974) proposed a similar modification to the low-energy region of the Bethe expression but their treatment leads to rather complex formulae for the stopping power.)

Rewriting equation (7.3) as a function of U gives

$$\frac{1}{S} = \int_{U_0}^{1} \frac{Q}{\mathrm{d}U/\mathrm{d}\rho s}\, \mathrm{d}U .$$

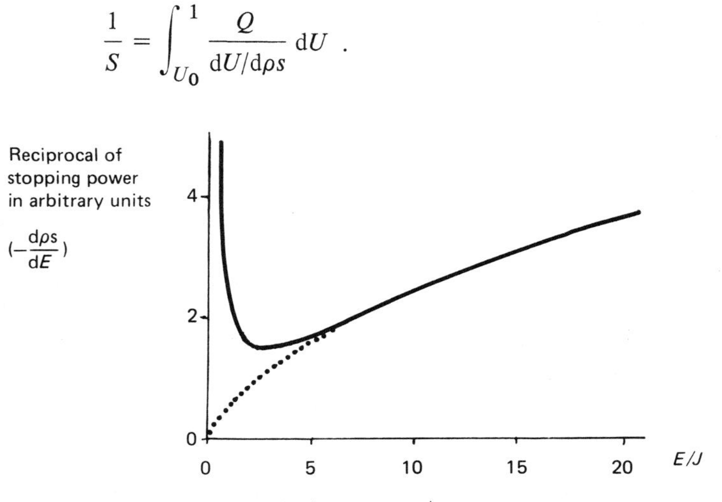

Fig. 7.1 – $\mathrm{d}\rho s/\mathrm{d}E$ versus E/J. Solid line, Bethe and Ashkin (1953); dotted line, Love *et al.* (1978a).

Substituting for Q using equation (7.6) and integrating leads to

$$\frac{1}{S} = \text{constant}\frac{1.4 \times 10^{-6} E_c^2}{\sum (c_i Z_i/A_i)}\left[U_0 \ln U_0 + (1 - U_0)\right]$$

$$\times \left[1 + 16.05\left(\frac{J}{E_c}\right)^{1/2} \frac{U_0^{1/2} \ln U_0 + 2(1 - U_0^{1/2})}{U_0 \ln U_0 + (1 - U_0)}\right] .$$

Some of these factors are common to both specimen and standard and hence cancel

$$\frac{1}{S} = \left[1 + 16.05\left(\frac{J}{E_c}\right)^{1/2} \frac{U_0^{1/2} \ln U_0 + 2(1 - U_0^{1/2})}{U_0 \ln U_0 + (1 - U_0)}\right] \Big/ \sum \frac{c_i Z_i}{A_i} .$$

This can be further simplified since Love demonstrated that

$$\frac{U_0^{1/2} \ln U_0 + 2(1 - U_0^{1/2})}{U_0 \ln U_0 + (1 - U_0)} \approx \left(\frac{U_0^{1/2} - 1}{U_0 - 1}\right)^{1.07}$$

over the range of U_0 encountered in electron-probe microanalysis. The final expression is thus

$$\frac{1}{S} = \left[1 + 16.05\left(\frac{J}{E_c}\right)^{1/2} \left(\frac{U_0^{1/2} - 1}{U_0 - 1}\right)^{1.07}\right] \Big/ \sum \frac{c_i Z_i}{A_i} .$$

The authors use the Bloch formula for the mean excitation energy, J (eV) = $13.5 Z$.

7.2 ELECTRON BACKSCATTERING

Before proceeding to devise an expression for the backscatter factor (R) it is useful to discuss the way in which electron backscattering is affected by target composition and electron energy.

As stated earlier the incident electrons do not lose all their energy in the target since large-angle (elastic) scattering events cause some electrons to approach the free surface with sufficient energy to escape. The fraction of incident electrons which emerge is termed the backscatter coefficient (η), and since the average angular deviation in an elastic collision increases with atomic number so does the backscatter coefficient. The experimental relationship between backscatter coefficient and atomic number is illustrated in Fig. 7.2 for electrons incident perpendicularly to the target. The curve shows an initial rapid rise of η with atomic number after which it slowly approaches a limiting value of ≈ 0.55.

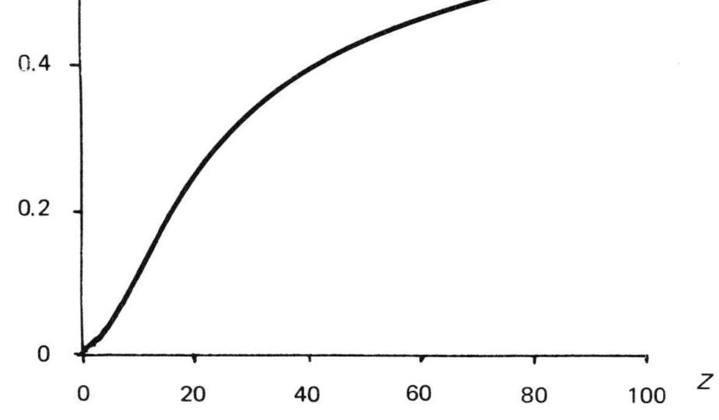

Fig. 7.2 – The relationship between backscatter coefficient (η) and atomic number (Z).

Backscatter coefficients have been measured by many workers (for example, Bishop, 1966b; Heinrich, 1966b; Darlington, 1975) and their values are in substantial agreement, although Heinrich (1968a) disputed the above simple monotonic increase with atomic number since he found small but significant discontinuities which correlated with variations in Z/A. The backscatter coefficient is almost independent of the incident electron energy for elements with atomic numbers close to 50. However, at lower atomic numbers η decreases with increase in incident electron energy whilst the reverse is true for $Z > 50$. These trends apply over the whole energy range of interest in electron-probe microanalysis (Darlington, 1975), although below 5 keV the backscatter coefficient depends in a complex way on the structural, chemical and electronic state of the surface (Bishop, 1966b; Darlington, 1971).

If the electron beam is not perpendicular to the target surface a greater fraction of the incident electrons is backscattered since the angular deviation required for primary electrons to escape from the surface is smaller. Elastic scattering occurs randomly as an electron penetrates into a solid and hence not all the backscattered electrons will escape after having travelled the same distance. Thus not only will they have suffered a different number of elastic scattering events but also they will have lost various amounts of energy by inelastic collisions. With a target of high atomic number, elastic scattering is strong and many of the primary electrons will be backscattered before they have travelled far. Consequently, they escape with little energy loss and their energy distribution will peak sharply at a value not much below the incident electron energy. With low atomic number materials, elastic scattering is weak and electrons traverse, on average, a considerable distance before re-emerging from the target. Hence the energy distribution will be less strongly peaked and have a mean energy which tends to one half the incident electron energy as the atomic number decreases. Some typical backscattered electron energy distributions are shown in Fig. 7.3 (Darlington, 1971), where the electron beam is perpendicular to the target surface and the electron-energy spectrometer accepts electrons back-scattered at 45° to the target surface. The data are presented as a function of the reduced unit $W = E/E_0$, since they are then almost independent of the incident electron energy. The area under each curve represents the fraction (η) of electrons which are backscattered, that is

$$\eta = \int_0^1 \frac{d\eta}{dW} \, dW \; .$$

If the acceptance direction of the spectrometer is moved away from the surface normal, the peak in the spectrum shifts to higher energies; a similar effect is noted as the incident beam direction is moved away from the perpendicular, both effects being due to the smaller nett deviation suffered by incident electrons before being backscattered.

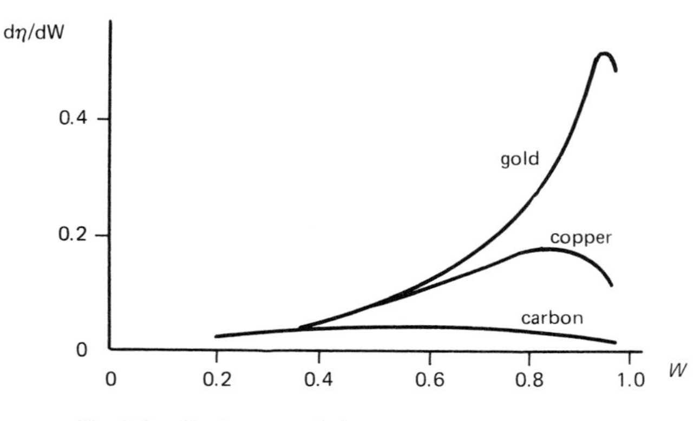

Fig. 7.3 – Backscattered electron energy distributions for carbon, copper and gold (after Bishop, 1966b).

Available information (Bishop, 1966b) indicates that that for multi-element targets,

$$\eta = \sum c_i \eta_i \; , \tag{7.11}$$

where η_i is the backscattering coefficient of the ith element. However, it is unlikely that such a simple additive relationship applies to the dependence of the energy distribution on composition, and computer simulations of electron behaviour suggest (Love *et al.*, 1978a) that the backscattered electron energy distribution for a multi-element target is similar to that of a pure element with the same backscatter coefficient.

7.3 BACKSCATTER FACTOR

The backscatter factor, R, is defined as the ratio of the number (n_1) of ionisations actually generated in the specimen to the number (n) that would have been produced if no electrons had escaped. If n_2 is the number of ionisations lost due to backscattering then

$$R = \frac{n_1}{n} = \frac{n - n_2}{n} = 1 - \frac{n_2}{n} . \tag{7.12}$$

There are few direct experimental measurements of the backscatter factor apart from those of Castaing and Derian (1966). These workers placed above the specimen a thin film which contained a small hole to allow through the electron beam, and then measured the x-ray intensity generated in the film by electrons backscattered from the specimen. After various corrections had been applied to the recorded x-ray intensity, values of $(1 - R)$ were obtained for several accelerating voltages; for example at 30 kV values of 0.157 for copper and 0.302 for gold were reported. However, most workers prefer a different and less direct approach to determine R which involves using measured back-scattered electron energy distributions (see Fig. 7.3) because such data are more readily available (Kulenkampff and Spyra, 1954; Kanter, 1957; Bishop, 1966a, 1966b, 1966c; Darlington, 1971).

The way in which $d\eta/dW$ can be used to establish the magnitude of R is described below. Consider a single backscattered electron with energy E $(= WE_0)$. This would have caused

$$c_A \frac{N}{A} \int_{WE_0}^{E_c} \frac{Q}{dE/d\rho s} \, dE$$

ionisations had it remained in the target. Hence for a fraction (η) of electrons backscattered, the number of ionisations lost is given by

$$n_2 = c_A \frac{N}{A} \int_{W_0}^{1} \frac{d\eta}{dW} \int_{WE_0}^{E_c} \frac{Q}{dE/d\rho s} \, dE \, dW ,$$

where $W_0 = E_c/E_0$.

Substituting in equation (7.12) for n (see equation (7.2)) and n_2 gives

$$R = 1 - \frac{\displaystyle\int_{W_0}^{1} \frac{d\eta}{dW} \int_{WE_0}^{E_c} \frac{Q}{dE/d\rho s} \, dE \, dW .}{\displaystyle\int_{E_0}^{E_c} \frac{Q}{dE/d\rho s} \, dE .}$$

Equations (7.5) and (7.6) are used for $\mathrm{d}E/\mathrm{d}\rho s$ and Q respectively, and the value for R determined by numerical integration.

A number of authors (Duncumb and Reed, 1968; Bishop, 1968; for example) have evaluated R in this way and there is a fair measure of agreement between their published values and the direct experimental measurements of Castaing and Derian. Some typical results expressed as a function of atomic number and overvoltage ratio are shown in Fig. 7.4. The shape of the curves illustrates the effect of increased backscattering with higher atomic number. The greater ionisation loss seen at high overvoltage ratios is due to the increased fraction of electrons backscattered with energies in excess of the critical excitation energy. Duncumb and Reed give a table of R values for pure elements which cover a range of overvoltage ratios. They present the data in terms of W (that is, $1/U$) since the variation in R for a given Z is approximately linear with W which greatly facilitates interpolation. Also the values have been fitted to polynomial equations which allow evaluation for any element at any overvoltage ratio (Duncumb, cited in Beaman and Isasi, 1972).

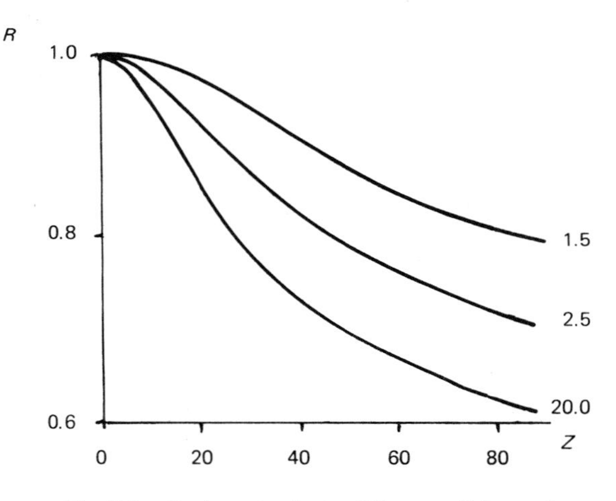

Fig. 7.4 – Backscatter factor (R) versus Z for various overvoltage ratios (after Duncumb and Reed, 1968).

Another formula for R is due to Springer (1966) who chose

$$R = a_0 + a_1 U_0 + a_2 U_0^2 + a_3 U_0^3 + a_4 U_0^4 \quad ,$$

where the coefficients a are each expressed as further polynomials in Z, that is

$$a_i = b_{0i} + b_{1i} Z + b_{2i} Z^2 + b_{3i} Z^3 + b_{4i} Z^4 \quad .$$

Six fourth-order polynomial equations with twenty five b coefficients are required to obtain R (Table 7.1).

Table 7.1 – Coefficients for Springer's polynomials.

i	b_{0i}	b_{1i}	b_{2i}	b_{3i}	b_{4i}
0	1.0088×10^{2}	-7.6070×10^{-1}	-3.5702×10^{-3}	1.6329×10^{-4}	-9.6521×10^{-7}
1	-6.1134×10^{-1}	6.0271×10^{-1}	1.6222×10^{-3}	-4.5936×10^{-4}	2.5267×10^{-6}
2	-9.1447×10^{-1}	2.9326	-1.7636×10^{-1}	2.8558×10^{-3}	-1.3294×10^{-5}
3	-7.9753×10^{-1}	-4.6855	2.9116×10^{-1}	-4.6797×10^{-3}	2.1597×10^{-5}
4	1.3735	1.9015	-1.2703×10^{-1}	2.1144×10^{-3}	-9.8423×10^{-6}

It is usually assumed that, since the backscatter coefficient (η) is a simple additive function of composition (equation (7.11)), R may be expressed in a similar manner, that is

$$R = \sum c_i R_i \ .$$

However, there is no theoretical justification for this and, although it appears to give satisfactory results in much microanalysis work, significant errors may be experienced if the specimen and standard differ widely in mean atomic number.

The reliability of backscatter factor data is determined to a large extent by the accuracy of the electron energy distributions from which they are derived. Systematic errors associated with the expression for $\mathrm{d}E/\mathrm{d}\rho s$ and Q (equations (7.5) and (7.6)) are also present. Relatively few sets of results exist for backscattered electron energy distributions and in the main they are confined to a handful of elements, an incident electron energy of 20 keV, electrons incident along the target surface normal and backscattered electrons emerging at 45° to the specimen surface. In practical analysis, however, accelerating voltages may vary between 5 keV and 40 keV, and the angle of incidence from 20° to 75°. Such effects combine to reduce the versatility of R values derived from experimental $\mathrm{d}\eta/\mathrm{d}W$ data. As an alternative, the backscatter factor has been estimated using Monte Carlo calculations (Chapter 12), where the behaviour of electrons within the target is computed using known laws of electron scattering. Provided that the results from such calculations agree with available experimental information the method can be used with some confidence to predict data for widely ranging input conditions. Following the early Monte Carlo calculations of Green (1963a) and Bishop (1966a), Love *et al.* (1978a) derived a comprehensive series of R values and then used a semi-empirical approach to obtain analytical expressions for R as a function of the backscatter coefficient (η) and overvoltage ratio, viz.

$$R = 1 - \eta[I(U_0) + \eta G(U_0)]^{1.67} \tag{7.13}$$

where $I(U_0)$ and $G(U_0)$ are functions only of the overvoltage ratio:

$$I(U_0) = 0.33148 \ln U_0 + 0.05596 (\ln U_0)^2 - 0.06339 (\ln U_0)^3 + 0.00947 (\ln U_0)^4$$

and

$$G(U_0) = \frac{1}{U_0}[2.87898 \ln U_0 - 1.51307 (\ln U_0)^2 + 0.81312 (\ln U_0)^3 - 0.08241 (\ln U_0)^4] .$$

In order to use equation (7.13) it is necessary to know η and, for normal electron incidence, Love and Scott (1978) gave

$$\eta = \eta_{20}\left[1 + \frac{H(Z)}{\eta_{20}} \ln\left(\frac{E_0}{20}\right)\right],$$

where η_{20} is the backscatter coefficient at an incident electron energy of 20 keV and $H(Z)$ is a function of Z only,

$$\eta_{20} = (-52.3791 + 150.48371Z - 1.67373Z^2 + 0.00716Z^3) \times 10^4$$

and

$$\frac{H(Z)}{\eta_{20}} = (-1112.8 + 30.289Z - 0.15498Z^2) \times 10^4 .$$

It should be noted that equation (7.13) is valid for any target composition, incident electron energy and angle of incidence of electrons provided that the η value appropriate to these parameters is inserted. For inclined specimens the relationship of Darlington (1975) should be used

$$\eta(\alpha) = 0.891\left(\frac{\eta(0)}{0.891}\right)^{\cos\alpha},$$

where α is the angle between the specimen surface normal and the incident electron beam.

The authors recommend that the backscatter coefficient for a multi-element target is calculated using equation (7.11).

7.4 COMPARISON OF ATOMIC NUMBER CORRECTIONS

The atomic number correction is unique in that a particular value is not assigned to a target and it has significance only when the ratio of intensities of a specific radiation from two targets is considered. The question of which proposed correction method works best is not perhaps obvious. Some methods require complex calculations, although this is not usually a great objection when high-speed computers are available. Accuracy and range of applicability are more important features.

The Duncumb–Reed method makes many approximations but the final form of the S factor is simple and the simplifications cause only minor changes to the final correction factor. The backscatter factor may be obtained from tables or it can be calculated from a fifth-order polynomial requiring solution of 30 equations, a straightforward procedure which can be readily programmed on a computer. The Philibert–Tixier approach is a rigorous evaluation of equation (7.3) and, consequently, the only sources of error are those inherent in the equations for Q and $\mathrm{d}E/\mathrm{d}\rho s$, the limitations of which have been discussed above. However, the final expression is complex and evaluation of the logarithmic integrals tedious. The method of Love, Cox and Scott uses a modified Bethe expression and provides a form for equation (7.3) which is easily integrated to give a simple analytical expression. These authors adopt also a different approach to the evaluation of R which is easier to implement than the polynomial expressions of Duncumb *et al.* or Springer. Another feature of their method is its ready extension to situations where the target is tilted with respect to the electron beam.

The input parameters required in all three methods are Z, A, E_0, E_c and J, the first four of which have been defined with considerable accuracy. The mean excitation energy (J) is, however, less easily established although early theoretical work (Bloch, 1933) suggested that J could be equated to $13.5Z$. Relatively few experimental values are available and all have been measured using high energy (~MeV) particles (for example, protons). Such experiments yielded J/Z values of ~11.5 for heavier ($Z > 30$) elements and suggested that J/Z decreased for atomic numbers less than 30; this led Berger and Seltzer to conclude that they were best represented by equation (7.8). The J values of Duncumb and Reed differ markedly from others in the low atomic number region, due possibly to the approximate formula for S used by these authors. It is clear that J is not a quantity which can be defined exactly in the energy range of interest in electron-probe microanalysis and for this reason many workers have adopted the simplest equations for J (that is, $J/Z = 11.5$ or 13.5) in the belief that they are probably no less wrong at low energies than the more complex expressions of Berger and Seltzer or of Duncumb and Reed. Fortunately, the expressions for S are not very sensitive to errors in the J value except in certain extreme cases. For example, if we take the Duncumb and Reed expression

$$S = \frac{Z}{A} \ln\left(1.166 \frac{E_0 + E_c}{2J}\right)$$

and differentiate it with respect to J we get

$$\frac{\partial S}{\partial J} = -\frac{Z}{A}\frac{1}{J} ;$$

therefore

$$\frac{\partial S}{S} = -\frac{\partial J}{J \ln\left[1.166 (E_0 + E_c)/2J\right]} .$$

Taking $E_0 = 20\,\text{keV}$, $E_c = 9\,\text{keV}$ and $J \sim 0.4\,\text{keV}$ as typical microanalysis conditions gives

$$\frac{\partial S}{S} = -0.27 \frac{\partial J}{J} .$$

Thus a 10% error in J changes S by only 2.7%. Furthermore, since a ratio of S for specimen and standard is used in the atomic number correction it is likely that this error is even further reduced. Only when E_0 approaches J do errors in J have a large effect on S. A worse case might be analysis of a high atomic number target at low incident electron energy; for example, for a gold target (Au Mα radiation) at 5 keV, $E_0 + E_c \sim 7\,\text{keV}$ and $J \sim 1\,\text{keV}$, then

$$\frac{\partial S}{S} \sim -0.7 \frac{\partial J}{J} .$$

Such conditions are however, only rarely used in microanalysis.

Apart from the early papers of Martin and Poole (1971) and Beaman and Isasi (1972) few systematic comparisons of atomic number corrections have been carried out, although the consensus of opinion suggests there is little to choose between them (see also section 10.2).

8

X-ray absorption correction

G. LOVE

The absorption correction is often the largest of the three factors employed in the ZAF approach and therefore the precision with which it can be calculated stongly influences the overall reliability of corrected microanalysis data.

In this chapter a number of absorption models are discussed and input parameters, such as mass absorption coefficients, are examined. The depth of treatment devoted to each model is not the same because (a) methods which are based upon the physics of electron-induced x-ray emission generally warrant a more detailed study than those which are purely empirical and (b), some models are of limited usefulness since the authors do not provide sufficient information for them to be employed over a range of experimental conditions. No attempt has been made here to compare and contrast the performance of absorption corrections but in Chapter 10 some of the more promising methods are combined with appropriate atomic number and fluorescence corrections for assessment purposes.

8.1 GRAPHICAL METHODS

8.1.1 $f(\chi)$ Curves

One of the earliest methods of calculating the absorption factor $f(\chi)$ (see section 2.5 for definition) involved a graphical procedure. From the tracer experiments of Castaing and Descamps (1955), values of $f(\chi)$ were calculated as a function of χ and the results presented graphically as in Fig. 8.1. From the figure it may be deduced that, for an electron energy of 20 keV and 35° take-off angle, $f(\chi)$ for aluminium is 0.8 ($\mu/\rho_{\mathrm{Al}}^{\mathrm{Al}} = 386$), that is 20% of the x-rays are absorbed. Unfortunately the absorption correction is not usually established as easily as this because graphs are available only for a few experimental conditions and interpolation is therefore necessary. Additional experimental results of Green (1963b) extended the range of applicability of the method, especially since he was able

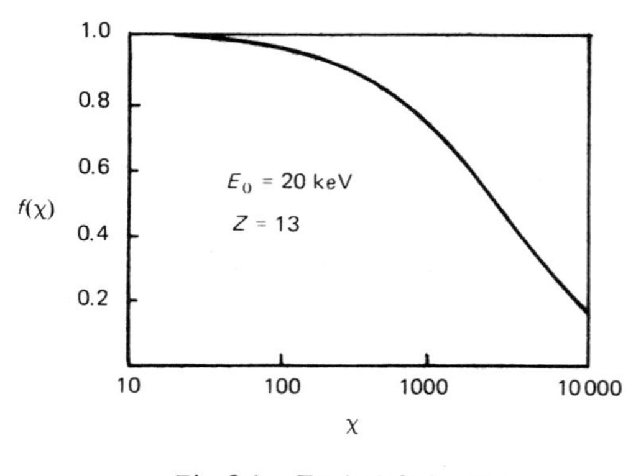

Fig. 8.1 – Typical $f(\chi)$ curve.

to show that $f(\chi)$ was not strongly dependent upon Z and that a series of 'universal' curves of $f(\chi)$ versus χ could be plotted for fixed values of $E_0 - E_c$ (Fig. 8.2).

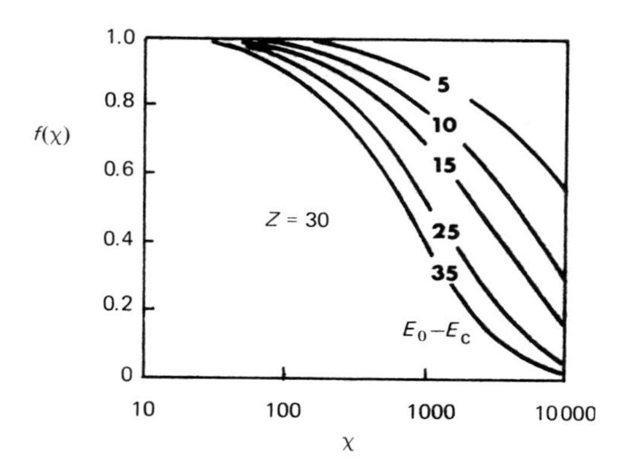

Fig. 8.2 – Set of 'universal' $f(\chi)$ curves (after Green, 1963).

8.1.2 Andersen–Wittry

The objective of condensing $f(\chi)$ data onto a single graph was achieved by Andersen and Wittry (1968) who expressed the absorption correction in terms of two parameters, χ and an effective mean mass depth of x-ray generation, $\overline{\rho z}^*$. Initially the mean mass depth was derived using the Monte Carlo results of Bishop (1966c) but after comparison with experimental $f(\chi)$ data it was given as

$$\overline{\rho z}^* \propto \frac{A}{Z^{1.33}} \left[\frac{E_0^{1.8}}{1.8 \ln (174 E_0/Z)} - \frac{E_c^{1.8}}{1.8 \ln (174 E_c/Z)} \right] .$$

By plotting $f(\chi)$ against $\chi\overline{\rho z}^*$ a universal curve (Fig. 8.3) was constructed which enabled the absorption correction to be determined for a range of experimental situations. However when $f(\chi)$ was less than 0.5 it was not possible to fit all the experimental results to a single line and the curve was therefore split into two, the upper portion (BC) being used for elements with atomic numbers greater than 26 and the lower (BD) for lighter elements. Unfortunately the vertical scale which Andersen and Wittry chose to adopt made their method particularly prone to reading errors. Much later Rao-Sahib and Wittry (1974) overcame this objection by deriving a fifth-order polynomial to describe the lower of the two curves.

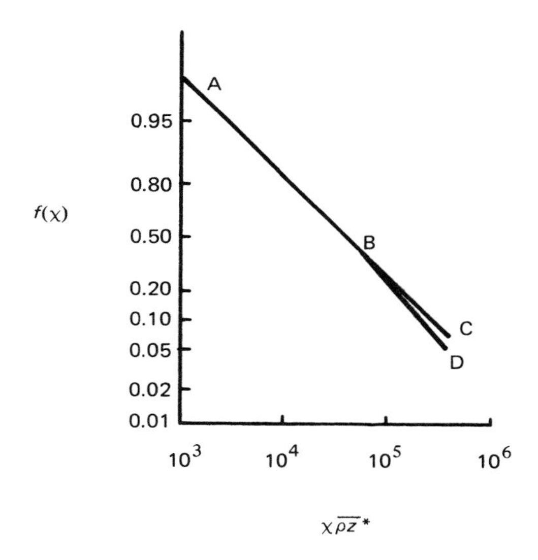

Fig. 8.3 – Universal $f(\chi)$ curve of Andersen and Wittry.

The Andersen–Wittry model has been tested by Love and Scott (1980) and shown to give rise to substantial errors when $f(\chi)$ is less than 0.6, that is, when absorption corrections are large, $f(\chi)$ is not well represented simply as a function of the product of χ and $\overline{\rho z}^*$. Because of these limitations the method will not be considered further.

8.2 ANALYTICAL EXPRESSIONS OF PHILIBERT

Graphical methods are generally unpopular because of the errors involved in extracting information from the curves and the time-consuming nature of the process, especially when several iterations are required. It is much more desirable if $f(\chi)$ can be described by an appropriate formula and in 1963 Philibert derived an equation which is still widely used today (Philibert, 1963).

8.2.1 Derivation

An expression was obtained for the normalised x-ray depth distribution, $\phi(\rho z)$, by considering the physical processes involved in electron-induced x-ray emission. In principle, the atomic number and absorption effects may then be dealt with as a single entity because the emitted x-ray intensity, $F(\chi)$, can be derived directly

$$F(\chi) = \epsilon \int_0^\infty \phi(\rho z) \exp(-\chi\rho z)\, \mathrm{d}\rho z \quad . \tag{8.1}$$

The term ϵ takes into account x-ray collection efficiency but henceforth it will be omitted as it cancels out when ratios are taken using the x-ray emission from a standard. The form of $\phi(\rho z)$ may be established by considering the amount of ionisation occurring in a layer of mass thickness $\Delta\rho z$ at a mass depth ρz in the specimen. This will depend upon:

(a) the number of electrons travelling in the forward direction reaching ρz in the sample;
(b) the number of electrons backscattered from deeper layers which pass through $\Delta\rho z$ a second time;
(c) the average path length of the electrons in the layer.

Philibert assumed that the number of electrons, $n_{\rho z}$, reaching ρz in the sample could be described using the attenuation law of Lenard and Becker (1972)

$$n_{\rho z} = n_0 \exp(-\sigma\rho z) \tag{8.2}$$

where n_0 is the number of electrons incident on the specimen surface and σ is a constant (Lenard's constant) for a particular incident electron energy. The number of electrons passing through $\Delta\rho z$ may be represented by $rn_{\rho z}$ which is composed of $n_{\rho z}$ travelling in the forward and $(r-1)n_{\rho z}$ in the reverse direction. The path length of the ith electron in the layer will be $(\Delta\rho z)/\cos\alpha_i$, α_i being the angle that the electron trajectory makes with the normal to the specimen surface. Then the effective average path length per electron reaching mass depth ρz may be described by $R_{\rho z}\Delta\rho z$, where

$$R_{\rho z}\Delta\rho z = \frac{\Delta\rho z}{n_{\rho z}} \sum_{i=1}^{i=rn_{\rho z}} \frac{1}{\cos\alpha_i} \quad .$$

In order to establish the general form of $R_{\rho z}$ it is necessary to determine its value at $\rho z = 0$ and $\rho z = \rho z_\mathrm{d}$, where ρz_d is the depth at which the electrons are completely diffused.

For electrons incident normally on the surface of the specimen $1/\cos\alpha = 1$ but some electrons will be backscattered so that the value of R_0 will be greater than unity. At greater depths in the specimen the value of R increases until the electrons are completely diffused. When the diffusion depth is reached there will be as many electrons travelling in the forward direction through $\Delta\rho z$ as those backscattered through it and hence $r = 2$. If any electron direction is assumed to be equally probable then the average value of α is 60°; thus R_∞, the limiting value of $R_{\rho z}$, is 4. Philibert makes the assumption that the transition from R_0 to R_∞ can be described by an exponential law as follows

$$R_{\rho z} = R_\infty - (R_\infty - R_0) \exp(-\rho z/h') \ . \qquad (8.3)$$

The value of h' is adjusted such that at the diffusion depth (ρz_d), $R_{\rho z}$ differs by only 0.1 from its limiting value, that is

$$\exp(-\rho z_d/h') = 0.1$$

and therefore

$$h' = 0.434 \rho z_d \ . \qquad (8.4)$$

The diffusion depth is determined using the scattering law of Bothe (1929) which takes the form

$$\alpha_p = \frac{400}{E_0}\left(\frac{Z^2}{A}\rho z\right)^{1/2} ,$$

where α_p is the most probable scattering angle. When $\alpha_p = 45°$ the electrons must be completely diffused and we may write

$$\rho z_d = \frac{\pi^2}{16}\left(\frac{E_0}{400}\right)^2 \frac{A}{Z^2} \ .$$

Substituting ρz_d into equation (8.4) leads to

$$h' = 1.67 \times 10^{-6} \frac{A}{Z^2} E_0^2 \ . \qquad (8.5)$$

The complete expression for the x-ray distribution with depth in the sample normalised with respect to a thin film of the same composition (as described in section 2.3) may now be constructed.

The effective path length of all electrons in a layer $\Delta\rho z$ at mass depth ρz in the specimen is equal to $R_{\rho z} n_{\rho z} \Delta_{\rho z}$ and the number of characteristic x-rays (ΔI) generated in the layer by these electrons will be given by

$$\Delta I = R_{\rho z} n_{\rho z} \frac{N}{A} Q(E) \omega \Delta\rho z \ ,$$

where N is Avogadro's number, $Q(E)$ is the ionisation cross-section per atom and ω is the fluorescence yield. The x-ray intensity (ΔI_0) produced in an isolated thin film is equal to $n_0(N/A)Q(E_0)\omega\Delta\rho z$. Hence

$$\phi(\rho z) = \frac{\Delta I}{\Delta I_0} = R_{\rho z}\frac{n_{\rho z}}{n_0}\;\frac{Q(E)}{Q(E_0)}\ .$$

The ionisation cross-section is a function of electron energy (see section 2.3), and the energy possessed by a particular electron in a layer $\Delta\rho z$ at mass depth ρz in the specimen will be dependent upon the amount of inelastic scattering it suffered before passing through the layer. In practice the electrons will have a range of energies so that no single value of Q will be appropriate. It is necessary therefore to introduce some approximation and Philibert chose to make Q independent of electron energy. In this case

$$\phi(\rho z) = R_{\rho z}\frac{n_{\rho z}}{n_0}$$

and by substituting for $n_{\rho z}$ and $R_{\rho z}$ (equations (8.2) and (8.3)) we obtain

$$\phi(\rho z) = \exp(-\sigma\rho z)\,[R_\infty - (R_\infty - R_0)\exp(-\rho z/h')]\ . \tag{8.6}$$

The intensity of x-ray emission may now be calculated using equation (8.1) to give

$$F(\chi) = R_\infty\left[\frac{1}{\sigma+\chi} - \frac{1-R_0/R_\infty}{\sigma+\chi+1/h'}\right]$$

which may rewritten as

$$F(\chi) = 4\left[\frac{1}{\sigma+\chi} - \frac{1-\phi(0)/4}{\chi+\sigma(1+h)/h}\right]\ , \tag{8.7}$$

where $\phi(0)$ is the surface ionisation function and is equivalent to R_0. The parameter h is equal to $\sigma h'$. Hence from equation (8.5)

$$h = 1.67\times 10^{-6}\,\frac{A}{Z^2}\,\sigma E_0^2\ .$$

Attenuation experiments of Lenard and Becker (1927) indicated that for electron energies greater than 20 keV, σ was inversely proportional to E_0^2 and thus h = const. A/Z^2. Philibert showed that the expression for $F(\chi)$ appeared quite promising when compared with experimental results of Castaing and Descamps (1955) but unfortunately in 1963 the variation of $\phi(0)$ with Z, E_0 and U_0 was not understood and hence the formula could not be generally applied.

Nevertheless, an absorption correction can be easily calculated from equation (8.7) since $f(\chi) = F(\chi)/F(0)$, that is,

$$f(\chi) = \frac{1 + \phi(0)h\chi/[(4 + \phi(0)h)\sigma]}{(1 + \chi/\sigma)\left(1 + \dfrac{h}{1+h}\chi/\sigma\right)} \quad . \tag{8.8}$$

Equation (8.8) will henceforth be referred to as the rigorous Philibert absorption correction to distinguish it from his simplified procedure. The simplification involved setting $\phi(0) = 0$ and $h = 1.2A/Z^2$, which Philibert justified on the grounds that $f(\chi)$ was not particularly sensitive to the exact form of either $\phi(0)$ or h. Thus the above expression was reduced to

$$f(\chi) = \left[(1 + \chi/\sigma)\left(1 + \frac{h}{1+h}\chi/\sigma\right)\right]^{-1} \quad . \tag{8.9}$$

8.2.2 Selection of σ and h values for the simplified model

There has been some debate about the most appropriate values of σ to insert in equation (8.9). Philibert employed the original values of Lenard but if these are used no account is taken of the fact that an electron is incapable of producing x-rays once its energy falls below the critical ionisation energy of the x-ray line being measured. A modified formula for σ was proposed by Duncumb and Shields (1966) to compensate. This was based upon experimental results (Cosslett and Thomas, 1964c) which showed that the amount of electron penetration into a solid is proportional to $E_0^{1.5}$. Duncumb and Shields therefore felt that the range over which the electrons could generate characteristic radiation would be proportional to $(E_0^{1.5} - E_c^{1.5})$. A new constant in the formula for σ was now required and this was found by fitting equation (8.9) to the $f(\chi)$ curves of Castaing and Descamps (1955) and Green (1963b). The final form of the equation was

$$\sigma = \frac{2.39 \times 10^5}{E_0^{1.5} - E_c^{1.5}} \quad .$$

Subsequently Heinrich (1967) re-examined the question of the energy dependence of σ. He began by re-writing equation (8.9) as

$$\frac{1}{f(\chi)} = 1 + \left(\frac{1+2h}{1+h}\right)\frac{\chi}{\sigma} + \frac{h}{(1+h)}\frac{\chi^2}{\sigma^2} = 1 + \alpha_1\gamma\chi + \alpha_2\gamma^2\chi^2 \quad . \tag{8.10}$$

The constant in the σ term and the compositional dependence (h) of the absorption correction are incorporated in α_1 and α_2, while γ takes into account energy dependence, that is $\gamma = E_0^n - E_c^n$. Heinrich used Green's $f(\chi)$ results on aluminium to produce a series of plots of $1/[f(\chi)]$ versus χ (one for each voltage). From equation (8.10) it follows that, providing absorption is not large, these will have gradients of $\alpha_1 \gamma$. Next a graph was constructed of log $\alpha_1 \gamma$ versus log E_0. The reason for selecting the Kα x-ray absorption data from an element such as aluminium now becomes obvious; the value of E_c is small so that the gradient of the graph will be equal to n. Heinrich found that $n = 1.65$ fitted the experimental $f(\chi)$ results quite well and therefore proposed that

$$\sigma = \frac{4.5 \times 10^5}{E_0^{1.65} - E_c^{1.65}} \ .$$

Duncumb *et al.* (1969) tested this equation and confirmed that improved accuracy was obtained by its adoption. Most absorption corrections are currently carried out using equation (8.9) and incorporate Heinrich's expression for σ and $h = 1.2\,A/Z^2$.

Because h is composition-dependent it must be averaged when dealing with multi-element specimens. Many methods have been used but the most frequently employed is weight averaging according to

$$h = \sum c_i h_i \ .$$

Love *et al.* (1975) have pointed out that on theoretical grounds it is more correct to use atomic fractions (a_i) as suggested by Martin and Poole (1971). In this case

$$h = \frac{B \sum a_i A_i}{\sum a_i (Z_i)^2}$$

where B is a constant and equals 1.2 if Heinrich's formula is adopted. However, in practice the method of averaging does not significantly affect $f(\chi)$ unless the absorption correction is large.

8.2.3 Selection of σ, h and $\phi(0)$ values for the rigorous model

It was recognised that where absorption effects were likely to be large the simplified Philibert treatment (with $\phi(0) = 0$) would be inappropriate. Duncumb and Melford (1966a) in their analysis of carbides by direct measurement of the light element adopted, therefore, the more rigorous expression (equation (8.8)). Because approximations had been made in deriving this equation they optimised the σ and h formulae by adjusting them to give the best fit to experimental $f(\chi)$ data of Castaing (1960) and Green (1963b). The equations obtained were

$$h = 4.5 \frac{A}{Z^2}$$

and

$$\sigma = \frac{2.54 \times 10^5}{E_0^{1.5} - E_c^{1.5}} \ ;$$

values of $\phi(0)$ were determined from the Monte Carlo calculations of Bishop (1965) and also by direct measurement.

Tests of the rigorous Philibert absorption correction incorporating the modifications described above, showed that it did not account satisfactorily for the observed intensity changes with kilovoltage of carbon K x-rays from a diamond sample. However, when used to predict the ratio of carbon K-emission from a specimen (silicon carbide) and standard (diamond) a measure of agreement was obtained with experimental results. These findings suggest that although the magnitude of the correction is in error some degree of compensation will be achieved where standards are employed.

8.3 METHODS BASED UPON THE PHILIBERT APPROACH

Since the inception of Philibert's absorption correction models there have been a number of modifications proposed. Some have involved simplifications to enable the correction to be carried out by hand or to speed computer processing of large amounts of data. Others have incorporated refinements intended to improve the range of applicability of the method.

8.3.1 Heinrich's simplifications

Heinrich *et al.* (1972) produced a modified form of Philibert's formula (equation (8.9)) which contained no atomic number dependence and it was subsequently employed in the computer program FRAME B (Yakowitz *et al.*, 1973). It takes the form

$$\frac{1}{f(\chi)} = 1 + 3 \times 10^{-6}(E_0^{1.65} - E_c^{1.65})\chi + 4.5 \times 10^{-13}(E_0^{1.65} - E_c^{1.65})^2\chi^2 \ .$$

The argument the authors advanced for introducing the expression was that the dependence of $f(\chi)$ upon target composition was not proven, having reached this conclusion after studying available $f(\chi)$ curves and analysing well characterised specimens. However, in spite of the fact that the parameter h has been eliminated from the equation, very little computer processing time has been saved because the value of χ is still a function of the elemental weight concentrations and $f(\chi)$ must therefore be determined by an iterative procedure. Furthermore it is obvious from Monte Carlo studies that the shape of the $\phi(\rho z)$ curve changes with atomic number and these changes will affect the magnitude of the absorption correction. The fact that no atomic number dependence was found in the original study carried out by Heinrich suggests that not enough reliable $f(\chi)$ data were available to establish the trend.

8.3.2 Reuter

Reuter (1972) re-examined the physical principles upon which the Philibert model was constructed and employed equation (8.6) as a basis for development. New empirical expressions for R_∞, h' and $n_{\rho z}$ were derived using the experimental data of Cosslett and Thomas (1964a, 1964b, 1964c, 1965) on electron transmission through thin metal foils.

Cosslett and Thomas showed that the angular distribution of electrons in the diffusion region closely obeys a $\cos^2\alpha$ law and hence the mean value of $1/\cos\alpha_i$ for electrons travelling in the forward direction is

$$\frac{\int_0^{\pi/2} [\cos^2\alpha\, 2\pi \sin\alpha/\cos\alpha]\, d\alpha}{\int_0^{\pi/2} \cos^2\alpha\, 2\pi \sin\alpha\, d\alpha} = 1.5 \; .$$

Under diffusion conditions, where there will be as many electrons travelling in the backward as the forward direction, R_∞ takes the value 3.

Rather than following Philibert and using Bothe's law based upon the transmission of fast electrons through thin foils to obtain the most probable scattering angle (α_p), Reuter employed the data of Cosslett and Thomas for 20–30 keV electrons. This gave

$$\alpha_p^2 = 1.2 \times 10^4 \frac{Z^{1.5}}{AE_0} \rho z \; .$$

In all targets $\alpha_p \to 38°$ when the depth of complete diffusion has been reached and

$$\rho z_d = 3.67 \times 10^{-5} E_0 A / Z^{1.5} \; .$$

Using the criterion proposed by Philibert that at the diffusion depth $R_{\rho z}$ is within 10% of the limiting value (R_∞), a new value of h' can now be established

$$\begin{aligned} h' &= 0.434\, \rho z_d \\ &= 1.59 \times 10^{-5} E_0 A / Z^{1.5} \; . \end{aligned}$$

The results of Cosslett and Thomas (1964b) indicate that if the depth of diffusion is not exceeded, electron transmission decreases linearly with depth in the specimen. Reuter therefore developed the following expression,

$$n_{\rho z} = n_0 (1 - 4 \times 10^4 Z^{0.5} \rho z / E_0^{1.7})$$

to describe experimental results when ρz is less than ρz_d. At greater depths the exponential transmission law (equation (8.2)) has been adopted with σ set equal to $4.5 \times 10^5 / E_0^{1.65}$. The precise depth at which the transition from one equation to the other takes place is established by prior determination of the crossover point of the two curves.

The formula used to represent the surface ionisation function is

$$\phi(0) = 1 + 2.8(1 - 0.9/U_0)\eta \tag{8.11}$$

where $U_0 = E_0/E_c$ and η, the backscatter factor, is given by

$$\eta = -0.0254 + 0.016Z - 1.86 \times 10^{-4}Z^2 + 8.31 \times 10^{-7}Z^3 \ .$$

Reuter has demonstrated that the equation for $\phi(0)$ fits his experimental data very well and is also in reasonable agreement with the calculations of Duncumb and Melford (1966b).

The ionisation cross-section adopted is $Q = (1/U^{0.7})\ln U$, which was empirically determined by fitting to experimental data of Clark (1935) and Green (1962).

Now the absorption correction can be evaluated by inserting the new expression for $\phi(\rho z)$ in

$$f(\chi) = \frac{\int_0^\infty \phi(\rho z) \exp(-\chi \rho z) \, \mathrm{d}\rho z}{\int_0^\infty \phi(\rho z) \, \mathrm{d}\rho z}$$

although, because of the greater complexity of formulae employed, numerical integration is required to obtain a value of $f(\chi)$.

8.3.3 Ruste–Zeller

Ruste and Zeller (1977) have proposed new h and σ formulae for incorporation into the rigorous absorption correction of Philibert (equation (8.9)). From an examination of the experimental data of Cosslett and Thomas (1964b) and Bishop (1967) they concluded that σ should be modified since, as Reuter (1972) pointed out, electron transmission is not well represented by an exponential law. However Ruste and Zeller prefer to retain this general form while increasing the flexibility of the Philibert treatment by making σ atomic number dependent. This permitted a somewhat better correspondence with the experimental electron transmission data than could be achieved using the formula of Duncumb and Melford (see section 8.2.3). The σ term may be written as

$$\sigma = \frac{\sigma_0 \times 10^5}{E_0^n - E_c^n} \ .$$

Ruste and Zeller chose to represent both σ_0 and n by equations of the type

$$X = X_\infty + a \exp(-bZ^p)$$

where a, b, p and X_∞ are constants. More specifically

$$\sigma_0 = 4.5 + 8.95 \exp\left(-\frac{Z^{2.5}}{500}\right)$$

and

$$n = 1.65 + 1.08 \exp\left(-\frac{Z^2}{100}\right) .$$

This means that for specimens composed of high atomic number elements $\sigma_0 \rightarrow 4.5$ and $n \rightarrow 1.65$, which are of course the values proposed by Heinrich for use in the *simplified* absorption correction of Philibert (equation 8.9)).

The formula for h given by Ruste and Zeller incorporates an energy dependent term

$$h = \frac{0.45\,\sigma_0}{E_0^{(n-1.5)}} \frac{A}{Z^2}$$

and for $Z > 24$ and $E_0 = 30$ keV, h approximates to $1.2A/Z^2$.

Athough h is averaged for multi-element systems according to $\Sigma c_i h_i$, σ_0 and n are not averaged at all despite the fact that they are now dependent upon atomic number. Ruste and Zeller have indicated that they take values approximate for the element and radiation being measured.

8.4 METHODS BASED ON $\phi(\rho z)$ CURVE FITTING

When the x-ray depth distribution is known, it is possible to treat the atomic number and absorption corrections as a single entity provided that the equation describing $\phi(\rho z)$ is a good fit to the true x-ray distribution with depth. If there is not a good match, the equation for $\phi(\rho z)$ may still provide a reasonably accurate absorption correction but the atomic number correction is then best calculated using a separate treatment.

A set of accurate $\phi(\rho z)$ curves is, of course, an essential prerequisite and in Table 8.1 is a list of experimental $\phi(\rho z)$ data currently available. Most of the experimental $\phi(\rho z)$ curves have been produced using the tracer method developed by Castaing and Descamps described in section 2.3. However, another way in which $\phi(\rho z)$ curves can be determined is by use of the wedge technique (Schmitz *et al.*, 1969). A wedge of element A is formed on top of element B (as in Fig. 8.4), the choice of element B in the specimen being governed by similar considerations to those controlling the choice of the tracer in the tracer method. Electron-probe measurements are carried out at successive positions along the line X–Y and the intensity of characteristic emission from element A is recorded. These data are normalised with reference to the intensity of x-ray emission (I_∞) from a region of the wedge where the thickness of element A exceeds the depth of x-ray generation. If the angle of the wedge (δ) is known accurately the thickness (z) of

Table 8.1 – Experimentally determined $\phi(\rho z)$ data.

System	Tracer	Voltage (keV)	Analysed radiation	Angle of incidence of e. beam	Reference
Cu–Zn	Zn	29	Zn Kα	80°	Castaing & Descamps (1955)
Au–Bi	Bi	29	Bi Lα	80°	
Al–Cu	Cu	29	Cu Kα	80°	
Al–Bi	Bi	29	Bi Lα	80°	
Al–Cr	Cr	29	Cr Kα	80°	
Al–Cu	Cu	29	Cu Kα	80°	
Al–Mg	Mg	10, 15, 20	Mg Kα	90°	Castaing & Henoc (1966)
Al–Mg	Mg	25, 29	Mg Kα	90°	
Cu–Zn	Zn	14.2, 15.9	Zn Kα	90°	Shimizu, Murata &
Cu–Zn	Zn	19.0, 21.1	Zn Kα	90°	Shinoda (1966)
Cu–Zn	Zn	23.8, 26.5	Zn Kα	90°	Shimizu, Kishimoto,
Cu–Zn	Zn	28.5, 31.7	Zn Kα	90°	Shirai, Murata, Shinoda &
Cu–Zn	Zn	33.1, 37.0	Zn Kα	90°	Miura (1966)
Cu–Zn	Zn	38.0, 42.3	Zn Kα	90°	Shinoda (1966)
Cu–Zn	Zn	13.4, 18.2	Zn Kα	60°	Brown (1966)
Cu–Zn	Zn	23.1, 27.6	Zn Kα	60°	Brown (1969)
Ti–V	V	17, 20	V Kα	90°	Vignes & Dez (1968)
Ti–V	V	25, 29	V Kα	90°	
Pb–Bi	Bi	29, 33	Bi Lα	90°	
Cu	Wedge	20, 25.5	Cu Kα	90°	Schmitz, Ryder &
Cu	Wedge	30	Cu Kα	90°	Pitsch (1969)
Cu–Zn	Zn	12, 15, 20	Zn Kα	90°	Durr, Hofer, Schulz &
Cu–Zn	Zn	25, 30	Zn Kα	90°	Wittmaack (1971)
C	Wedge	5, 10, 15	C Kα	90°	Furuno & Izui (1971)
Al–Si	Si	15, 20, 25, 30	Si Kα	90°, $62\frac{1}{2}$°	Brown & Parobek (1972)
Al–Zn	Zn	15, 20, 25, 30	Zn Kα	90°, $62\frac{1}{2}$°	Brown & Parobek (1973)
Al–Cd	Cd	15, 20, 25, 30	Cd Lα	90°, $62\frac{1}{2}$°	
Al–Bi	Bi	20, 25, 30	Bi Lα	90°, $62\frac{1}{2}$°	
Cu–Si	Si	15, 20, 25, 30	Si Kα	90°, $62\frac{1}{2}$°	
Cu–Zn	Zn	15, 20, 25, 30	Zn Kα	90°, $62\frac{1}{2}$°	
Cu–Cd	Cd	15, 20, 25, 30	Cd Lα	90°, $62\frac{1}{2}$°	
Cu–Bi	Bi	20, 25, 30	Bi Lα	90°, $62\frac{1}{2}$°	
Ag–Si	Si	15, 20, 25, 30	Si Kα	90°, $62\frac{1}{2}$°	
Ag–Zn	Zn	15, 20, 25, 30	Zn Kα	90°, $62\frac{1}{2}$°	
Ag–Cd	Cd	15, 20, 25, 30	Cd Lα	90°, $62\frac{1}{2}$°	
Ag–Bi	Bi	20, 25, 30	Bi Lα	90°, $62\frac{1}{2}$°	
Au–Si	Si	15, 20, 25, 30	Si Kα	90°, $62\frac{1}{2}$°	
Au–Zn	Zn	15, 20, 25, 30	Zn Kα	90°, $62\frac{1}{2}$°	
Au–Cd	Cd	15, 20, 25, 30	Cd Lα	90°, $62\frac{1}{2}$°	
Au–Bi	Bi	20, 25, 30	Bi Lα	90°, $62\frac{1}{2}$°	

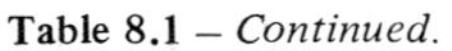

Table 8.1 – *Continued.*

System	Tracer	Voltage (keV)	Analysed radiation	Angle of incidence of e. beam	Reference
Al–Si	Si	6, 8, 10	Si Kα	90°	Brown & Parobek (1974)
Ni–Si	Si	6, 8, 10	Si Kα	90°	Brown & Parobek (1976)
Ag–Si	Si	6, 8, 10	Si Kα	90°	
Au–Si	Si	6, 8, 10, 12	Si Kα	90°	
Al–Cu	Cu	12, 15	Cu Kα	90°	
Ni–Cu	Cu	12, 15	Cu Kα	90°	
Ag–Cu	Cu	12, 15	Cu Kα	90°	
Ag–Zn	Zn	15, 20, 25, 30	Zn Kα	90°	
Al–Zn	Zn	15, 20, 25, 30	Zn Kα	90°	Parobek & Brown (1974)
Cu–Zn	Zn	15, 20, 25, 30	Zn Kα	90°	
Au–Zn	Zn	15, 20, 25, 30	Zn Kα	90°	
Systems listed by Brown & Parobek (1972 & 1973)		20–30		Range of tilt not specified	Robinson & Brown (1978)
C–Al	Al	18.1	Al Kα	90°	Weisweiler (1975)

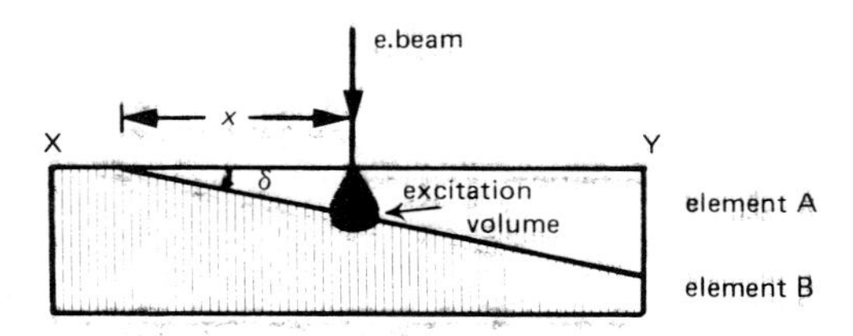

Fig. 8.4 – Determination of x-ray depth distributions using the wedge technique.

element A can be calculated at each point since $z = x \tan \delta$. Now if $I_{\rho z}$ represents the x-ray intensity from a region of thickness z and density ρ, we may write

$$\frac{I_{\rho z}}{I_\infty} = \frac{\int_0^{\rho z} \phi(\rho z) \exp(-\chi \rho z) \mathrm{d}\rho z}{\int_0^{\infty} \phi(\rho z) \exp(-\chi \rho z) \mathrm{d}\rho z} , \tag{8.12}$$

that is,

$$\phi(\rho z) \exp(-\chi \rho z) = \left[\int_0^{\infty} \phi(\rho z) \exp(-\chi \rho z) \mathrm{d}\rho z \right] \frac{\mathrm{d}}{\mathrm{d}\rho z} \left(\frac{I_{\rho z}}{I_\infty} \right)$$

and

$$\phi(\rho z) = F(\chi)\left[\frac{\mathrm{d}}{\mathrm{d}\rho z}\left(\frac{I_{\rho z}}{I_\infty}\right)\right]\exp(\chi\rho z) \; .$$

Hence from the gradient of the $I_{\rho z}/I_\infty$ versus ρz curve, values of $\phi(\rho z)/F(\chi)$ are established. The magnitude of $F(\chi)$ can be determined indirectly because, as $\rho z \to 0$,

$$F(\chi) \to \phi(0) \Big/ \left(\frac{\mathrm{d}}{\mathrm{d}\rho z}\left(\frac{I_0}{I_\infty}\right)\right) .$$

Therefore by using the $\phi(0)$ results of Duncumb and Melford (1966b), Reuter (1972) or Love *et al.* (1978b), x-ray depth distributions may be readily calculated.

8.4.1 Offset Gaussian

It has long been recognised that the shape of a $\phi(\rho z)$ curve resembles a Gaussian profile with the peak displaced along the mass depth axis. Wittry (1958) proposed that $\phi(\rho z)$ could be represented by

$$\phi(\rho z) = A\exp\left[-\left(\frac{\rho z - \rho z_0}{\Delta\rho z}\right)^2\right] ,$$

where ρz_0 describes the position of the peak and $\Delta\rho z$ the half-width. In principle the method appears to be a good one although $f(\chi)$ can be calculated only by numerical integration and the symmetrical Gaussian function given above cannot be adjusted to give simultaneously the correct peak position, half width and $\phi(0)$ value. These limitations are not serious and Kyser (1972) has suggested the introduction of an additional term in the equation for $\phi(\rho z)$ which would overcome the latter objection and allow the correct $\phi(0)$ value to be obtained. The major difficulty with this and other early curve-fitting methods (for example, Criss and Birks, 1966) is that the authors provide only sufficient data to fit a few $\phi(\rho z)$ curves and the models are therefore not generally applicable.

The first curve-fitting models which could be used in practical situations (Buchner and Pitsch, 1971; Parobek and Brown, 1978) are described in the following section.

8.4.2 Parobek–Brown

The formula used by Parobek and Brown (1978) to represent x-ray depth distributions is similar to that proposed some years earlier by Buchner and Pitsch (1971) and its development is most readily understood by first referring to the work of the latter authors.

Buchner and Pitsch obtained values of $I_{\rho z}/I_\infty$ (see equation (8.12) for definition of parameters), using the wedge technique, and then corrected them for absorption effects to establish the generated intensity ratio $I'_{\rho z}/I'_\infty$, from

$$\mathrm{d}(I_{\rho z}) = \mathrm{d}(I'_{\rho z})\exp(-\chi\rho z) \; ,$$

which may be written as

$$I'_{\rho z} = \int_0^{\rho z} \left(\frac{\mathrm{d}I_{\rho z}}{\mathrm{d}\rho z}\right) \exp(\chi \rho z)\, \mathrm{d}\rho z \;.$$

Now a normalised $\phi(\rho z)$ curve, $\phi'(\rho z)$, can be calculated since

$$\phi'(\rho z) = \frac{\phi(\rho z)}{\int_0^{\infty} \phi(\rho z)\mathrm{d}\rho z} = \frac{\mathrm{d}}{\mathrm{d}\rho z}\left(\frac{I'_{\rho z}}{I'_{\infty}}\right) .$$

Examination of experimental data (see for example Fig. 8.5) shows that $I'_{\rho z}/I'_{\infty}$ is well represented by an equation of the form $1 - \exp(-a\rho z)^n$ and least-squares

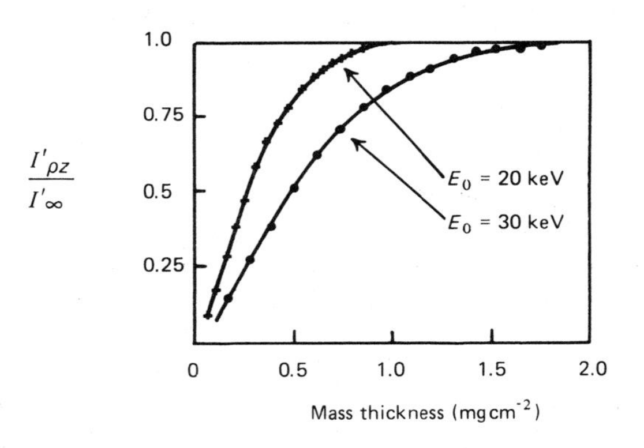

Fig. 8.5 – Plot of $I'_{\rho z}/I'_{\infty}$ versus mass thickness, where $I'_{\rho z}$ is the intensity of generated x-rays from a surface film of thickness ρz and I'_{∞} is the intensity generated from a bulk specimen of the same composition as the film. Experimental data from a nickel coating on cobalt fitted using equation of the form $1 - \exp(-a\rho z)^n$.

analysis is used to determine the optimum values of both n and a. The absorption correction is given by

$$f(\chi) = \int_0^{\infty} \frac{\mathrm{d}}{\mathrm{d}\rho z}\left(\frac{I'_{\rho z}}{I'_{\infty}}\right) \exp(-\chi\rho z)\,\mathrm{d}\rho z$$

$$= \int_0^{\infty} [a^n n(\rho z)^{n-1} \exp(-(a\rho z)^n)] \exp(-\chi \rho z)\,\mathrm{d}\rho z \;. \qquad (8.13)$$

The integral cannot be solved analytically and must, therefore, be determined by numerical integration or by using the graphical data provided by Buchner and

Wepner (1971). The main problem with the Buchner and Pitsch model is that $\phi'(\rho z)$ (given by the square-bracketed term in equation (8.13)) is zero when $\rho z = 0$ and thus the model will possess the same limitations as the simplified Philibert procedure.

Parobek and Brown (1978) have used an almost identical expression for $\phi(\rho z)$ and have described the x-ray depth distribution by the equation $\phi(R) = Dkn(KR)^{n-1} \exp(-(KR)^n)$, where $R = \rho z + \rho z_0$. Introduction of the term ρz_0 means that the x-ray depth distribution has effectively been displaced along the mass depth axis by ρz_0 and hence $\phi(0)$ is no longer zero. In order to establish the appropriate values for D, K, n and ρz_0 Parobek and Brown matched the equation to their own experimental $\phi(\rho z)$ curves using a least-squares fitting procedure. Their original $\phi(\rho z)$ results, obtained using the tracer technique, covered the electron energy range 6 to 15 keV so that the analytical equations listed below are applicable only in this region.

$$D = (3.1245E_c)^{1.135} (E_0 - E_c)^P \exp(0.0115Z) \ ,$$

where

$$P = \left(\frac{1.84}{E_c}\right)^{0.3045} (1 + 0.7417 \exp(-0.00325Z) \ ;$$

$$K = BF(E_0 - E_c)^{-m} \ ,$$

where

$$m = \left(\frac{3.873}{E_c}\right)^{0.3526} \ ,$$

$$B = 0.188E_c^{-1.149}$$

and

$$F = 1 + 0.0595Z^{0.7} \exp(-0.000294Z^2) \ ;$$

$$n = 1.95AZ^{-1.32}$$

and

$$\rho z_0 = 43.0E_c^{0.106} E_0 Z^{-1.12} (1 + \exp(-0.088Z)) \ .$$

Subsequently a second, similar, set of equations was produced by Brown and Robinson (1979) which were suitable for electron energies in the keV range $15 \leqslant E_0 \leqslant 30$.

From the $\phi(R)$ curve the magnitude of the combined atomic number and absorption correction is determined by integrating the product of $\phi(R)$ and the x-ray attenuation term $\exp(-\chi\rho z)$. Multi-element specimens are dealt with by using weight-averaged atomic numbers in the equations for D, P, F, n and ρz_0.

8.4.3 Bishop

It is obviously desirable that analytical formulae should describe the $\phi(\rho z)$ curves as well as possible and the methods referred to earlier in this section attempt to do this. However, in general, the equations increase in complexity as the fit is improved and it often becomes necessary to use numerical integration to evaluate $f(\chi)$. On occasions this may prove unsatisfactory, for example during on-line processing of results, and approximate representations of $\phi(\rho z)$ might be preferable if they enable quantitative data to be acquired more rapidly. Such a model has been proposed by Bishop (1974). The x-ray distribution with depth is taken to be rectangular (or square if the axes are plotted in reduced coordinates), see Fig. 8.6. A constant intensity of x-ray generation (H) with depth is assumed until a value of twice the mean depth of x-ray generation is reached, whereupon the intensity falls abruptly to zero.

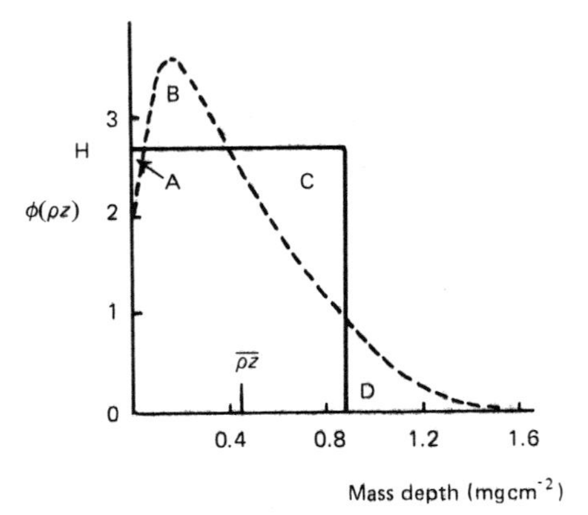

Fig. 8.6 – X-ray depth distributions for Au Lα radiation at 30 keV. Dashed line, Monte Carlo calculation; solid line, approximation of Bishop.

The formula for $f(\chi)$ may be calculated directly from equation (2.1) but the limit of integration is now $2\overline{\rho z}$

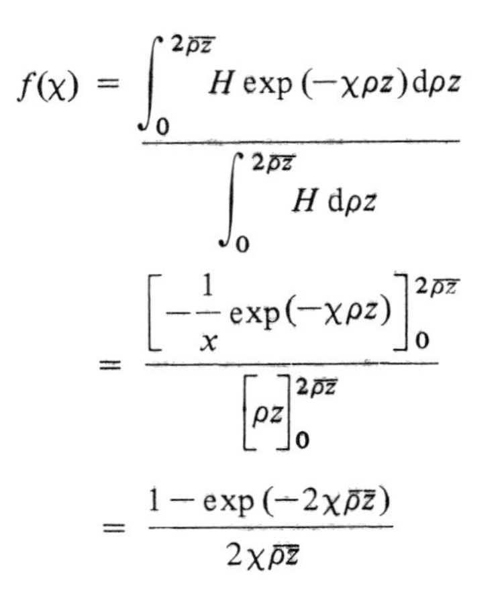

$$f(\chi) = \frac{\int_0^{2\overline{\rho z}} H \exp(-\chi\rho z)\,\mathrm{d}\rho z}{\int_0^{2\overline{\rho z}} H\,\mathrm{d}\rho z}$$

$$= \frac{\left[-\frac{1}{x}\exp(-\chi\rho z)\right]_0^{2\overline{\rho z}}}{\left[\rho z\right]_0^{2\overline{\rho z}}}$$

$$= \frac{1-\exp(-2\chi\overline{\rho z})}{2\chi\overline{\rho z}}$$

Thus the magnitude of the absorption correction is determined by the values of χ and $\overline{\rho z}$, the height of the rectangular distribution (H) being irrelevant. Bishop suggested that the mean depth could be calculated form the simplified Philibert model in which case $\overline{\rho z} = (1 + 2h)/[(1 + h)\sigma]$.

Subsequently Love and Scott (1978) derived equations for the mean depth from Monte Carlo calculations and wrote

$$\overline{\rho z} = \rho s_m \frac{(D_0 + D_1\eta + D_2\eta^2) \ln U_0}{F_0 + F_1\eta + F_2\eta^2 + \ln U_0} ,$$

where ρs_m, the electron range, was expressed by

$$\rho s_m = \frac{A}{Z} (0.787 \times 10^{-5} J^{0.5} E_0^{1.5} + 0.735 \times 10^{-6} E_0^2) .$$

The constants D_0, D_1, D_2 and F_0, F_1, F_2 are given in Table 8.2. The mean excitation energy for a multi-element specimen ($\bar{J}$) is averaged according to

$$\ln \bar{J} = \sum \frac{c_i Z_i}{A_i} \ln J_i \Big/ \sum \frac{c_i Z_i}{A_i}$$

(see section 7.1.2) but Z, A and η are weight-averaged in the usual manner.

Table 8.2 – Coefficients for use in the Bishop absorption correction (Love and Scott, 1978).

D_0	0.49269	F_0	0.70256
D_1	−1.09870	F_1	−1.09865
D_2	0.78557	F_2	1.00460

8.4.4 Thin film

This model (Duncumb and Melford, 1966b) can be regarded as a rather special case of curve fitting but, because the method is intended to operate successfully only when absorption is very large a detailed knowledge of the x-ray depth distribution is not required.

The method depends upon the fact that when $\chi \to \infty$ the emitted x-rays originate from the surface region of the specimen. If the overvoltage is high, which will generally be the case in light element analysis, $\phi(\rho z)$ is only a slowly changing function with mass depth. Therefore to a first approximation it may be considered constant and equal to $\phi(0)$, the value at the surface. Hence we may write

$$F(\chi) = \int_0^\infty \phi(0) \exp(-\chi\rho z)$$

$$= \frac{\phi(0)}{\chi} .$$

The model provides a combined atomic number and absorption correction and few input parameters are required.

Duncumb and Melford (1966a) derived values of $\phi(0)$ using backscattered electron energy distributions of Bishop (1966c) and employed the model in one of the earliest attempts at quantitative analysis of light elements. Carbon measurements were made on carbides of titanium. tungsten and boron using a silicon carbide specimen as a standard and good results were obtained over a wide range of kilovoltage (10–40 kV). However, the method is applicable only in cases where the value of χ in both specimen and standard is high. This normally rules out the use of pure element standards because self-absorption coefficients are not usually sufficiently large.

From oxygen x-ray measurements obtained on a well-characterised set of oxides Love *et al.* (1974a) suggested that the thin film model could be applied with some confidence provided that $\chi/\sigma > 3$ in both specimen and standard. However, on the basis of Monte Carlo results, Bishop (1974) proposed $f(\chi) <$ 0.01 (that is, $\chi/\sigma > 50$) as the criterion which, unfortunately, is almost impossible to satisfy in practical microanalysis work. While it is evident that Bishop's criterion is strictly correct, errors in analysis will generally be less than 5% when standards are used provided that $\chi/\sigma > 3$.

As one might expect the model is not used a great deal in practice because the experimental conditions required for successful operation (high kV and/or low x-ray take-off angle) are exactly the opposite of those needed to achieve good detection sensitivity.

8.5 SPECIMENS INCLINED TO THE ELECTRON BEAM

Throughout this chapter it has been assumed that the electron beam is incident normally on the specimen surface, the configuration adopted in most purpose-built electron-probe microanalysers. However, in certain types of scanning electron microscope it may be necessary to tilt the specimen towards the x-ray spectrometer in order to carry out x-ray measurements. Furthermore, inclined specimen geometry may be desirable on occasions in a conventional electron-probe microanalyser to improve the sensitivity of analysis of thin surface films and/or to optimise light-element detection efficiency. It is unfortunate, therefore, that there is no absorption correction theory which is capable of dealing adequately with electron beams at non-normal incidence.

Figures 8.7(a) and 8.7(b) show schematically the situations for normal and non-normal incidence respectively. If one makes the assumption that the x-ray profile in the direction of the incident electron beam is the same for both specimen geometries then

$$\chi_\beta = (\mu/\rho)\,\rho x \sin\beta \operatorname{cosec}\psi$$

or

$$\chi_\beta = \chi_{90} \sin\beta \ . \tag{8.14}$$

Green (1964) tested this relationship by comparing it with his experimental results obtained using copper Kα radiation at $\beta = 90°$ and $45°$ and for electron beam energies of 49.7 and 30.2 keV. He found that at both energies the predicted values of χ_β were too low. Reed (1975b) pointed out that correct compensation could be achieved at 30.2 keV if a multiplying factor of $1.12 \sin\beta$ was employed but this would not appear to be generally applicable. From Monte Carlo studies undertaken at different angles of electron incidence Bishop (1965) proposed that

$$\chi_\beta = \chi_{90}(1 - 0.5 \cos^2\beta) \tag{8.15}$$

but later, after studying more extensive Monte Carlo data, he came to the conclusion that no satisfactory formula for χ_β exists (Bishop, 1968).

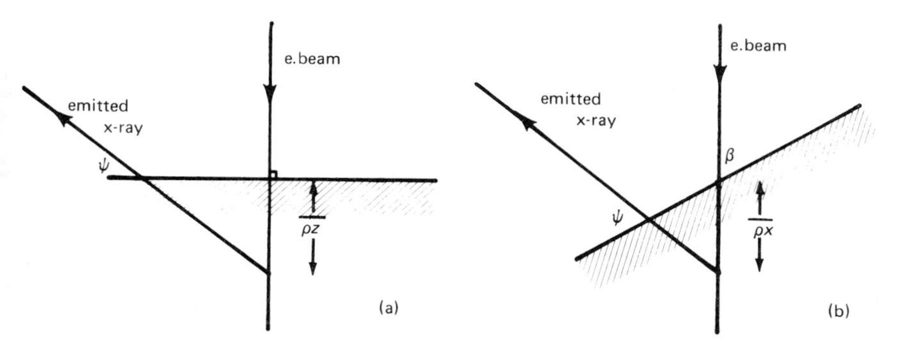

Fig. 8.7 – X-ray emission from specimens with electron beam (a) normal to the surface and (b) inclined at an angle β.

The reasons for this are not hard to find. When the angle of incidence is altered not only will the mean depth of x-ray generation be affected but also the shape of the x-ray depth distribution. The peak in the $\phi(\rho z)$ curve increases and moves closer to the surface of the specimen as the angle of incidence (β) is lowered. Therefore the assumption made in deriving equation (8.14) is not valid and the formula is prone to error. Furthermore, because the shape of the $\phi(\rho z)$ has been fundamentally changed after tilting, it is unlikely that the true x-ray profile can be obtained by a simple modification to the $\phi(\rho z)$ distribution determined at normal incidence.

When the mean depth largely controls the magnitude of the absorption correction (for $f(\chi) > 0.6$) it may be possible to establish an empirical factor but formulation of a tilt correction which will be effective when χ is large is likely to prove extremely difficult.

Ultimately the problem can only be resolved by a systematic study using Monte Carlo programs in conjunction with experimental data, and some fundamental work in this area has recently been undertaken (Newbury and Myklebust, 1981; Brown, 1981). However for the time being it is probably best to use equation (8.15) which, as Reed (1975b) points out, should give reasonable accuracy for moderate absorption because any errors in $f(\chi)$ tend to cancel when taking measurements from specimen and standard.

8.6 INPUT PARAMETERS

The accuracy of the absorption correction is influenced not only by the effectiveness of the model itself but also by the reliability of the input parameters which are inserted into the equations. The major parameters are:

(a) mass absorption coefficients;
(b) x-ray take-off angle (ψ);
(c) incident electron energy (E_0).

Yakowitz and Heinrich (1968) have investigated the effect that errors in each of the above input parameters have upon the value of $f(\chi)$ as computed by the simplified Philibert absorption correction. Whilst their analysis is strictly appropriate only for the Philibert model it is likely that similar results would be obtained from a study of other absorption methods and their findings are therefore worth noting. Let us consider each of the three input parameters in turn. If the only error is in the mass absorption coefficient value then the relative error in the absorption correction is given by

$$\left(\frac{\Delta f(\chi)}{f(\chi)}\right)_{E_0,\psi} = -B\frac{\Delta\mu/\rho}{\mu/\rho} \tag{8.16}$$

where $B \cong \chi/\sigma$.

If the only error is in ψ then

$$\left(\frac{\Delta f(\chi)}{f(\chi)}\right)_{E_0,(\mu/\rho)} = B\cot\psi\Delta\psi \tag{8.17}$$

and finally if μ/ρ and ψ are correct

$$\left(\frac{\Delta f(\chi)}{f(\chi)}\right)_{\psi,(\mu/\rho)} = -B\left(\frac{1.5\,\Delta E_0}{E_0 - E_c U_0^{-0.5}}\right) . \qquad (8.18)$$

Errors in mass absorption coefficients are beyond the control of the micro-analyst and will be of the order of several per cent (more will be said about this in section 8.7). Taking as a typical value $\Delta(\mu/\rho)/(\mu/\rho) = 0.05$ it is evident from equation (8.16) that if $\Delta f(\chi)/f(\chi)$ is to be less than 0.01, then χ/σ must be less than 0.2. Since χ is usually fixed it may be necessary to lower the incident electron energy to achieve this objective.

Provided that care is taken in mounting the specimen the error in ψ should not exceed 2°. Examination of equation (8.17) indicates that with an x-ray take-off angle of 40°, χ/σ must be kept below 0.24 if the error in $f(\chi)$ is not to exceed 1%. For a low take-off angle (20°) instrument the situation is more serious and the criterion becomes $\chi/\sigma < 0.1$.

The effect of a kilovoltage error upon $f(\chi)$ is more difficult to quantify from equation (8.18) since the bracketed term increases and χ/σ decreases when E_0 is lowered. A graph of $\Delta f(\chi)/f(\chi)$ versus E_0 (Fig. 8.8) reveals that the χ/σ term dominates and errors are minimised by reducing the energy of the incident electron beam. The data plotted in the graph correspond to a 1 kV measurement error but, whilst this is not uncommon on commercial equipment, the kilo-

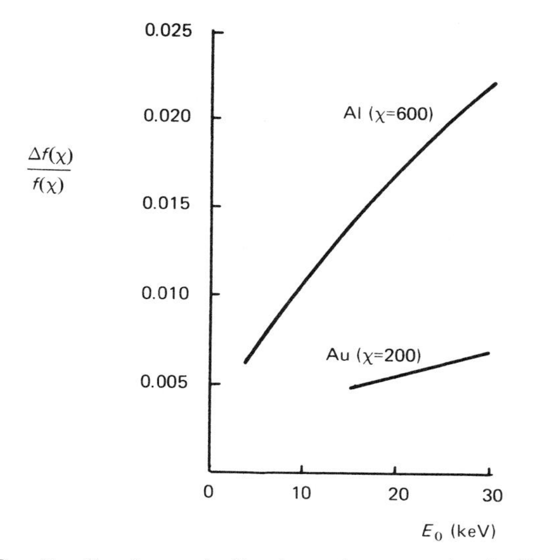

Fig. 8.8 – Fractional error in the absorption correction $[\Delta f(\chi)/f(\chi)]$ plotted as a function of incident electron energy E_0; measurement error, ΔE_0, assumed to be 1 keV.

voltage meter can be accurately calibrated using the methods described in section 6.3.2.

With care both the kilovoltage and the take-off angle errors may be reduced to well below 1%. However, errors in the mass absorption coefficients provide a fundamental limitation to the accuracy of $f(\chi)$ and Heinrich recommends keeping $f(\chi) > 0.8$ so that the relative error in the absorption correction is below 1%.

8.7 MASS ABSORPTION COEFFICIENTS

The mass absorption coefficients of many elements have been experimentally determined only for a few wavelengths and for some elements there are no results at all. Fortunately interpolation methods may be employed to establish tables of coefficients from these limited data because it has been found that, between absorption edges, the mass absorption coefficient of an element varies smoothly with wavelength and varies regularly with atomic number of the absorber for a given wavelength.

8.7.1 Coefficients for the wavelength range $1\text{Å} < \lambda < 10\text{Å}$

The theoretically derived expression for the mass absorption coefficient proposed by Kramers (1923) (see section 2.2.4) predicts the variation with atomic number described above. However, from a comparison with experimentally determined μ/ρ results it is evident that the exponents in Kramers' expression are not exact integers nor are their values independent of wavelength and absorber. It follows therefore that the equation is unsuitable for the purpose of interpolation and instead the empirical expression $\mu/\rho = C\lambda^n$ is often employed. Leroux (1961) adopted this approach and made the assumption that the value of n between a particular set of absorption edges was identical for all absorbers. This permitted him to calculate a series of values for C and n. It was found, as expected, that values of C were a function of both the absorber and the set of absorption edges bounding λ. The C and n data were tabulated and from them a set of internally consistent mass absorption coefficients was constructed for elements with $Z > 2$ and for $0.18\text{Å} < \lambda < 10\text{Å}$.

Heinrich (1966a) re-examined the experimental data base employed by Leroux and concluded, from a careful study, that some values were unreliable; he rejected these but included some new experimental results of his own. Heinrich then re-optimised C and n for the narrower range of wavelengths of interest in microprobe analysis and showed that n was not independent of atomic number of the absorber as assumed by Leroux. As well as listing values of C and n, he provided a table of mass absorption coefficients for $K\alpha$, $K\beta$, $L\alpha$, $L\beta$, $M\alpha$ and $M\beta$ lines covering the wavelength range $0.7\text{Å} \leqslant \lambda < 12\text{Å}$ and elements with $3 \leqslant Z \leqslant 94$. (Table 8.3 gives a typical set of μ/ρ values taken from the compilation of Heinrich).

Table 8.3 – Mass absorption coefficients for Kα lines (after Heinrich, 1966a).

Absorber		Emitter																			
Z		Na	Mg	Al	Si	P	S	Cl	Ar	K	Ca	Sc	Ti	V	Cr	Mn	Fe	Co	Ni	Cu	Zn
6	C	1534	905	557	357	235	160	111	79	57	42	32	24	18	14	11	9	7	6	5	4
7	N	2450	1448	893	573	378	258	179	128	93	68	51	39	30	23	18	14	12	9	8	6
8	O	3613	2139	1322	849	562	383	267	190	138	102	76	58	44	35	27	22	17	14	11	9
9	F	5169	3066	1898	1221	809	552	386	275	200	148	111	84	65	50	40	31	25	20	17	14
10	Ne	6967	4140	2567	1654	1098	750	524	375	272	201	151	115	89	69	54	43	35	28	23	19
11	Na	571†	5409	3359	2168	1441	986	690	494	359	266	200	152	117	91	72	57	46	37	30	25
12	Mg	770	464	4377	2825	1877	1285	899	643	468	346	260	198	153	119	94	75	60	48	40	32
13	Al	1021	615	386	3493	2325	1593	1117	800	583	432	325	247	191	149	117	93	75	61	50	41
14	Si	1333	802	503	328	2840	1949	1368	981	716	531	400	304	235	184	145	116	93	75	61	50
15	P	1696	1021	641	417	280	2371	1664	1193	870	645	486	370	286	224	176	141	113	91	75	61
16	S	2103	1266	794	518	347	239	1966	1411	1031	765	577	440	340	266	210	167	135	109	89	73
17	Cl	2578	1552	974	635	425	294	207	1657	1210	898	677	516	399	312	247	197	158	128	105	86
18	Ar	3132	1886	1183	771	517	357	252	181	1390	1033	779	595	461	361	285	227	183	149	121	100
19	K	3729	2245	1409	918	615	425	300	216	158	1190	898	685	530	415	328	262	211	171	140	115
20	Ca	4413	2657	1667	1086	728	502	354	255	187	139	1011	772	599	469	371	296	239	194	159	131
21	Sc	5183	3120	1958	1276	855	590	416	300	220	164	124	879	681	534	422	337	272	221	181	149
22	Ti	6057	3464	2288	1491	999	690	486	350	257	191	145	111	86	597	473	378	304	247	203	167
23	V	6939	4178	2621	1708	1145	790	557	402	294	219	166	127	98	77	531	424	342	278	228	188
24	Cr	7943	4782	3001	1955	1311	904	638	460	337	251	190	145	113	88	70	474	382	311	254	210
25	Mn	9042	5444	3416	2225	1492	1030	726	523	383	286	216	165	128	101	80	64	423	344	282	232
26	Fe	10167	6121	3841	2502	1677	1158	816	588	431	321	243	186	144	113	89	71	58	380	311	256
27	Co	11465	6902	4331	2822	1892	1305	921	663	486	362	274	209	162	127	101	81	65	53	341	281
28	Ni	12806	7710	4838	3152	2113	1458	1028	741	543	404	306	234	181	142	113	90	73	59	48	309
29	Cu	12165	8569	5377	3503	2348	1621	1143	824	604	450	340	260	202	158	125	100	81	66	54	44
30	Zn	9691	9507	5965	3886	2605	1798	1268	914	670	499	377	288	224	175	139	111	89	73	60	49

† No value given by Heinrich: this value obtained by extrapolation.

Data for absorption of Kα lines in oxygen have been modified according to the recommendation of Springer and Nolan (1976).

However, because of the absence of experimental data, no mass absorption coefficients were given for wavelengths between the M_{IV} and M_V absorption edges and beyond the N_I edge. Heinrich's tabulations have proved extremely useful and have been widely used by electron-probe microanalysts.

With the increasing use of computers to carry out ZAF corrections it became desirable to use an analytical expression to represent the atomic number dependence of C and n values between absorption edges, thus avoiding the need to store large blocks of data in the computer or to enter the individual mass absorption coefficients by hand. Yakowitz *et al.* (1973) used a least-squares fitting procedure to obtain equations for log n and log C in terms of a second-order polynomial in log Z. Springer and Nolan (1976) have examined the equations for C and n given by Yakowitz *et al.* and also the expressions developed by Colby (1968) which represent C and n as polynomials in Z. From a comparison with the original data provided by Heinrich they concluded that the formulae of Colby were preferable provided that low atomic number elements and data between the M and N absorption edges were excluded. During their assessment Springer and Nolan found some minor discrepancies in the coefficients of the sixth-order polynomials employed by Colby and recalculated these; the new values are given in Table 8.4. It should be stressed that use of the equations will not give any improvement over the mass absorption coefficients of Heinrich; they merely provide a convenient method of condensing information.

When calculating mass absorption coefficients using empirical equations it is vital that absorption edge energies and line energies are accurately defined otherwise incorrect C and n values may be assigned. For convenience these energies are usually calculated in a similar manner to mass absorption coefficients and Springer and Nolan (1976) have used an equation of the form $E = \exp(A_0 + A_1 \ln Z + A_2(\ln Z)^2 + A_3(\ln Z)^3)$. Values of coefficients for the K, L and M series of absorption edges and for the principal x-ray emission lines are given in Table 8.5. Calculated energies agree with values tabulated by White and Johnson (1970) to within 10 eV in almost all cases and frequently to within 5 eV.

Recently Thinh and Leroux (1979) have compiled a new set of mass absorption coefficients using an empirical formula $\mu/\rho = CE_c\lambda^n$, where E_c is the lower of the two absorption edges enclosing the energy of the absorbed radiation. The authors have shown that by adopting this approach a unique value of C may be ascribed to each element and also that their equation provides a very good fit to experimentally determined data. Tables of the mass absorption coefficients for atomic numbers between 1 and 94 and for radiations between 1 and 40 keV are commercially available (Leroux and Thinh, 1977) and values have been computed for the Kα, Kβ_1, Lα_1, Lβ_1 and Mα_1 emission lines. In principle, the method of Thinh and Leroux should allow better cross-correlation of available experimental results but it would be useful if the C and n values could be expressed analytically so that the data could be more easily incorporated in computer programs.

Table 8.4 – Expressions for mass absorption coefficients (after Springer and Nolan, 1976).

$$\mu/\rho = c.\lambda^n$$

$$c = A_0 + A_1.Z + A_2.Z^2 + A_3.Z^3 + A_4.Z^4 + A_5.Z^5 + A_6.Z^6$$

$$n = A_0 + A_1.Z + A_2.Z^2$$

μ/ρ: absorption coefficient Z: atomic number λ: wavelength, Å

c,n	Z	A_0	A_1	A_2	A_3	A_4	A_5	A_6
c_K	11–42	–40.6242	11.0687	–1.25603	0.084617	–2.71262E–3	4.61653E–5	–3.18732E–7
c_{KL}	1–92	–1.31911	0.316461	–0.035114	2.41261E–3	–3.92914E–5	3.91759E–7	–1.6333E–9
c_{LM}	30–92	162.151	–19.7529	0.979945	–2.52389E–2	3.61003E–4	–2.67888E–6	8.12823E–9
c_{MN}	60–92	–18475.2	1515.58	–51.5749	0.931841	–9.42334E–3	5.05609E–5	–1.12416E–7
n_K	11–42	2.8601	–6.96196E–3	5.67643E–5				
n_{KL}	42–92	2.72812	1.82246E–3	–4.55921E–5				

Table 8.5 – Expressions for emission and critical excitation energies (after Springer and Nolan, 1976).

$$E = \exp(A_0 + A_1 . \ln Z + A_2 . (\ln Z)^2 + A_3 . (\ln Z)^3)$$

Z: atomic number E: Energy in keV

	Z[†]	A_0	A_1	A_2	A_3
Line					
Kβ	4–92	−6.6101	3.48039	−0.373102	3.42585E−2
Kα	4–92	−5.49184	2.54555	−0.128429	1.20661E−2
Lβ	4–92	−21.4703	12.7049	−2.65398	0.228538
Lα	4–92	−13.2015	5.92457	−0.790156	5.60193E−2
Mβ	45–92	−38.6556	20.1655	−3.66665	0.255776
Mα	45–92	6.66542	−11.2046	3.58209	−0.303821
Edge					
K	4–92	−6.74556	3.65218	−0.442671	4.35567E−2
L_1	4–61	−15.8443	8.16606	−1.42787	0.121083
L_2	4–61	−21.1589	11.8839	−2.32253	0.194394
L_3	4–61	−20.0371	10.8106	−1.97776	0.156716
M_1	45–92	191.314	−126.755	27.4254	−1.91602
M_2	45–92	505.443	−352.372	81.3256	−6.20098
M_3	45–92	186.278	−130.174	29.8087	−2.2247
M_4	45–92	232.328	−165.694	38.7299	−2.96003
M_5	45–92	96.4983	−71.3325	16.9044	−1.2797
L_1	62–92	−30.4478	19.7531	−4.44501	0.379591
L_2	62–92	−23.818	15.5315	−3.41493	0.29695
L_3	62–92	−1.48449	−1.67038	0.840482	−5.73591E−2

† Although the formulae were developed for the energy range 1–20 keV, their application may be extended to elements with greater or smaller characteristic energies, as indicated.

It is possible to derive mass absorption coefficients from theoretical studies of photon–atom interactions. Veigele (1973) has produced a set of atomic photoelectric cross-sections for photon energies of 0.1 keV to 1.5 MeV and for elements $1 \leqslant Z \leqslant 94$. At energies between 1 and 10 keV the theoretical results agree with the experimental data to within 5 or 10% but in the energy range 0.1 to 1 keV discrepancies of between 10% and 200% have been observed (Hubbell and Veigele, 1976). It would appear that as yet the theory is not sufficiently well developed to provide accurate mass absorption coefficients for all the x-ray energies of importance to the microanalyst.

8.7.2 Coefficients for the wavelength range ($10\,Å < \lambda < 120\,Å$)

In the compilation of Heinrich (1966a) no mass absorption coefficients were listed for K lines from elements below $Z = 11$, although some values could be derived for x-rays with energies greater than the K edge of the absorber using the C and n data (Table 8.6). Leroux and Thinh (1977) also give no data for $Z < 11$ if the energy of the x-ray line is below the energy of the K absorption edge. Reasons for this are the paucity of experimental data and the inconsistencies evident in results which are available (Weisweiler, 1970).

Table 8.6 – Values of c_k and n_k for low atomic number elements (after Springer and Nolan, 1976).

Z	3	4	5	6	7	8	9	0
c_k	0.135	0.350	0.740	1.35	2.21	3.34†	4.90	6.77
n_k	2.88	2.86	2.85	2.84	2.83	2.82	2.81	2.80

† Heinrich's c_k value for oxygen (3.8) has been replaced since it is in error. The value of 3.34 is consistent with other tabulated data.

The method commonly used for hard x-rays is to measure their attenuation through thin foils but there are considerable practical difficulties in extending it to softer x-rays. The foil must be uniformly thin and for adequate mechanical strength is often supported upon a plastic film which also absorbs the x-rays. Hence the thickness of the plastic support and the evaporated metal film must be reliably determined and the attenuation of the x-rays in the plastic established. The experimental difficulties encountered when carrying out such measurements are likely to result in substantial errors.

Reflection techniques have been used (Lukirskii *et al.*, 1964; Ershov, 1967), mass absorption coefficients being determined from the measured coefficient of reflectance. The method is highly sensitive to the smoothness of the material, to any decrease in density near the surface and to surface oxidation. Values obtained by reflectance and transmission methods show poor agreement. Usually discrepancies of 20 to 50% exist between the two sets of values, the measurements from the transmission method being generally deemed the more reliable.

An interesting method for determining soft x-ray mass absorption coefficients involves the use of a proton probe (Lurio *et al.*, 1977). A substrate was selected which would emit the x-ray line of interest and part was coated with an absorbing film of known mass thickness, ρt. The proton beam was directed at each region in turn and the x-ray intensities from the coated (I_F) and uncoated (I_0) portions recorded. Mass absorption coefficients were calculated using

$$I_F = I_0 \exp\left(-\mu/\rho \, . \, \rho t \operatorname{cosec} \psi\right)$$

that is

$$\left(\frac{\mu}{\rho}\right) = \frac{1}{\rho t \operatorname{cosec} \psi} \ln\left(\frac{I_0}{I_F}\right)$$

where ψ is the x-ray take-off angle. The above formula can be applied only if there is negligible energy loss and scattering of protons in the coating. However, if the coating thickness does not greatly exceed 1000 Å these conditions can be satisfied and the loss of x-ray intensity due to the reduction of proton energy in passing through the coating is less than 1%. The only problems with the technique are the accurate measurement of mass thickness of the coating and the production of a coating which does not contain significant impurities. Mass absorption coefficients for carbon Kα (Lurio *et al.*, 1977) and boron Kα (Lurio and Reuter, 1977) in a range of metallic elements have been determined using the method and an experimental accuracy of between 5 and 10% is claimed.

Some mass absorption coefficients have been derived from microanalysis measurements on specimens of known composition (Manzione and Fornwalt, 1965; Kohlhass and Scheiding 1970). The coefficients were adjusted so that after a ZAF correction had been performed the correct weight concentrations were obtained. Such a method is almost always unsatisfactory because the calculated mass absorption coefficients will reflect any inadequacy in the correction procedure employed.

Measurements on the attenuation of x-rays in gas mixtures have been performed by Henke *et al.* (1967) where, by controlling the pressure of the gas, the effective mass thickness of the absorber may be established with considerable precision. Experimental results were claimed to be accurate to within 10% and, after carrying out interpolation, Henke suggested an accuracy of 3% for the listed values. A more complete set of soft x-ray mass absorption coefficients based upon the earlier experiments has subsequently been published by Henke and Ebisu (1974) and these probably represent the most reliable and internally consistent tabulation of data available at the present time.

8.7.3 Extended fine structure

Although equations of the type adopted by Leroux (1961) are generally satisfactory for calculating mass absorption coefficients, they break down when x-ray energies are within a few hundred electron volts of the high energy side of an absorption edge. In this region the mass absorption coefficient does not vary smoothly with energy and instead exhibits oscillatory behaviour which is known as Kronig structure. The effect occurs most strongly in materials possessing a high-symmetry crystal lattice such as face-centred cubic or hexagonal-close-packed structures and it is therefore observable in most metals and many oxides. The phenomenon may be explained by treating the ejected photoelectron as a plane wave travelling in the periodic field of the crystal. Mass absorption coefficient minima are observed when a significant number of directions are forbidden because the wave undergoes Bragg diffraction rather than transmission. Maxima occur at wavelengths for which no directions are forbidden.

Kronig structure may cause the mass absorption coefficient to differ significantly from the value obtained from empirical equations and it is therefore desirable to avoid, whenever possible, using an x-ray line which is within 200 eV of an absorption edge.

9

Fluorescence corrections

S. J. B. REED

9.1 FLUORESCENCE EXCITATION

In electron-probe microanalysis characteristic x-rays are excited principally by electron bombardment, but secondary excitation (fluorescence) also occurs as a result of inner-shell ionisation caused by the absorption of primary x-rays within the sample. For a given element, fluorescence is excited only by x-rays of energy greater than the critical excitation energy (E_c) of the particular shell. For example, iron K-radiation is excited by x-rays of energy greater than 7.11 keV, including any characteristic peaks present (for example, Ni Kα at 7.47 keV), and the relevant part of the continuum (see Fig. 9.1). Both characteristic and continuum fluorescence, but especially the former, are composition-dependent: hence a correction is required to allow for the difference between specimen and standard.

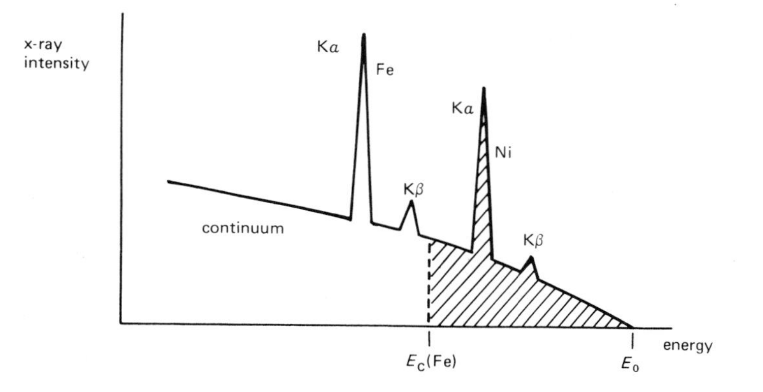

Fig. 9.1 – Spectrum of Ni–Fe alloy; shaded area indicates x-rays of sufficient energy to excite Fe K radiation.

Owing to the energy condition, $E > E_c$, excitation of the Kα line of one element by the K radiation of another (K–K fluorescence) occurs only when the exciting element is of greater atomic number, as in the case of nickel ($Z = 28$) exciting iron ($Z = 26$) cited above. For $Z > 21$ an atomic number difference of 2 is the minimum necessary for excitation by the Kα line, though excitation by the Kβ line alone occurs when the difference is only 1. For $Z \leqslant 21$ excitation by both Kα and Kβ lines takes place with an atomic number difference of only 1. In specimens containing heavy elements excitation by L lines is also possible. If an L line is used for analysis, fluorescence excitation by either K or L radiation (K–L and L–L fluorescence respectively) may occur provided the energy criterion is satisfied.

9.2 FLUORESCENCE EXCITATION EFFICIENCY

The first step towards obtaining a formula for fluorescence corrections is to derive an expression for the efficiency (ϵ) with which fluorescence radiation of the element of interest (A) is generated by primary x-rays of a given energy, E. For a multi-element specimen the fraction of the absorbed primary radiation which is absorbed by element A is

$$c_A \left(\frac{\mu}{\rho}\right)_E^A \Bigg/ \left(\frac{\mu}{\rho}\right)_E$$

where c_A is the mass concentration of A, $(\mu/\rho)_E^A$ is the mass absorption coefficient of A for x-rays of energy E and $(\mu/\rho)_E$ is that of the sample as a whole for such x-rays. However, not all of the resulting inner-shell vacancies are in the relevant shell (usually the K shell). On the low-energy side of the K absorption edge, absorption is attributable to the L, M, etc. shells, and the edge itself arises from additional absorption caused by the K shell. The size of this step can be expressed in terms of the 'absorption edge jump ratio' (r) which is defined as the ratio of μ/ρ on the high-energy side to that on the low-energy side. The fraction of μ/ρ on the high-energy side which is attributable to the K shell is thus equal to $(r - 1)/r$. Since the energy-dependence of the contributions to μ/ρ of different shells is the same (to a fairly close approximation), this expression is applicable anywhere on the high-energy side of the edge (see Fig. 9.2). The above expression must therefore be multiplied by $(r_K(A) - 1)/r_K(A)$ in order to obtain the rate of production of vacancies in the K-shell of element A. Finally, it is also necessary to multiply by the fluorescence yield, $\omega_K(A)$, this being the fraction of K-shell ionisations which result in the emission of an x-ray photon rather than an Auger electron. Hence:

$$\epsilon = c_A \frac{(\mu/\rho)_E^A}{(\mu/\rho)_E} \frac{r_K(A) - 1}{r_K(A)} \omega_K(A) \; . \tag{9.1}$$

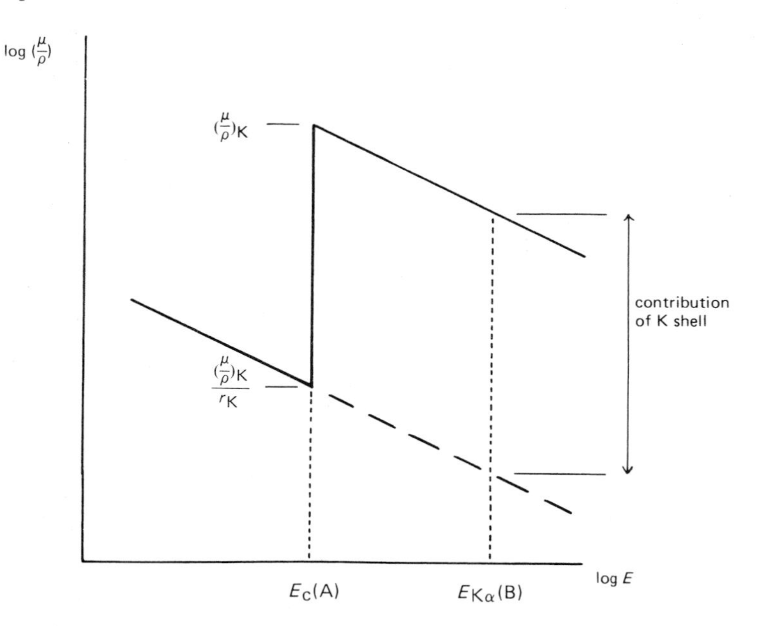

Fig. 9.2 – Mass absorption coefficient of element A as a function of energy showing effect of K absorption edge superimposed on contribution of other shells (L, M, etc.).

9.3 CHARACTERISTIC FLUORESCENCE

We now consider the case where fluorescence in A is excited by the characteristic radiation from another element, B, following the approach of Castaing (1951). In order to derive this correction it is necessary to relate the characteristic fluorescence intensity (I_F) to the primary x-ray intensity (I_A) emitted by element A. The measured intensity is equal to $I_A + I_F$ and thus the correction factor is $1/(1+\gamma)$, where $\gamma = I_F/I_A$.

In considering fluorescence excitation the simplest geometrical assumption is that the primary x-rays are produced at a point on the surface, so that half the primary radiation is absorbed in the sample. Applying equation (9.1) for the fluorescence excitation efficiency, we have:

$$\frac{I_F}{I_A} = 0.5 c_A \frac{(\mu/\rho)_B^A}{(\mu/\rho)_B} \frac{r_K(A)-1}{r_K(A)} \omega_K(A) \frac{I_B}{I_A} \quad . \tag{9.2}$$

The Green and Cosslett (1961) formula for characteristic x-ray intensity gives:

$$\frac{I_B}{I_A} = \frac{c_B \omega_K(B) A_A (U_B - 1)^{1.67}}{c_A \omega_K(A) A_B (U_A - 1)^{1.67}}$$

where U_A and U_B refer to the overvoltage ratios. Substituting this expression in equation (9.2), we have:

$$\frac{I_F}{I_A} = 0.5c_B \frac{(\mu/\rho)_B^A}{(\mu/\rho)_B} \frac{r_K(A)-1}{r_K(A)} \omega_K(B) \frac{A_A}{A_B} \left(\frac{U_B-1}{U_A-1}\right)^{1.67} \tag{9.3}$$

which is the basic equation for characteristic fluorescence corrections, neglecting absorption of the fluorescence radiation and the finite size of the primary x-ray source.

It is instructive to study the role of the various terms in equation (9.3). Sample composition has an effect through c_B and $(\mu/\rho)_B$. For example, consider the case of Ni–Fe alloys (A = Fe, B = Ni). If $c_A \rightarrow 0$ and $c_B \rightarrow 1$, then $c_B(\mu/\rho)_B^A/(\mu/\rho)_B =$ 380/59 = 6.44, which illustrates the strong enhancement of fluorescence due to differential absorption of Ni K radiation by iron. Now for a 50–50 alloy ($c_A = c_B = 0.5$), $c_B(\mu/\rho)_B^A/(\mu/\rho)_B = 0.5 \times 380/220 = 0.86$. Comparison with the previous value demonstrates the marked non-linearity of fluorescence intensity with respect to concentration.

The absorption edge jump ratio is only a slowly varying function of atomic number (see Table 9.1); for iron the value of $(r_K - 1)/r_K$ is 0.88. On the other hand, the fluorescence yield (ω_K) increases rapidly with Z (see Table 9.2) and plays a dominant role in determining the size of the correction. For nickel, $\omega_K = 0.39$: this is some ten times greater than for light elements such as sodium which therefore are not subject to large fluorescence corrections.

Table 9.1 – Values of $(r-1)/r$ for the K shell where r is the absorption edge jump ratio.

Z	$(r-1)/r$	Z	$(r-1)/r$
11	0.947	23	0.885
12	0.934	24	0.882
13	0.928	25	0.879
14	0.923	26	0.877
15	0.919	27	0.873
16	0.914	28	0.871
17	0.910	29	0.870
18	0.905	30	0.867
19	0.901	31	0.865
20	0.896	32	0.863
21	0.892	33	0.862
22	0.887		

Table 9.2 – Values of the K-shell fluorescence yield (ω_K).

Z	ω_K	Z	ω_K
11	0.021	24	0.260
12	0.028	25	0.291
13	0.037	26	0.323
14	0.048	27	0.356
15	0.061	28	0.389
16	0.075	29	0.422
17	0.091	30	0.454
18	0.108	31	0.486
19	0.128	32	0.517
20	0.151	33	0.549
21	0.177	34	0.576
22	0.202	35	0.604
23	0.231		

Table 9.3 – The fluorescence parameter (j) for K–K fluorescence.

Z	j	Z	j
11	0.013	23	0.120
12	0.016	24	0.133
13	0.021	25	0.146
14	0.026	26	0.162
15	0.033	27	0.171
16	0.038	28	0.178
17	0.044	29	0.193
18	0.059	30	0.202
19	0.066	31	0.221
20	0.071	32	0.228
21	0.085	33	0.244
22	0.106		

The terms in equation (9.3), which depend only on the atomic numbers of the exciting and excited elements (A and B) can be combined in a single variable:

$$j(\mathrm{A}) = 0.5\frac{r_{\mathrm{K}}(\mathrm{A})-1}{r_{\mathrm{K}}(\mathrm{A})}\omega_{\mathrm{K}}(\mathrm{B})\frac{A_{\mathrm{A}}}{A_{\mathrm{B}}} \qquad (9.4)$$

which to a fair approximation is a function of Z_A only (Reed, 1965). Neglecting the dependence of $j(A)$ on Z_B gives rise only to small absolute errors, since the size of the correction decreases rapidly as $(Z_B - Z_A)$ increases. Values of $j(A)$ suitable for use in the simplified form of correction calculation are given in Table 9.3.

The remaining terms in equation (9.3) are dependent mainly upon the difference in atomic number between A and B. Thus, both the atomic weight ratio A_A/A_B and the expression $[(U_B - 1)/(U_A - 1)]^{1.67}$ cause the fluorescence intensity to decrease with increasing atomic number difference. In the case of nickel and iron where the difference is only 2,

$$\frac{A_A}{A_B}\left(\frac{U_B-1}{U_A-1}\right)^{1.67} = 0.65$$

for a typical accelerating voltage of 25 kV. (This term is only weakly dependent on accelerating voltage.) Combining all the contributions for the 50–50 Ni–Fe alloy we have:

$$\frac{I_F}{I_A} = 0.5 \times 0.86 \times 0.88 \times 0.39 \times 0.65 = 0.096 \; .$$

The above calculation thus predicts that the generated fluorescence intensity will be 9.6% of the primary Fe Kα intensity, and the correction factor is $1/1.096 = 0.91$.

9.3.1 The effect of the Kβ line

So far it has been assumed that the exciting radiation is monochromatic and consists solely of the Kα line of the exciting element B. With a typical case of K–K fluorescence such as Ni–Fe, the ‘B’ radiation includes about 10% of Kβ radiation which differs significantly in wavelength from that of the Kα. However, in the absence of absorption edges between the Kα and Kβ lines, the difference in $(\mu/\rho)^A_B/(\mu/\rho)_B$ is negligible owing to the fact that the *ratio* of the mass absorption coefficients is practically independent of wavelength. With a Ni–Co–Fe alloy, the Co K absorption edge lies between the Ni Kα and Kβ lines and in principle could have a significant effect. However, since the contribution of the Kβ line to the total fluorescence intensity is small, the effect on the correction will generally be negligible.

Another possibility is that only the Kβ line has sufficient energy to excite fluorescence (for example, Co Kβ exciting iron). Here the procedure for calculating the correction follows exactly the same lines as for Kα excitation except that the fluorescence intensity is multiplied by the intensity ratio Kβ/(Kα + Kβ) (for B radiation). This factor is 0.10 for cobalt and is of the same order for other elements of similar atomic number. The size of the correction is thus reduced by a factor of about 10 and is therefore always small, though not necessarily completely negligible.

9.3.2 Fluorescence in ternary and higher-order compounds

Thus far it has been assumed that fluorescence is excited by only one element but in compounds containing three or more elements the situation may be more complex: for example in a Ni–Fe–Cr alloy, chromium is excited by both iron and nickel. In such cases I_F/I_A can be calculated separately for each exciting element and the individual contributions summed when calculating γ. Thus, for a 50–50 Ni–Fe alloy containing a trace of chromium we have, for iron exciting chromium, $I_F/I_A = 0.260$, and for nickel exciting chromium $I_F/I_A = 0.048$; hence $\gamma = 0.260 + 0.048 = 0.308$, and the correction factor $1/(1 + \gamma)$ is equal to 0.765. (These figures are calculated for an accelerating voltage of 25 kV.) Note that nickel is much less effective than iron in exciting chromium because iron absorbs Ni K radiation strongly and greatly reduces its relative effect on the chromium.

This procedure neglects the second-order effect, whereby the intensity of the Fe K radiation which excites chromium is itself enhanced owing to the fluorescence excited by nickel. As shown previously, in a 50–50 Ni–Fe alloy the iron intensity is increased by about 10% due to nickel fluorescence. It is therefore appropriate to multiply I_F/I_A for iron exciting chromium by 1.10, changing the correction factor from 0.765 to 0.776. This is of marginal significance and usually the second-order effect is neglected.

9.4 THE ABSORPTION TERM

Up to this point in the discussion absorption of the fluorescence radiation in emerging from the specimen has been neglected. Although the mass absorption coefficient of the specimen is the same for both fluorescence and primary radiation (since they have the same wavelength) absorption affects the ratio I_F/I_A owing to the difference in the depth at which they are produced. In order to derive the absorption factor for the fluorescence radiation it is necessary to carry out the integration set out below, which leads to a relatively simple analytical expression.

We start with the assumption that all primary x-ray production originates from a point 'O' on the surface (see Fig. 9.3). The solid angle between the cones defined by angles ϕ and $\phi + \mathrm{d}\phi$ is $2\pi \sin\phi \, \mathrm{d}\phi$ and, since characteristic x-rays are isotropically distributed over a solid angle of 4π, the intensity of the radiation emitted between these angles is equal to

$$\frac{1}{2} \sin\phi \, \mathrm{d}\phi \, I_B$$

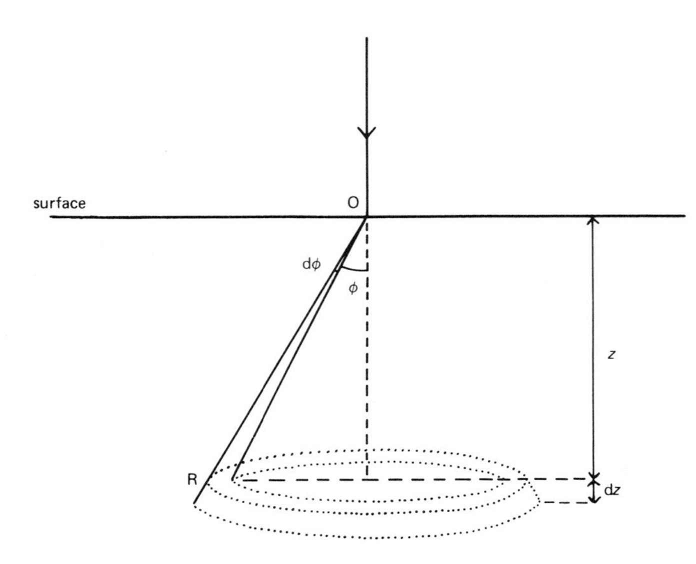

Fig. 9.3 – Geometry of fluorescence absorption calculation (see text).

where I_B is the total primary intensity emitted by element B. Now, consider an annular ring lying between depths z and $z + dz$. In travelling from the source at O to point R on the ring the x-rays are attenuated by the factor

$$\exp\left[-\left(\frac{\mu}{\rho}\right)_B \frac{1}{\cos\phi}\rho z\right] ,$$

where $(\mu/\rho)_B$ is the mass absorption coefficient of the sample for B radiation. Thus the intensity at R is:

$$\frac{1}{2}\sin\phi\, d\phi \exp\left[-\left(\frac{\mu}{\rho}\right)_B \frac{1}{\cos\phi}\rho z\right] I_B \ .$$

The path length of the x-rays in the ring is $dz/\cos\phi$, and the fraction of the x-rays arriving at R which are absorbed in this distance is $(\mu/\rho)_B\ (1/\cos\phi)d\rho z$. Hence the intensity absorbed in the ring is

$$\frac{1}{2}\sin\phi\, d\phi\left(\frac{\mu}{\rho}\right)_B \frac{1}{\cos\phi}\, d\rho z \exp\left[-\left(\frac{\mu}{\rho}\right)_B \frac{1}{\cos\phi}\rho z\right] I_B \ .$$

The absorbed primary radiation is converted to fluorescent 'A' radiation with an efficiency ϵ which can be derived from equation (9.1). The fluorescent 'A' radiation emerging from the sample suffers absorption which reduces the intensity by the factor: $\exp\{-(\mu/\rho)_A \operatorname{cosec}\psi\,\rho z\}$, where ψ is the x-ray take-off angle. The final fluorescence intensity is then given by

$$dI_F = \frac{1}{2}\sin\phi\, d\phi\left(\frac{\mu}{\rho}\right)_B \frac{1}{\cos\phi}\, d\rho z \exp\left[-\left(\frac{\mu}{\rho}\right)_B \frac{1}{\cos\phi}\rho z\right]$$

$$\times \exp\left[-\left(\frac{\mu}{\rho}\right)_A \operatorname{cosec}\psi\,\rho z\right]\epsilon\, I_B \,.$$

This expression can now be integrated to obtain the total fluorescence intensity

$$I_F = \frac{1}{2}\epsilon I_B\left(\frac{\mu}{\rho}\right)_B \int_0^{\pi/2} \tan\phi \int_0^{\infty} \exp\left[-\left\{\frac{(\mu/\rho)_B}{\cos\phi} + \left(\frac{\mu}{\rho}\right)_A \operatorname{cosec}\psi\right\}\rho z\right] d\rho z\, d\phi$$

$$= \frac{1}{2}\epsilon I_B\left(\frac{\mu}{\rho}\right)_B \int_0^{\pi/2} \frac{\tan\phi}{\dfrac{(\mu/\rho)_B}{\cos\phi} + \left(\dfrac{\mu}{\rho}\right)_A \operatorname{cosec}\psi}\, d\phi$$

$$= \frac{1}{2}\epsilon I_B \int_0^{\pi/2} \frac{\sin\phi}{1 + u\cos\phi}\, d\phi \,,$$

where

$$u = \frac{(\mu/\rho)_A \operatorname{cosec}\psi}{(\mu/\rho)_B}$$

Changing the variable, $t = 1 + u\cos\phi$, we have

$$\int_0^{\pi/2} \frac{\sin\phi}{1 + u\cos\phi}\, d\phi = \frac{1}{u}\int_1^{1+u} \frac{dt}{t} = \frac{\ln(1+u)}{u} \,.$$

The expression $[\ln(1 + u)]/u$ represents the absorption factor for fluorescence radiation. Typically u lies in the range 0.1–1 and the corresponding range of values of the absorption factor is 0.7–0.95.

So far it has been assumed that all primary x-ray production is concentrated at a point on the surface. In order to allow for the finite depth of primary x-ray production a further factor is required. Referring to Fig. 9.3: if the primary x-ray source is moved along OZ to a depth z_0, fluorescence radiation produced at any depth is subject to additional absorption, represented by the factor $\exp\{-(\mu/\rho)_A \operatorname{cosec}\psi\,\rho z_0\}$. In reality primary x-ray production takes place over a range of depths, described by the depth distribution function $\phi(\rho z)$. The total effect of the additional absorption can thus be represented by the integral:

$$\int_0^{\infty} \phi(\rho z) \exp\left[-\left(\frac{\mu}{\rho}\right)_A \operatorname{cosec}\psi\right] d\rho z \,,$$

which may be recognised as the factor $F(\chi)$ (see section 8.2). However, in this context the relevant $\phi(\rho z)$ function is that which applies to 'B' radiation, whereas the appropriate value of χ is that for 'A' radiation. Now, $\phi(\rho z)$ exhibits a dependence on critical excitation potential which is included in the absorption correction formula, but in fluorescence corrections no significant error arises from the assumption that $\phi(\rho z)$ is the same for 'A' as for 'B' radiation. The additional fluorescence absorption factor due to the finite depth of primary x-ray production therefore cancels out with the absorption factor for the primary radiation. Thus the ratio I_F/I_A as given by equation (9.3) is equal to the ratio of fluorescence to primary 'A' radiation as 'seen' by the spectrometer after emergence from the specimen.

9.4.1 The effect of finite source size

The integration in section 9.3 covers only the downward direction from the source. If the source has a finite extent along the z axis it becomes necessary also to consider the contribution of the upward component of the primary radiation ($\phi > \pi/2$ in Fig. 9.3). As shown by Castaing (1951) this leads to an additional term in the absorption expression, which becomes

$$\left\{\ln\left(\frac{1+u}{u}\right) + \ln\left(\frac{1+v}{v}\right)\right\}$$

where $v = \sigma/(\mu/\rho)_B$. The 'Lenard coefficient' σ is related to that which appears in the Philibert absorption correction formula (section 8.2) but the depth distribution assumed by Castaing is simplified to the form

$$\phi(\rho z) = \phi_0 \exp(-\sigma\rho z) .$$

The difference between this and more rigorous expressions is not of any practical significance. Since the simple exponential $\phi(\rho z)$ model differs from that upon which the absorption correction is based, it is not necessarily appropriate to use the same σ values. In fluorescence corrections the use of a σ function which includes a term dependent on the critical excitation potential is superfluous, and it is normal practice to use the values which depend only on the incident electron energy, E_0 (see Table 9.4). Returning to the example of the 50–50 Ni–Fe alloy, and assuming an accelerating voltage of 25 kV, we have $(\mu/\rho)_B = 220$ and $\sigma = 2550$; hence $v = 11.6$, $[\ln(1+v)]/v = 0.22$, and

$$\left\{\frac{\ln(1+u)}{u} + \frac{\ln(1+v)}{v}\right\} = 0.77 + 0.22 = 0.99$$

(for an x-ray take-off angle of 45°). Thus the contribution of the second term is relatively small and the correction is insensitive to σ and hence E_0.

Table 9.4 – Values of the Lenard coefficient (σ) for use in fluorescence corrections.

E_0(kV)	σ
10	8600
15	5300
20	3350
25	2300
30	1640

9.5 FLUORESCENCE INVOLVING L LINES

Fluorescence may be excited by the L lines of 'heavy' elements when these are present. The usual energy criterion applies: thus Fe Kα radiation is excited by the Lα radiation of elements of atomic number 69 (thulium) and upwards. Similarly, the minimum atomic number for L–L excitation of lanthanum ($Z = 57$) is 62 (samarium). K–L fluorescence (K radiation exciting L fluorescence) is also possible: for example, Mn Kα radiation has just sufficient energy to excite the L_{III} shell of lanthanum.

In calculating corrections for fluorescence involving L lines, the greater complexity of the L spectrum compared with the K spectrum and the existence of three sub-shells are complicating factors. Each sub-shell has its own excitation energy, fluorescence yield and absorption edge jump ratio; furthermore, the Lα line (Lα_1 and Lα_2 combined) contains only about 55% of the total intensity in the L spectrum (compared with 90% in the case of the Kα line). Ideally a relatively elaborate calculation is required but in the interests of simplicity it is usual to proceed as if there were only a single L shell and emission line (Reed, 1965). Thus, in equation (9.3) the overvoltage ratio U_0 is calculated for the L_{III} excitation energy, and the fluorescence yield ω(B) in the case of L–K and L–L fluorescence is assumed to be that for the L_{III} sub-shell (see Table 9.5). For K–L and L–L fluorescence the absorption edge jump ratio (r_A) is taken to be that for the L shell as a whole, which is the product of the jump ratios for the individual sub-shells (values of $(r_L - 1)/r_L$ are given in Table 9.6). Where L radiation is involved, mass absorption coefficients are those for the Lα line.

The expression for I_B/I_A substituted in equation (9.3) refers to K–K fluorescence only but the Green and Cosslett intensity formula is equally applicable to L radiation if suitable values are substituted for ω and U (as discussed above) and no modification is required for L–L fluorescence. However, since the constant of proportionality in the Green and Cosslett expression is different for K and L radiation, an additional factor is needed in equation (9.3) for K–L and L–K fluorescence, taking the values 0.24 and 4.2 respectively (Reed, 1965).

Table 9.5 – Values of the L fluorescence yield (ω_L) for the L_{III} sub-shell.

Z	ω_L	Z	ω_L
30	0.008	62	0.129
31	0.009	63	0.136
32	0.010	64	0.144
33	0.012	65	0.152
34	0.013	66	0.160
35	0.015	67	0.168
36	0.017	68	0.176
37	0.018	69	0.185
38	0.020	70	0.194
39	0.023	71	0.203
40	0.025	72	0.212
41	0.027	73	0.221
42	0.030	74	0.231
43	0.033	75	0.240
44	0.036	76	0.250
45	0.039	77	0.260
46	0.043	78	0.270
47	0.047	79	0.280
48	0.050	80	0.291
49	0.054	81	0.301
50	0.059	82	0.311
51	0.063	83	0.322
52	0.068	84	0.332
53	0.073	85	0.343
54	0.078	86	0.354
55	0.084	87	0.364
56	0.089	88	0.375
57	0.096	89	0.386
58	0.102	90	0.396
59	0.108	91	0.407
60	0.115	92	0.417
61	0.122		

Table 9.6 – Values of $(r-1)/r$ for the L shell where r is the total L-shell absorption edge ratio.

Z	$(r-1)/r$	Z	$(r-1)/r$
30	0.873	58	0.804
31	0.872	59	0.801
32	0.870	60	0.799
33	0.868	61	0.797
34	0.866	62	0.793
35	0.864	63	0.790
36	0.862	64	0.788
37	0.870	65	0.785
38	0.858	66	0.783
39	0.855	67	0.780
40	0.853	68	0.777
41	0.851	69	0.776
42	0.846	70	0.772
43	0.844	71	0.770
44	0.842	72	0.767
45	0.840	73	0.765
46	0.838	74	0.763
47	0.834	75	0.761
48	0.831	76	0.758
49	0.830	77	0.757
50	0.827	78	0.753
51	0.823	79	0.751
52	0.820	80	0.749
53	0.819	81	0.745
54	0.813	82	0.743
55	0.811	83	0.740
56	0.809		
57	0.806		

We now have the necessary data for calculating fluorescence corrections when L radiation is involved, bearing in mind that the accuracy will be somewhat worse than that of K–K fluorescence corrections in view of the more serious approximations entailed. Values of $j(A)$, the variable used in the simplified form of correction calculation (section 9.3), can be derived for fluorescence involving L lines, incorporating the above relative K–L intensity factors (see Tables 9.7 and 9.8).

Table 9.7 – Values of the fluorescence parameter (j) for L–K fluorescence.

Z	j	Z	j
11	0.006	23	0.078
12	0.007	24	0.084
13	0.010	25	0.099
14	0.011	26	0.108
15	0.017	27	0.122
16	0.021	28	0.129
17	0.026	29	0.149
18	0.034	30	0.161
19	0.037	31	0.181
20	0.045	32	0.204
21	0.055	33	0.211
22	0.065		

9.6 CONTINUUM FLUORESCENCE

The whole of the continuum lying above the critical excitation energy (E_c) of the element of interest is capable of exciting fluorescence; therefore in order to calculate the fluorescence correction it is necessary to integrate with respect to x-ray energy between E_0 and E_c. For this purpose an expression is required for the dependence of continuum intensity on energy, and this is provided by Kramers' law (Kramers, 1923; see also section 2.1.4)

$$I(E) = bZ\frac{E_0 - E}{E} ,$$

where b is Kramers' constant.

As shown by Green and Cosslett (1961), the number of continuum photons capable of exciting fluorescence for an element of critical excitation energy E_c, is given by the integral

$$bZ\int_{E_c}^{E_0}\frac{E_0 - E}{E}\,\mathrm{d}E ,$$

or

$$bZE_c\int_{1}^{U_0}\left(\frac{U_0}{U} - 1\right)\mathrm{d}U ,$$

Table 9.8 – Values for the fluorescence parameter (j) for L–L and K–L fluorescence.

Z	j		Z	j	
	L–L	K–L		L–L	K–L
30	0.004	0.006	57	0.048	0.074
31	0.004	0.009	58	0.050	0.075
32	0.005	0.009	59	0.052	0.082
33	0.005	0.011	60	0.055	0.083
34	0.006	0.012	61	0.059	0.087
35	0.007	0.015	62	0.063	0.090
36	0.007	0.015	63	0.066	0.100
37	0.008	0.018	64	0.069	0.103
38	0.010	0.019	65	0.072	0.104
39	0.010	0.022	66	0.076	0.106
40	0.012	0.023	67	0.078	0.112
41	0.013	0.025	68	0.082	0.113
42	0.015	0.026	69	0.085	0.114
43	0.016	0.028	70	0.090	0.116
44	0.017	0.029	71	0.093	0.117
45	0.019	0.035	72	0.097	0.122
46	0.021	0.036	73	0.100	0.123
47	0.023	0.042	74	0.103	0.129
48	0.024	0.044	75	0.106	0.130
49	0.025	0.047	76	0.115	0.131
50	0.028	0.048	77	0.119	0.133
51	0.032	0.053	78	0.118	0.139
52	0.034	0.055	79	0.121	
53	0.036	0.059	80	0.125	
54	0.039	0.060	81	0.129	
55	0.042	0.067	82	0.131	
56	0.047	0.069	83	0.135	

which is equal to $bZE_c(U_0 \ln U_0 - U_0 + 1)$. The dependence on U_0 is the same as for the electron-excited characteristic x-ray intensity (cf. equation (11.15)) and therefore the ratio of continuum fluorescence intensity to electron-excited characteristic radiation (I_f/I_A) is independent of E_0. Furthermore, since I_A is inversely proportional to Z and given the approximation $E_c \alpha Z^2$, it follows that for pure elements $I_f/I_A \; \alpha \; Z^4$ (see Fig. 9.4); hence continuum fluorescence increases rapidly in significance with increasing atomic number.

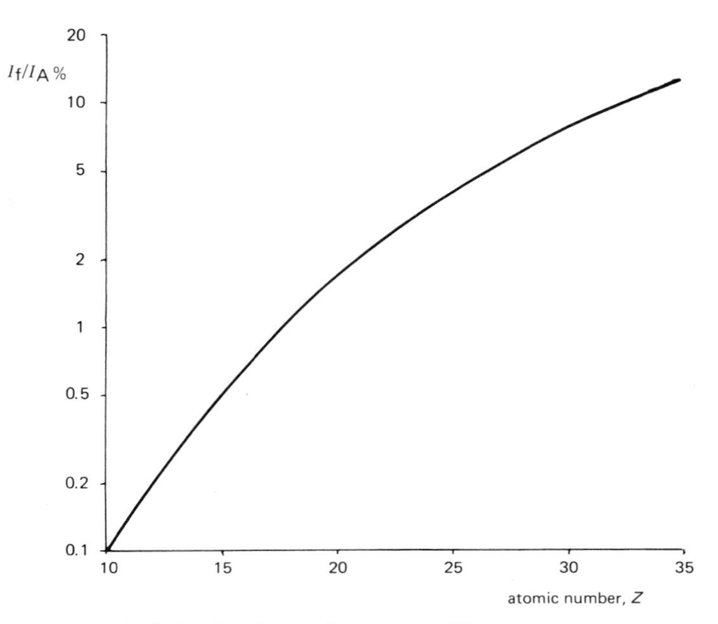

Fig. 9.4 – Continuum fluorescence (I_f) as a percentage of primary x-ray production (I_A) for pure elements.

For compounds we have to separate the factor Z which appears in Kramers' expression (and which now becomes $\bar{Z}$, the mean atomic number of the specimen) from these parameters which depend on Z_A, the atomic number of the emitting element. Neglecting absorption, the Green and Cosslett expression adapted for compounds (Springer, 1967) becomes

$$\frac{I_f}{I_A} = 4.34 \times 10^{-6} \frac{r_K(A) - 1}{r_K(A)} A_A \bar{Z} E_c(A) \frac{(\mu/\rho)_K^A}{(\mu/\rho)_K}$$

where $(\mu/\rho)_K$ is the mass absorption coefficient on the high energy side of the K absorption edge of element A, ($(\mu/\rho)_K^A$ refers to the pure element A and $(\mu/\rho)_K$ to the sample); values of $(\mu/\rho)_K^A$ are given in Table 9.9.

For Lα radiation the constant is 3.31×10^{-6},

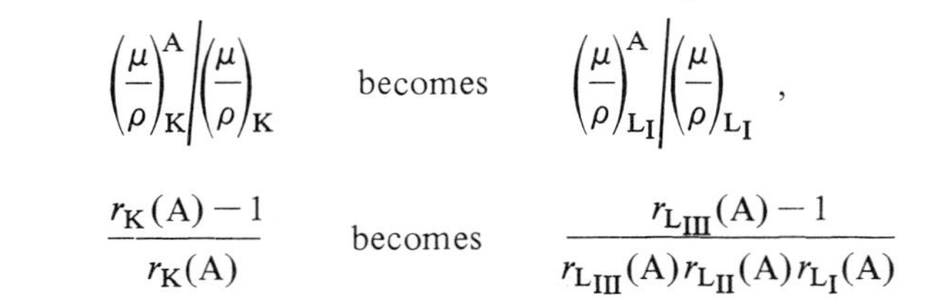

$$\left(\frac{\mu}{\rho}\right)_K^A \bigg/ \left(\frac{\mu}{\rho}\right)_K \quad \text{becomes} \quad \left(\frac{\mu}{\rho}\right)_{L_I}^A \bigg/ \left(\frac{\mu}{\rho}\right)_{L_I},$$

and

$$\frac{r_K(A) - 1}{r_K(A)} \quad \text{becomes} \quad \frac{r_{L_{III}}(A) - 1}{r_{L_{III}}(A) r_{L_{II}}(A) r_{L_I}(A)}$$

which is given to a sufficient accuracy by: $0.548 - 0.00231Z$ (Springer, 1967).

Table 9.9 – Values of $(\mu/\rho)_K^A$, the mass absorption coefficient of the absorbing element on the high-energy side of the K absorption edge of the emitting element, for use in calculating continuum fluorescence corrections (after Springer, 1967).

Absorber		Emitter																
Z		Mg	Al	Si	P	S	Cl	K	Ca	Ti	V	Cr	Mn	Fe	Co	Ni	Cu	Zn
6	C	800	483	300	195	130	89	44	32	17	10	–	–	–	–	–	–	–
7	N	1380	770	480	315	210	145	72	52	29	17	13	10	–	–	–	–	–
8	O	2180	1300	820	530	350	244	122	88	49	37	29	22	17	14	11	–	–
9	F	2670	1610	1010	665	442	310	155	133	63	48	37	29	21	18	14	12	–
11	Na	4850	2920	1830	1200	805	555	278	202	114	88	68	53	42	33	27	22	18
12	Mg	6260	3800	2400	1580	1050	730	367	266	150	111	89	70	56	44	35	29	23
13	Al	550	4680	2950	1950	1310	910	460	335	188	146	113	89	70	56	45	37	30
14	Si	715	435	3620	2370	1600	1120	560	412	231	179	138	108	86	68	56	44	37
15	P	920	555	350	2850	1930	1350	685	500	283	220	170	133	106	85	69	56	46
16	S	1140	695	438	290	2290	1580	810	590	332	260	200	157	125	99	81	66	54
17	Cl	1390	845	540	360	241	1810	935	690	390	300	235	184	147	117	95	78	63
19	K	2025	1230	775	515	347	242	1220	910	520	398	310	233	193	155	127	102	84
20	Ca	2380	1450	920	610	412	288	146	1010	580	450	350	275	218	175	143	117	96
22	Ti	3280	1980	1270	835	565	395	203	148	744	580	447	355	284	226	183	151	123
23	V	3750	2300	1450	960	645	450	230	168	96	630	500	390	315	253	205	168	139
24	Cr	4220	2600	1650	1090	740	520	265	195	112	87	550	440	350	280	230	187	153
25	Mn	4820	2950	1880	1250	840	560	290	220	125	97	76	484	385	306	250	206	169
26	Fe	5450	3300	2100	1400	950	660	336	250	140	110.	85	66	421	314	280	227	187
27	Co	6150	3750	2400	1580	1060	740	380	278	158	122	96	75	60	374	308	248	205
28	Ni	6800	4150	2650	1760	1180	835	423	313	177	138	107	84	68	54	336	275	227
29	Cu	7600	4600	2970	1960	1310	920	470	345	198	152	118	93	74	59	48	315	258
30	Zn	8600	5200	3300	2180	1460	1020	520	382	217	166	138	102	83	66	53	43	277

Assuming primary x-ray production takes place at the surface of the specimen, the absorption factor for fluorescence excited by continuum radiation of a particular energy is the same as that already derived for characteristic fluorescence, viz. $\ln(1+u)/u$, where $u = (\mu/\rho)_A \operatorname{cosec}\psi/(\mu/\rho)_c$, in which $(\mu/\rho)_c$ is the mass absorption coefficient for the continuum radiation. This, however, varies with energy, necessitating an integration to determine the overall absorption factor. In the absence of absorption edges, (μ/ρ) for all elements varies as E^{-n}, where $n = 3$ to a reasonably close approximation and we may write $(\mu/\rho)_c = (\mu/\rho)_K U^{-3}$.

Substituting for $(\mu/\rho)_c$ in the above equation gives

$$u = \left(\frac{\mu}{\rho}\right)_A U^3 \operatorname{cosec}\psi \bigg/ \left(\frac{\mu}{\rho}\right)_K \qquad \text{or} \qquad u = wU^3 \ ,$$

where $w = (\mu/\rho)_A \operatorname{cosec}\psi/(\mu/\rho)_K$. This makes the absorption expression integrable (Henoc, 1962), but the resultant formula is unduly complicated from a practical viewpoint. Springer (1967), however, showed that the absorption factor for continuum fluorescence could be approximated to $[\ln(1+U_0)]/U_0$ in view of the small size of the correction. The complete continuum fluorescence expression is thus:

$$\frac{I_f}{I_A} = 4.34 \times 10^{-6} \frac{r_K(A) - 1}{r_K(A)} A_A \bar{Z} G \ , \tag{9.5}$$

where

$$G = E_c(A) \frac{(\mu/\rho)_K^A}{(\mu/\rho)_K} \frac{\ln(1 + wU_0)}{w U_0} \ .$$

If absorption edges exist between E_c and E_0 due to the presence of other elements, $(\mu/\rho)_K^A/(\mu/\rho)_K$ and w are different between each pair of edges. It is therefore necessary to integrate G in equation (9.5) over each such region, with a weighting factor for the continuum intensity (Springer, 1972), giving

$$G = E_0 \sum \frac{(\mu/\rho)_{K,n}^A}{(\mu/\rho)_{K,n}} \left[(\ln U_{0,n} - 1 + U_{0,n}^{-1}) \frac{\ln(1 + w_n U_{0,n})}{w_n U_{0,n}} \right.$$
$$\left. -(\ln U_{0,n+1} - 1 + U_{0,n+1}^{-1}) \frac{\ln(1 + w_{n+1} U_{0,n+1})}{w_{n+1} U_{0,n+1}} \right]$$

in which the summation includes all energy intervals between $E_c(A)$ and E_0.

The size of the continuum fluorescence correction depends on the difference between I_f/I_A for specimen and standard. The most obvious variable is $\bar{Z}$, which represents the atomic number dependence of the continuum intensity. However, the term $(\mu/\rho)_K^A/(\mu/\rho)_K$ is sometimes very significant and can give rise to a substantial correction even when the difference in $\bar{Z}$ is small, for example when element A absorbs x-rays in the relevant energy range much more strongly than do the other elements present, that is $(\mu/\rho)_K^A \gg (\mu/\rho)_K$. This occurs in the case of a trace of copper in aluminium when using pure copper as the standard. Here a large differential absorption effect enhances the continuum fluorescence excitation of the copper by a factor which outweighs the atomic number difference, resulting in a *downwards* correction to the apparent copper concentration – the opposite to that which would be expected from consideration of the continuum intensity alone.

9.7 BOUNDARY EFFECTS

The discussion of fluorescence so far has been based on the assumption that the excited region is homogeneous. Since the directly analysed volume (the region penetrated by the electron beam) typically has dimensions of the order of 1 μm, and the exciting x-rays may penetrate much further than this, there is a significant likelihood of the assumption of homogeneity being invalid, in which case the calculated fluorescence correction will be incorrect. An obvious example is the analysis of a point close to a boundary between two phases of different composition, of which the most extreme example is the case of a particle of phase 1 just large enough to contain the incident electrons with most of the fluorescence excitation taking place in the surrounding phase (2), as shown in Fig. 9.5. Equation (9.3) can be modified to cover this situation by substituting $c_{B1}(c_{A2}/c_{A1})$ for c_B (where subscripts 1, 2 refer to phases 1, 2) and using mass

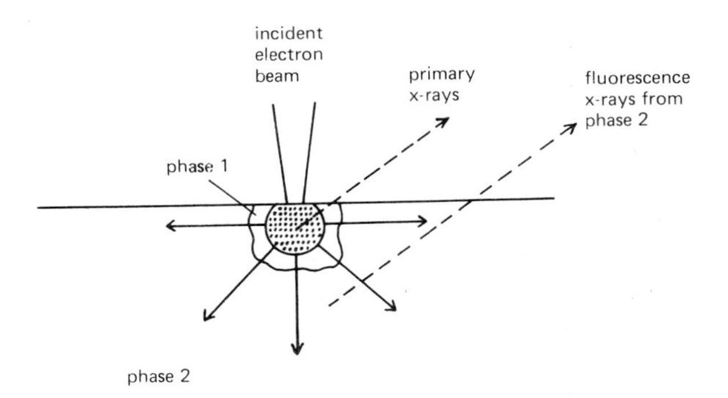

Fig. 9.5 – Analysis of small particle of phase 1, with fluorescence taking place in surrounding phase 2; shading represents region of primary excitation.

absorption coefficients which apply to phase 2, in which case I_F/I_A becomes the ratio of the fluorescence intensity excited in phase 2 to the electron-excited intensity that would be obtained with the beam on phase 2. The greatest effect occurs when $c_{B1} \to 1$ and $c_{A2} \to 1$, which gives $I_F/I_A \to \infty$. However, it is more useful to express the fluorescence intensity as a fraction of the primary intensity (I_{A0}) obtained from pure A, rather than that from phase 2.

$$\frac{I_F}{I_{A0}} = 0.5\, c_{B1} c_{A2} \frac{(\mu/\rho)_B^A}{(\mu/\rho)_{B2}} \frac{r_K(A)-1}{r_K(A)} \omega_K(B) \frac{A_A}{A_B} \left(\frac{U_B - 1}{U_A - 1}\right)^{1.67} \{\ldots\} \tag{9.6}$$

where $\{\ldots\}$ represents the absorption term, which is unchanged (see section 9.4). For the case of pure iron (phase 2) excited by pure nickel (phase 1), we find: $I_F/I_{A0} = 0.13$. Thus the maximum apparent concentration of iron in a small particle of pure nickel is 13% and boundary fluorescence can evidently be very significant.

It is also of interest to estimate the effect of continuum fluorescence in the same circumstances, namely a small particle of phase 1 surrounded by phase 2. Adapting equation (9.5) and expressing the fluorescence intensity relative to I_{A0}, as above, we have:

$$\frac{I_f}{I_{A0}} = 4.34 \times 10^{-6} c_{A2} \frac{r_K(A)-1}{r_K(A)} A_A \bar{Z}_1 E_c(A)$$

$$\times \frac{(\mu/\rho)_K^A}{(\mu/\rho)_{K2}} \frac{\ln(1 + w_2 U_0)}{w_2 U_0}\,. \tag{9.7}$$

An appropriate example of continuum fluorescence is when phase 1 is pure iron and phase 2 is pure nickel, so that nickel fluorescence is excited only by the continuum. Substituting appropriate values in equation (9.7) we find $I_f/I_{A0} = 0.033$, that is, the maximum apparent nickel concentration in pure iron is 3.3%, which, though not negligible, is a factor of 4 smaller than the equivalent figure for Ni–Fe characteristic fluorescence considered above.

So far discussion has covered the 'worst case' situation, which serves to indicate the upper limit to the analytical errors which can occur due to fluorescence across phase boundaries. Quantitative estimation of such effects is possible only for simple geometrical configurations, another example of which is that of an infinite vertical plane interface between two phases (Fig. 9.6). In the limiting case of an infinitely narrow beam located in phase 1 at the boundary with phase 2, equations (9.6) and (9.7) are applicable if an extra factor of 0.5 is included to allow for the different geometry. As the beam is moved away from the boundary, the fluorescence intensity decreases approximately exponentially, as shown experimentally with artificial boundaries (for example, Reed and Long, 1963). Thus the intensity at a distance x from the boundary is given by:

$$I_A(x) = I_A(0) \exp(-\tau x) \tag{9.8}$$

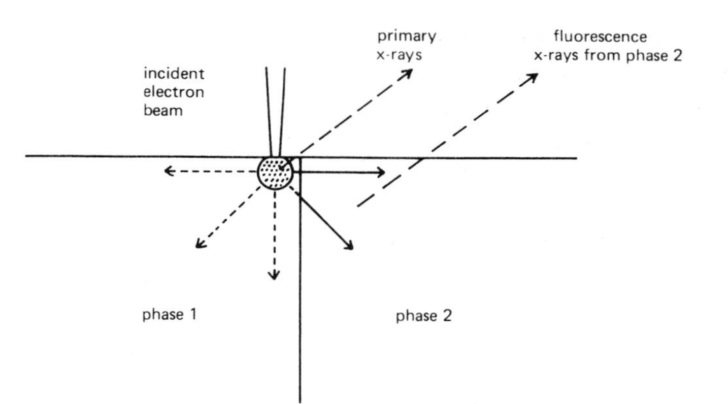

Fig. 9.6 – Analysis at the interface between phases 1 and 2, with fluorescence excitation across the boundary; shading represents region of primary excitation.

The size of the coefficient τ determines how far from the boundary the fluorescence effect remains significant.

In the case of characteristic fluorescence, the exciting radiation is emitted isotropically. Hence neglecting the effect of the finite source size, which is unimportant, the lateral distribution of the fluorescence radiation generated in the specimen may be equated with the depth distribution. It can be shown that the absorption factor derived in section 9.4 can be closely approximated by an expression based on an exponential depth distribution function corresponding to equation (9.8), which may be used for boundary fluorescence calculations, putting $\tau = 2.4(\mu/\rho)_{B1}\rho_1$.

In the case of a boundary between pure nickel (phase 1) and pure iron (phase 2), $\tau = 1250\,\text{cm}^{-1}$, the characteristic fluorescence intensity of iron excited by nickel falls to $1/e$ (37%) of its boundary value at a distance of 8μm from the boundary, and 1% at a distance of 37μm.

Continuum radiation is not emitted isotropically, but the random nature of electron trajectories arising from scattering ensures that the overall anisotropy is small when considering solid targets. Thus a similar approach can be used for continuum boundary fluorescence and the corresponding expression for τ is: $2.4\,U_0^{-1}(\mu/\rho)_{K1}\rho_1$.

In the case of excitation of nickel by the continuum from iron, $\tau = 5910\,\text{cm}^{-1}$, and the distance at which the intensity falls to $1/e$ is 1.7μm, while at 8μm it has fallen to 1%. Thus continuum fluorescence effects generally extend over considerably smaller distances, owing to the relatively strong absorption of the exciting continuum radiation in phase 1 (iron in this case).

9.8 FLUORESCENCE IN THIN FILMS

The analysis of thin specimens in TEM or STEM instruments is treated in Chapter 13. However, it seems appropriate to discuss the question of fluorescence in such specimens here. Fluorescence in thin (unsupported) films is sometimes assumed to be negligible on account of the relatively short absorption path for the exciting radiation. The following derivation allows thin-film fluorescence effects to be calculated quantitatively.

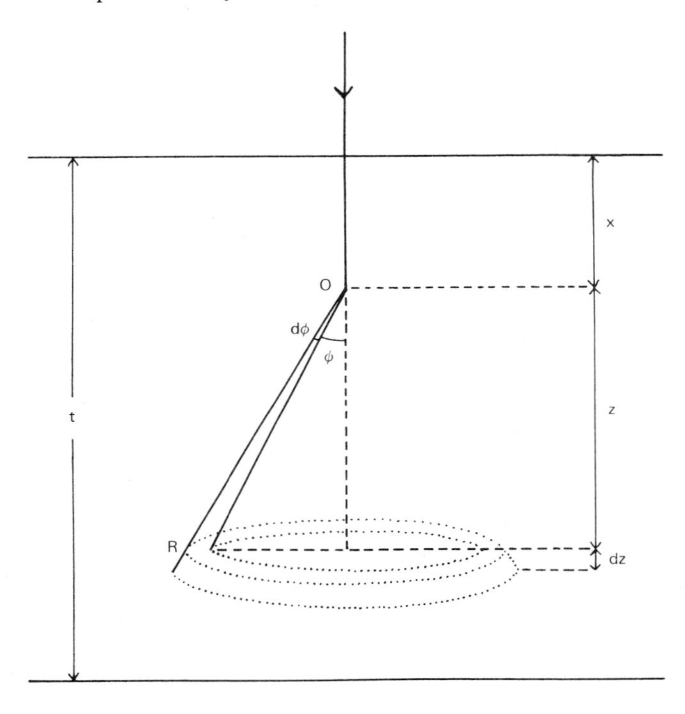

Fig. 9.7 – Geometry of thin-film fluorescence calculation (see text).

If the total intensity (I_B) of the exciting radiation from element B is produced uniformly throughout the thickness (t) of the film (which is true in practice to a sufficiently close approximation), then the intensity emitted per unit distance is I_B/t. Let us consider exciting radiation emitted in the forward direction from point O, as shown in Fig. 9.7. The geometry is the same as for the thick specimen derivation (section 9.4). Adapting the previous expression and neglecting the effect of absorption of the emerging fluorescence radiation, we have:

$$\mathrm{d}I_F = \frac{1}{2}\sin\phi\,\mathrm{d}\phi\left(\frac{\mu}{\rho}\right)_B \sec\phi\,\rho\,\mathrm{d}z\,\exp\left[-\left(\frac{\mu}{\rho}\right)_B \sec\phi\,\rho z\right]\epsilon\frac{I_B}{t}\,\mathrm{d}x\ .$$

The fluorescence intensity generated by 'B' radiation emitted from point O in the forward direction is thus given by

$$I_F(x) = \frac{1}{2}\epsilon\left(\frac{\mu}{\rho}\right)_B \frac{\rho}{t} I_B \left\{\int_0^{\pi/2}\int_0^x \tan\phi \exp\left[-\left(\frac{\mu}{\rho}\right)_B \sec\phi\rho z\right] dz\, d\phi\right\} dx$$

$$= \frac{1}{2}\epsilon\frac{I_B}{t}\left\{\int_0^{\pi/2} -\sin\phi\left(\exp\left[-\left(\frac{\mu}{\rho}\right)_B \sec\phi\rho x\right] - 1\right) d\phi\right\} dx \ .$$

Now, following Cox *et al.* (1979), we change the variable from x to a $(=(\mu/\rho)_B \rho x)$, hence:

$$I_F(a) = \frac{1}{2}\epsilon\frac{I_B}{(\mu/\rho)_B \rho t}\{1 - f(a)\}\, da$$

where

$$f(a) = \int_0^{\pi/2} \sin\phi \exp(-a\sec\phi)\, d\phi$$

$$= \exp(-a) + a\left(0.577 + \ln a - a + \frac{a^2}{2\times 2!}\ \cdots\right)$$

and $1 - f(a) = 0.423a - a\ln a$

(neglecting terms in a^2 and higher powers, which results in an error of less than 10% for $a < 0.3$). Hence:

$$I_F(a) = \frac{1}{2}\epsilon\frac{I_B}{a_t}(0.423a - a\ln a)\, da \ ,$$

where $a_t = (\mu/\rho)_B \rho t$.

We now integrate with respect to 'a' over the thickness of the film and multiply by two in order to take account of fluorescence excited by primary radiation emitted upwards (for which the same integration applies), thus:

$$I_F = \epsilon\frac{I_B}{a_t}\int_0^{a_t}(0.423a - a\ln a)\, da$$

$$= \frac{1}{2}\epsilon I_B a_t(0.923 - \ln a_t) \ .$$

The fluorescence conversion efficiency ϵ is given by equation (9.1), whereas

I_B/I_A is somewhat different in the thin-film case, being directly dependent on the ratio of the ionisation cross-sections (section 7.1).

$$\frac{Q_B}{Q_A} = \left(\frac{a \ln U}{E_c^2 U}\right)_B \bigg/ \left(\frac{a \ln U}{E_c^2 U}\right)_A = \frac{U_B \ln U_B}{U_A \ln U_A} ,$$

and

$$\frac{I_B}{I_A} = \frac{c_B \omega_K(B) A_A U_B \ln U_B}{c_A \omega_K(A) A_B U_A \ln U_A} .$$

Therefore

$$\frac{I_F}{I_A} = \frac{1}{2} c_B \left(\frac{\mu}{\rho}\right)_B^A \frac{r_K(A) - 1}{r_K(A)} \omega_K(B) \frac{A_A}{A_B} \frac{U_B \ln U_B}{U_A \ln U_A} \rho t (0.923 - \ln a_t) , \tag{9.9}$$

which is the same expression as given by Nockolds *et al.* (1980).

The subject of thin film fluorescence will be discussed further in section 13.2.8.

10

Evaluation of ZAF methods

G. LOVE

A stage has now been reached in the book when we can consider combining atomic number, absorption and fluorescence correction factors into a complete ZAF procedure. Although as pointed out in preceding chapters a number of correction models and variants thereon have appeared in the literature, it is not obvious which of these will prove the most suitable. Therefore in the present chapter some selected models will be examined in detail to identify areas of strength and weakness.

The evaluation will be confined to popular or promising approaches. For example, methods will be omitted which involve only slight modifications (such as adjusting the constants in the σ and h terms of the Philibert absorption correction), and no procedures will be included where authors provide insufficient information for their general application (as in the offset Gaussian method). An assessment of fluorescence models is not undertaken since (a) the characteristic fluorescence correction is generally less than 1% and very rarely exceeds 5% and (b) the continuum fluorescence contribution, although larger, tends to be similar in both specimen and standard so that again the nett effect is small (~1%), see for example Love and Scott (1981).

Following the process of selection we are left with essentially three atomic number corrections (Duncumb and Reed, 1968; Philibert and Tixier, 1968a; Love *et al.*, 1978a) and five absorption corrections (simplified Philibert, 1963; Reuter, 1972; Ruste and Zeller, 1977; Parobek and Brown, 1978; Love and Scott, 1978) which appear to merit further study. The ZAF procedures incorporating these are listed in Table 10.1 where, for ease of reference, the names associated with the absorption correction are used to distinguish between them. Where authors have expressed a preference for a particular combination of atomic number and absorption corrections that scheme has been adopted.

Table 10.1 – Details of ZAF correction models evaluated.

Model designation	Absorption correction	Atomic no. correction
Simplified Philibert	Simplified Philibert (1963) Heinrich's (1967) values for h and σ	Duncumb and Reed (1968)
Reuter	Reuter (1972)	Duncumb and Reed (1968)
Ruste–Zeller	Ruste and Zeller (1977)	Philibert and Tixier (1968a)
Parobek–Brown	Parobek and Brown (1978) Combined atomic number and absorption correction. Incorporates equations of Brown and Robinson (1979) for analysis carried out above 15 keV.	
Love–Scott	Love and Scott (1978)	Love *et al.* (1978a)

The formulae of Reed (1965) and Springer (1967) are employed to correct for characteristic and continuum fluorescence respectively. For elements with $Z > 11$, mass absorption coefficients are derived from the equations of Springer and Nolan (1976) and for oxygen, the values of Love *et al.* (1974a) are used.

10.1 EXPERIMENTAL MICROANALYSIS DATA

Clearly the most convincing way to assess any ZAF procedure is to see how well it performs when correcting microanalysis measurements obtained on specimens of known chemical composition; the wider the range of specimen compositions and microanalyser operating conditions included in the evaluation the more comprehensive the findings. Poole (1968) used such an approach, collecting together 229 microanalysis measurements for the purpose of comparing atomic number corrections. Subsequently Heinrich (1968b) examined Poole's data base and concluded that some of the measurements should be rejected as being unreliable. For example, he reasoned that where a set of data existed which contained a single variable, such as composition or probe voltage, the results should lie on a smooth monotonic curve and any significant deviation would indicate an erroneous measurement. A similar criterion was adopted by Love *et al.* (1976) when collecting 430 microanalysis results on binary alloys for the purpose of evaluating ZAF methods. This same data base is used in the assessment described in this chapter.

One way of presenting the corrected results is in the form of a histogram with horizontal axis labelled k'/k, where k refers, conventionally, to the measured intensity ratio for the element of interest and k' is the corresponding intensity ratio predicted by the ZAF method from a prior knowledge of the specimen composition. Now, if the experimental data are being corrected properly k'/k will be equal to unity and the spread of the results about this position gives a measure of the ZAF method's performance.

10.2 EVALUATION OF ATOMIC NUMBER CORRECTIONS

The most widely used atomic number corrections are those of Duncumb and Reed (1968, see section 7.1.1) and Philibert and Tixier (1968a, see section 7.1.2). Although workers have argued in favour of one or the other method for specific analyses (Walitski and Colby, 1969; Beaman, 1969), more comprehensive comparisons (Philibert, 1969; *Love et al.*, 1976) have concluded that there is little to choose between them despite the approximations made by Duncumb and Reed. In view of these findings the present assessment has been confined to the Duncumb and Reed method and the approach of Love *et al.* (1978a).

Atomic number corrections may be evaluated using histograms of corrected data and looking for any bias which occurs when a heavy element is being analysed in a light matrix, and vice versa. However, for the assessment to be meaningful it is important to select binary systems where absorption and fluorescence effects are relatively small. This leaves relatively few of the above 430 microanalysis results where both (a) a light element is analysed in a heavy matrix and (b) $f(\chi)$ exceeds 0.8. Consequently the percentage root mean square error (r.m.s.) is preferred here when describing the performance of atomic number corrections.

The respective atomic number corrections, when combined with an appropriate absorption correction (Love and Scott, 1978a), gave r.m.s. errors of 3.0% for the Duncumb and Reed method and 2.4% for that of Love *et al.*, which suggests that neither correction is introducing an error which is larger than that inherent in the measured intensity ratios (probably ~2%).

Since it has been established previously that fluorescence effects are generally small it may be concluded that an inadequate absorption correction must be primarily responsible for the large errors often found in microanalysis data. Therefore the remainder of this chapter is devoted to a study of this factor.

10.3 METHODS FOR EVALUATING THE ABSORPTION CORRECTION

Before using microanalysis measurements to assess absorption corrections it is important to establish how k'/k is influenced by errors in $f(\chi)$. It will be evident that $k'_A = c_A$ ZAF and the absorption correction (A) is given by $f(\chi)_A^{AB}/f(\chi)_A^A$. Now usually the larger source of error will be in $f(\chi)_A^{AB}$ because absorption in the specimen always tends to be greater than in pure (single element) standards (in each of the 430 systems studied pure standards were employed). Consequently if a model under-corrects for absorption, that is $f(\chi)$ is too high, one may expect $f(\chi)_A^{AB}/f(\chi)_A^A$ to be too large and hence $k'/k > 1$. Similarly k'/k will be less than unity in the event of over-correction.

No light element ($Z < 11$) results were included in the compilation of binary alloys and generally such information is lacking apart from some work published on oxides, carbides and borides by Ruste (1976) and on oxides by Love *et al.* (1974b). Here the latter set of data is used to provide information on a model's suitability for analysing light elements.

10.3.1 Ratios of $f(\chi)$ plotted as a function of χ

Different absorption corrections are perhaps more directly evaluated by comparing their predicted $f(\chi)$ values, $f(\chi)_{mod}$, with $f(\chi)$ data determined either from reliable $\phi(\rho z)$ curves or from Monte Carlo calculations. Love and Scott (1980) chose to employ Monte Carlo methods for the purpose of comparison because they provided a self-consistent set of data. However, for elements of high atomic number (gold and lead) some disagreement exists between individual sets of experimental results and also between these and Monte Carlo data (Love *et al.*, 1977a). Hence we confine the evaluation to the fairly representative series of elements – carbon, aluminium and copper.

Data are presented graphically as $f(\chi)_{mod}/f(\chi)_{MC}$ versus log χ, where $f(\chi)_{MC}$ is the magnitude of the absorption correction given by the Monte Carlo method. If the ratio of $f(\chi)$ values is significantly greater than unity then the absorption correction is too small ($f(\chi)_{mod}$ too large) and vice versa. This form of presentation was adopted by Bishop (1974) and permits the performance of a given model to be readily assessed for a wide range of absorption conditions.

10.3.2 Study of $\phi(\rho z)$ curves

This method of analysis complements the foregoing treatment by providing an insight into what may be at fault with a particular absorption correction model. Deficiencies are revealed by comparing $\phi(\rho z)$ distributions derived from the absorption model with similar data obtained by the Monte Carlo method. Sometimes the comparison of $\phi(\rho z)$ shape functions is rendered easier by normalising the x-ray depth distributions such that the horizontal axis is scaled in terms of the mean depth of x-ray generation and the vertical axis adjusted so that the area under each curve is the same.

10.4 REQUIREMENTS OF AN ABSORPTION CORRECTION

Before evaluating particular correction models it is useful to decide which features of the $\phi(\rho z)$ distribution are of greatest importance when calculating $f(\chi)$.

The first point to note is that the scaling factor along the vertical axis is not significant. For example the $\phi(\rho z)$ curve A in Fig. 10.1 will give exactly the same value of $f(\chi)$ as curve B no matter what the value of χ. This is because $f(\chi)$ contains the term $\phi(\rho z)$ in both numerator and denominator, viz.

$$f(\chi) = \frac{\int_0^\infty \phi(\rho z) \exp(-\chi \rho z) \mathrm{d}\rho z}{\int_0^\infty \phi(\rho z) \mathrm{d}\rho z} . \tag{10.1}$$

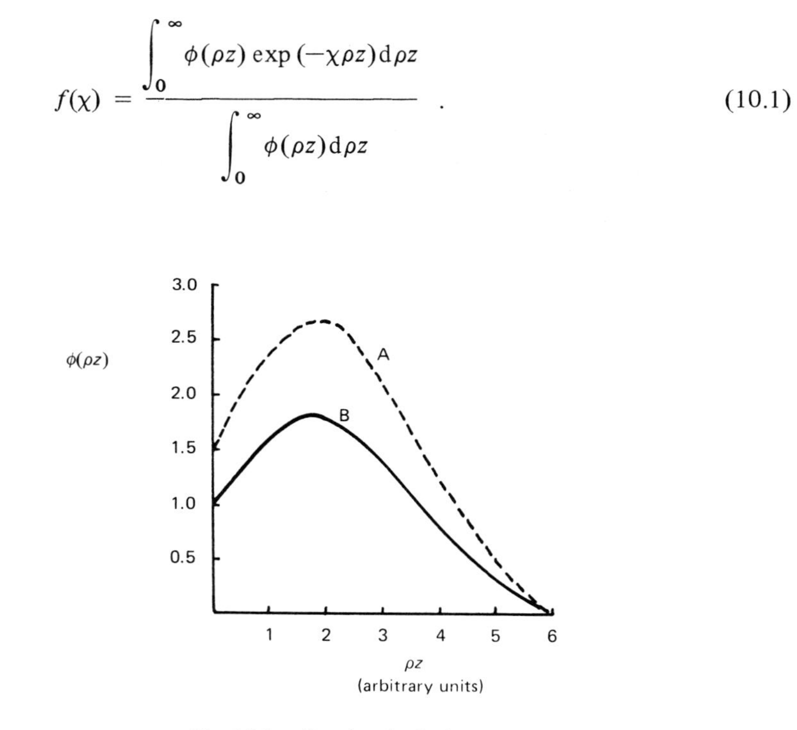

Fig. 10.1 – Simulated $\phi(\rho z)$ curves.

However, the scaling along the horizontal axis cannot be ignored. Indeed, when χ is small ($f(\chi) > 0.7$) the average depth at which x-rays are produced in the specimen becomes the single most important factor controlling the magnitude of $f(\chi)$ (Bishop, 1974). This may be demonstrated by expanding the exponential term in equation (10.1) which, to a first approximation, becomes

$$f(\chi) = \frac{\int_0^\infty \phi(\rho z)(1 - \chi \rho z) \mathrm{d}\rho z}{\int_0^\infty \phi(\rho z) \mathrm{d}\rho z}$$

$$= 1 - \chi \frac{\int_0^\infty \phi(\rho z)\rho z \, \mathrm{d}\rho z}{\int_0^\infty \phi(\rho z) \, \mathrm{d}\rho z}$$

$$= 1 - \chi \overline{\rho z} \ , \qquad (10.2)$$

where $\overline{\rho z}$ is the mean depth of x-ray production.

At the other extreme where absorption is very large, that is $f(\chi) < 0.1$, $f(\chi)$ depends strongly upon x-ray emission from the surface of the specimen. Here the thin film model (section 8.4.5) is appropriate and

$$F(\chi) = \frac{\phi(0)}{\chi} \quad .$$

Now by definition

$$f(\chi) = \frac{F(\chi)}{F(0)} \ ,$$

where

$$F(0) = \int_0^\infty \phi(\rho z) \, \mathrm{d}\rho z \ ,$$

and hence

$$f(\chi) = \frac{\phi(0)}{\chi \int_0^\infty \phi(\rho z) \, \mathrm{d}\rho z} \quad . \qquad (10.3)$$

Thus the actual shape of the $\phi(\rho z)$ curve is not important, $f(\chi)$ being determined by $\phi(0)$ and the area under the $\phi(\rho z)$ curve.

It is apparent from the discussion in section 8.4.4 that the thin film model strictly applies only when $f(\chi) < 0.01$, but that $\phi(0)$ becomes increasingly important as $f(\chi)$ falls below 0.1. Hence it follows that for $0.05 < f(\chi) < 0.7$, the shape of the $\phi(\rho z)$ curve plays an important role in determining the size of the absorption correction.

Having established which features of an x-ray depth distribution are important over particular ranges of χ, we are now in a position to explain why the absorption correction models perform in the way that they do.

10.5 SIMPLIFIED PHILIBERT METHOD

This correction method (Philibert, 1963; see sections 8.2.1 and 8.2.2) is the one most commonly used in quantitative analysis and is an essential component of a number of correction programs such as COR2 (Henoc *et al.*, 1973). Judging from its wide usage many people appear satisfied with the results it gives although some caution has been expressed concerning its range of application. The reservations appear more obvious when the simplified Philibert method is applied to the 430 microanalysis data compiled by Love *et al.* (1975, 1976). The resulting histogram (Fig. 10.2(a)) gives a skewed distribution with a greater number of results having $k'/k > 1$, indicating that the model has not provided a large enough correction factor. In almost all of the 430 systems the value of $f(\chi)$ exceeds 0.25 and consequently the skewness of the distribution may be explained by a tendency for the Philibert method to under-correct for absorption over this $f(\chi)$ range.

This view is confirmed from graphs of $f(\chi)_{SP}/f(\chi)_{MC}$ plotted as a function of log χ (Fig. 10.3(a)). They show significant under-correction (>5% relative) for absorption in aluminium at 20 KV when χ is between 1000 and 13 000 ($0.72 > f(\chi) > 0.1$) and in copper at 30 kV when χ is between ~1000 and 25 000 ($0.65 > f(\chi) > 0.05$).

An examination of the x-ray depth distribution for copper at 30 kV given by the simplified Philibert absorption correction (Fig. 10.4) shows a peak which is too high and too close to the specimen surface. This is why the model tends to under-correct for much of the $f(\chi)$ range. The over-correction which occurs when χ is very large (>20 000) results from the fact that the surface ionisation term (zero for this model) controls the magnitude of the correction. It should be noted that although the shape of the x-ray depth distribution approximates only roughly to that derived from the Monte Carlo model, the mean depth of x-ray generation is virtually the same, 5.7×10^{-4} g cm^{-2} compared with 5.5×10^{-4} g cm^{-2}. The values are similar because the over-extended tail of the simplified Philibert curve tends to compensate for poor representation of the peak, a feature which also explains why the absorption correction performs

Table 10.2 – Percentage root mean square errors obtained when using the correction procedures listed below.

Model	Analysed elements	
	$Z > 10$	$Z = 8$
Simplified Philibert	6.8	14.0
Reuter	8.0	35.0
Ruste–Zeller	7.1	9.2
Parobek–Brown	7.2	28.1
Love–Scott	5.4	5.6

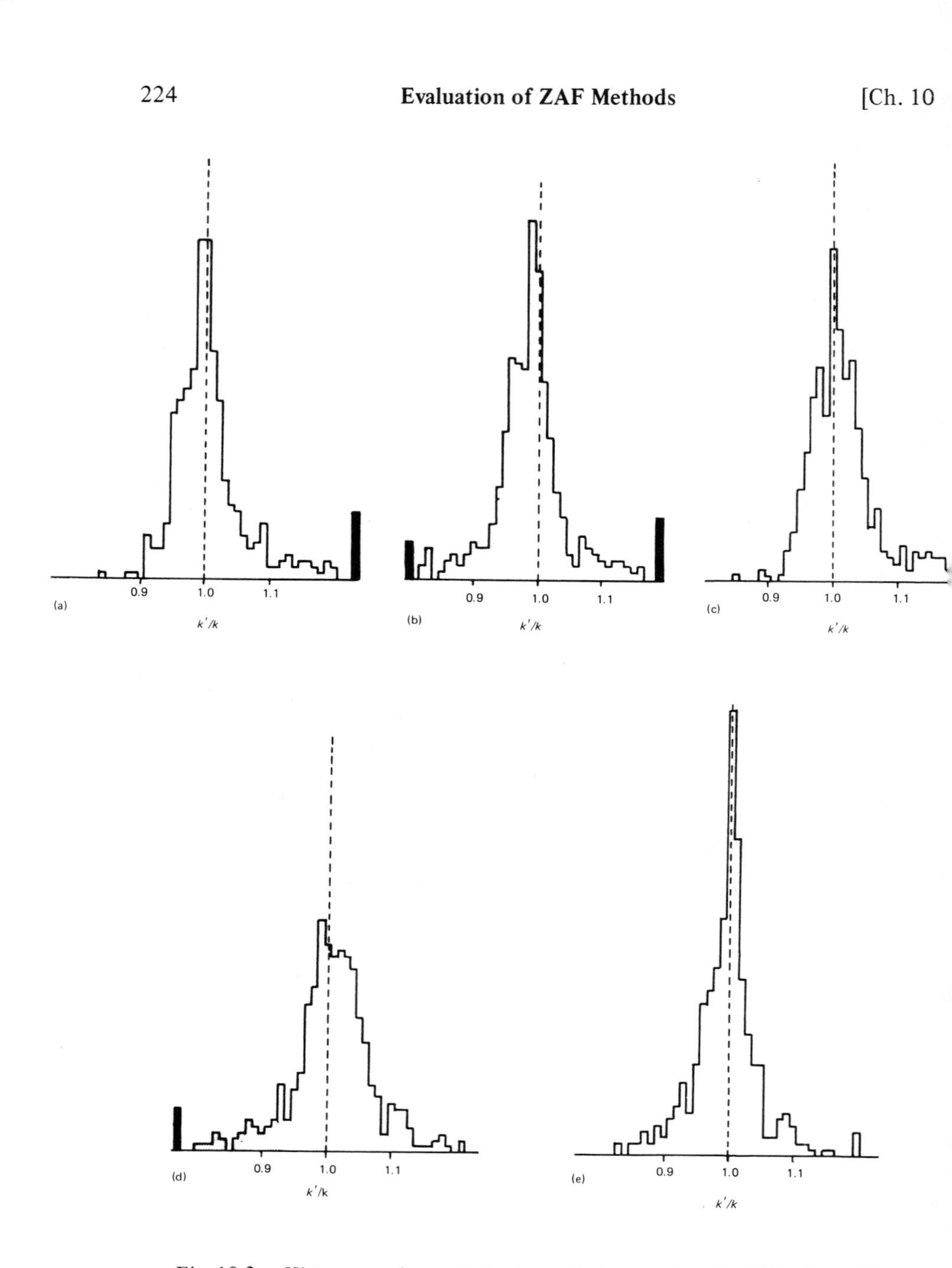

Fig. 10.2 – Histograms of corrected microanalysis data plotted as k'/k where k' is the intensity ratio predicted by the ZAF model and k is the measured value: (a) simplified Philibert, (b) Reuter, (c) Ruste–Zeller, (d) Parobek–Brown, (e) Love–Scott. Solid bars indicate errors in excess of 20%.

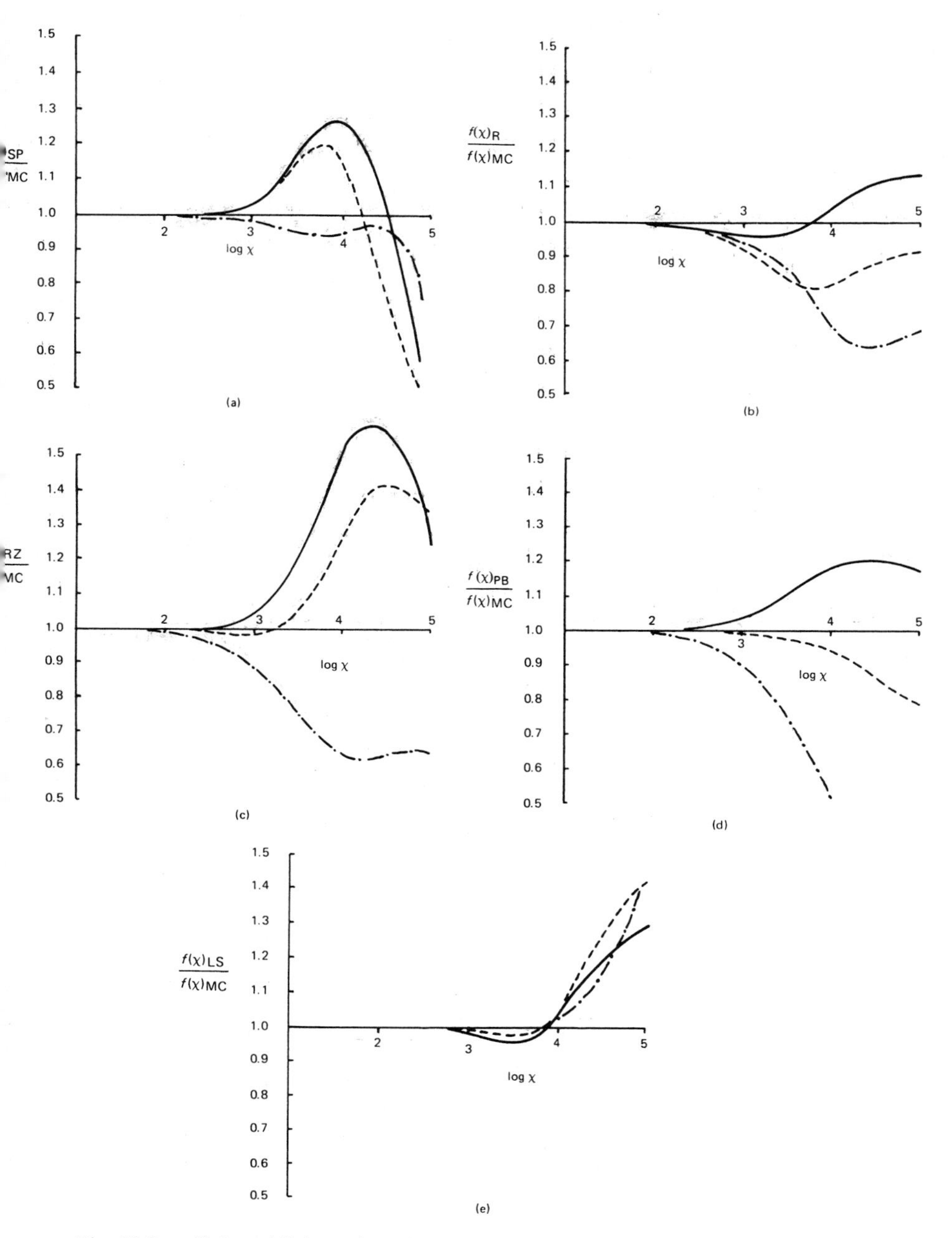

Fig. 10.3 – $f(\chi)_{mod}/f(\chi)_{MC}$ plotted as a function of log χ: —— Cu Kα 30 kV, - - - - Al Kα 20 kV and -·-·- C Kα 10 kV; (a) simplified Philibert, (b) Reuter, (c) Ruste–Zeller, (d) Parobek–Brown, (e) Love–Scott.

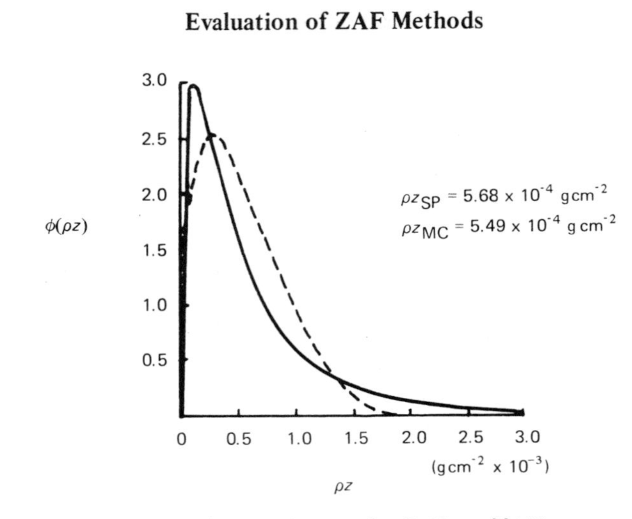

Fig. 10.4 – $\phi(\rho z)$ curves for Cu Kα at 30 kV; —— simplified Philibert, - - - Monte Carlo.

satisfactorily for copper at 10 kV when $f(\chi)$ exceeds 0.7. Similar reasoning may be advanced to explain the aluminium data shown in Fig. 10.3(a).

Because of the massive over-correction which takes place when χ is very large it is evident that the model is not well suited to light-element analysis. As expected, when applying the model to the oxide measurements mentioned earlier a large r.m.s. error is found (Table 10.2). However, it should be noted that the largest discrepancies arise at kilovoltages (25–30 kV) not normally used for soft x-ray analysis. Indeed, errors may be very much smaller at low kilovoltages, as can be inferred from the $f(\chi)$ plot for carbon at 10 kV (Fig.10.3(a)).

10.6 METHODS BASED UPON THE RIGOROUS PHILIBERT TREATMENT

In principle, the rigorous Philibert model (Philibert, 1963; see sections 8.2.1 and 8.2.3) should provide a more accurate correction than the simplifed version because x-ray emission from the surface is taken into account. Furthermore when absorption is large the expression for $f(\chi)$ (equation (8.9)) gives the correct limiting behaviour, viz.

$$f(\chi) = \frac{1 + \phi(0)h.\chi/[(4 + \phi(0)h)\sigma]}{(1 + \chi/\sigma)\left(1 + \dfrac{h}{1+h}\chi/\sigma\right)}$$

and for large χ where $\chi/\sigma \gg 1$,

$$f(\chi) \approx \frac{\phi(0)h}{4 + \phi(0)h}\left(\frac{1+h}{h}\right)\frac{\sigma}{\chi},$$

that is,

$$f(\chi) \propto \frac{\phi(0)}{\chi} \quad ,$$

(cf. equation (10.3)).

Love *et al.* (1974b, 1975) indicated that the method performed well for light elements, giving an r.m.s. error of 6% when applied to the oxide data but found, somewhat surprisingly, that it was not as good as the simplified approach for elements of higher atomic number. Love also showed that some improvement could be achieved by altering constants in the h and σ terms proposed by Duncumb and Melford (1966b) but concluded that, even after these modifications had been undertaken, performance was still inadequate.

10.6.1 Reuter

Reuter's correction method (Reuter, 1972; see section 8.3.2) was developed primarily for analysing thin films on substrates (see also section 13.1.2) and there appears little published information concerning its application to more conventional samples.

When the method is applied to the 430 systems the resulting histogram (Fig. 10.2(b)) is very broad and this is reflected in the percentage r.m.s. error of 8.8. There is also some over-correction with a much larger proportion of the results having $k'/k < 1$. The method nevertheless performs well on copper where errors do not exceed 5% until χ is above 10 000 (Fig. 10.3(b)), a restriction ($f(\chi)<0.09$) unlikely to be encountered in practical microanalysis. It would appear Reuter's correction is less satisfactory when dealing with lower atomic number elements such as aluminium where substantial errors arise when $\chi > 800$, a value frequently exceeded when analysing aluminium alloys. This is because the model predicts for aluminium a $\phi(\rho z)$ curve which has too low a peak and too extensive a tail (Fig. 10.5), features which combine to give a large overestimate of the mean depth of x-ray generation and a correspondingly small $f(\chi)$ value.

The performance of the Reuter method when applied to the oxygen results is extremely poor (Table 10.2). A study of the $f(\chi)$ plot for carbon (Fig. 10.3(b)) reveals that the model is unlikely to be suited to any light-element analysis because the value of $f(\chi)_R$ is found to be much too low throughout almost the entire χ range. The $\phi(\rho z)$ curve for carbon (Fig. 10.6) is helpful in explaining why this should be. It is noticeable that the mean depth of x-ray generation is vastly overestimated but, more importantly, the $\phi(\rho z)$ curve of Reuter is a most peculiar shape. A simple calculation reveals that the sharp cut-off, which occurs at a mass depth of $\sim 2.8 \times 10^{-4}$ g cm^{-1}, arises because the Bethe energy-loss law predicts that at this point the energy of the electrons is less than the value of E_c for carbon, that is virtually zero. This cannot be reconciled with the linear electron attenuation formula of Reuter which would indicate that some 45% of the incident electrons reach a depth of 2.8×10^{-4} g cm^{-1}. Clearly then this electron attenuation formula is not suitable for elements with $Z < 13$. In fact if the exponential law is used over the whole electron range the r.m.s. error for the oxygen data improves from 35% to 5.7%.

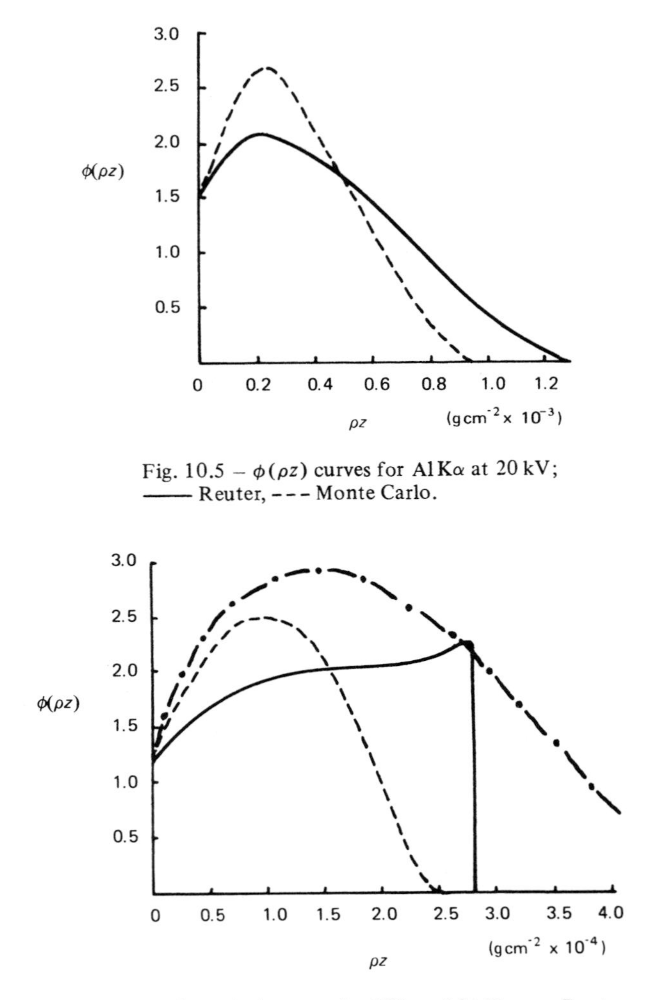

Fig. 10.5 – $\phi(\rho z)$ curves for Al Kα at 20 kV; —— Reuter, - - - Monte Carlo.

Fig. 10.6 – $\phi(\rho z)$ curves for C Kα at 10 kV; —— Reuter, - - - Monte Carlo, -·-·- Parobek–Brown.

10.6.2 Ruste–Zeller

The method of Ruste–Zeller (Ruste and Zeller, 1977; see section 8.3.3) is essentially the rigorous Philibert treatment modified to improve the quality of soft x-ray analysis. Satisfactory performance was claimed for light elements

(Ruste, 1976; Ruste *et al.*, 1978) although no information was available concerning the model's applicability for heavier elements.

The shape of the histogram (Fig. 10.2(c)) is very similar to that obtained when using the simplified Philibert model, which is not perhaps surprising because there is little difference between the models for elements with $Z > 26$. The r.m.s. error (7.1%) is marginally worse than that given by the simplified Philibert treatment (6.8%) and the bias in the histogram is more pronounced (cf. Fig. 10.2(c) and 10.2(a)).

Further evidence of similarity between the two models is obtained from the respective $f(\chi)$ data (cf. Fig. 10.3(c) and 10.3(a)). The Ruste–Zeller curve shows slightly greater deviations in the ratios of $f(\chi)_{\text{mod}}/f(\chi)_{\text{MC}}$, under-correcting for absorption in copper at 30 kV when $\chi > 1100$ ($f(\chi) < 0.6$) and in aluminium at 10 kV.

When the Ruste–Zeller $\phi(\rho z)$ curve for copper is constructed it is apparent that the general shape and the value of the mean depth of x-ray generation are almost identical to those given by the simplified Philibert model. The only difference is that the former method does not over-correct for absorption when χ is very large because the value of $\phi(0)$ is approximately correct. However, in spite of the introduction of a realistic value for $\phi(0)$ the Ruste–Zeller method is a little less satisfactory than the simplified Philibert treatment in dealing with elements of atomic number greater than 11. Hence it may be concluded that putting $\phi(0) = 0$ into a Philibert-type equation has the advantage of reducing the amount of under-correction in the medium-to-high χ range and a deleterious effect is experienced only when χ is extremely large.

The good light-element performance of the Ruste–Zeller method (9.2% r.m.s. error) when applied to the oxide systems is difficult to reconcile with the $f(\chi)$ plot for carbon at 10 kV†, see Fig. 10.3(c), which indicates substantial over-correction when χ exceeds 400 ($f(\chi) < 0.96$). However, if one assumes that the 10 kV oxygen $f(\chi)$ plot is similar to the carbon graph in Fig. 10.3(c), then it turns out that, fortuitously, almost all of the oxide systems studied fall on the horizontal part of the curve. The end result is that $f(\chi)$ values for both specimen and standard are over-corrected but the ratio of the two is approximately correct.

Given in Fig. 10.7 are the $\phi(\rho z)$ profiles for carbon at 10 kV calculated using the Ruste–Zeller and Monte Carlo models. It is evident that Ruste and Zeller have adjusted the equations so that the front end of the curve is modelled correctly but in so doing the mean depth has been made far too large. The equivalent set of normalised distributions in Fig. 10.8 indicates that the Ruste–Zeller distribution is essentially the same as that obtained from the simplified Philibert model although this is not immediately obvious from Fig. 10.7.

The conclusion which may be drawn is that the correct shape of the $\phi(\rho z)$

† An error was made in the $f(\chi)_{\text{RZ}}/f(\chi)_{\text{MC}}$ plot of carbon in the paper of Love and Scott (1981) and this favourably coloured the judgement made about the model at that time.

curve near the specimen surface and the correct value of the mean depth of x-ray generation cannot both be obtained using the equations of Ruste–Zeller. In addition, although these workers have fairly accurately modelled the front of the x-ray distribution with depth, the proper limiting behaviour as $\chi \rightarrow \infty$ (given by equation (10.3)) is not obtained because the area under the curve is much too large; this will seriously impair the light-element capability of the model.

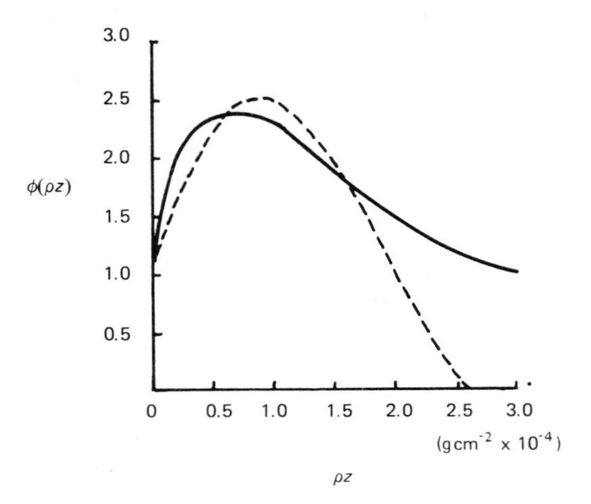

Fig. 10.7 – $\phi(\rho z)$ curves for C Kα at 10 kV; —— Ruste–Zeller, - - - Monte Carlo.

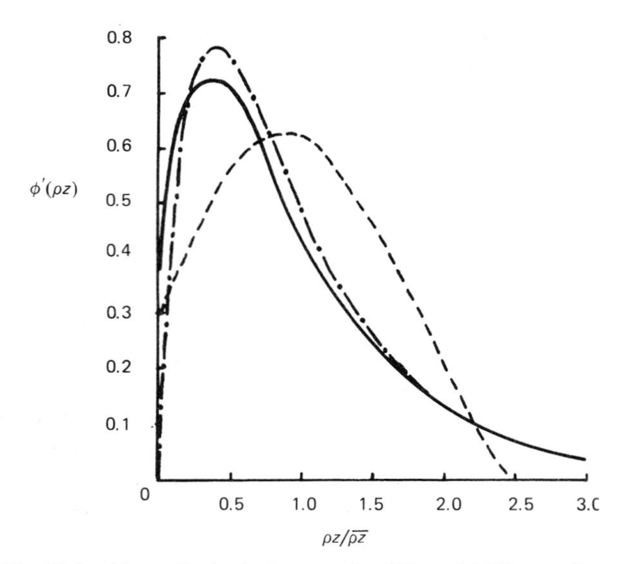

Fig. 10.8 – Normalised $\phi(\rho z)$ curves for C Kα at 10 kV; —— Ruste–Zeller, - - - Monte Carlo, -·-·- simplified Philibert.

10.7 METHODS BASED ON $\phi(\rho z)$ CURVE FITTING

10.7.1 Parobek–Brown

The method of Parobek and Brown (Parobek and Brown, 1978, see section 8.4.2) is an empirical one which attempts to describe $\phi(\rho z)$ curves precisely by means of complex equations. Exact fitting then permits atomic number and absorption effects to be dealt with together.

Brown *et al.* (1979) described its use for carbon analysis on a number of carbides and considered there was good agreement with the known concentrations when using an Fe_3C standard. Unfortunately it is difficult to comment on these results because although intensity ratios were given no corrected results were provided.

For elements with $Z > 11$ the correction does not appear to offer any improvement on the simplified Philibert method as evidenced by the histogram (Fig. 10.2(d)) which is broad and exhibits some bias towards under-correction (r.m.s. error 7.2%). However, more detailed examination of these results shows substantial over-correction will occur for data at high overvoltage ratios ($U_0 > 10$) as illustrated by the hatched portion of the histogram.

These findings are substantiated by the $f(\chi)$ graphs in Fig. 10.3(d). In the case of copper at 30 kV ($U_0 = 3.34$) under-correction takes place when χ exceeds ~1000 ($f(\chi) < 0.6$), but for aluminium at 20 kV ($U_0 = 12.8$) the model performs adequately up to χ ~10 000 ($f(\chi) > 0.14$) although now the tendency is towards over-correction.

Examination of a large number of $\phi(\rho z)$ curves indicates that the shape of the distribution is modelled fairly well by Parobek and Brown but the mean depth of x-ray generation is generally too small at low overvoltage ratios and too large when ratios are high. This is apparent from the $\phi(\rho z)$ curve for carbon at 10 kV (Fig. 10.6) where the mean depth is approximately twice the true value at an overvoltage ratio ~35. Not surprisingly the model does not work well for light element analysis where values of U_0 often exceed 10. This is borne out by the 28.1% r.m.s. error value obtained from the oxygen data and also the $f(\chi)$ curve of carbon which reveals substantial discrepancies when $\chi > 500$ ($f(\chi) < 0.95$). The good agreement claimed by Brown *et al.* on carbon referred to earlier is difficult to reconcile with these findings although Brown's corrected data will benefit from his use of the compound carbon standard, Fe_3C.

10.7.2 Love–Scott

As with the Parobek–Brown approach, the Love–Scott method is essentially empirical, but Love treats the atomic number (section 7.4.3) and absorption corrections (section 8.4.3) separately. There are few reports of its use in microanalysis apart from the authors' own work (Love and Scott 1978, 1980, 1981).

The histogram of the 430 systems corrected by the method is illustrated in Fig. 10.2(e). There is little evidence of any bias and the relative r.m.s. error is 5.4%.

The lack of bias and low error value are explained by reference to the $f(\chi)$ plots (Fig. 10.3(e)). For copper at 30 kV, agreement with the Monte Carlo data is very good and significant differences do not arise unless $\chi > 20\,000$ ($f(\chi) < 0.5$). Similarly for aluminium at 20 kV, discrepancies are less than 5% provided that $\chi < 10\,000$ ($f(\chi) > 0.15$).

At first sight this performance may seem surprising in view of the drastic simplification of the shape of the x-ray distribution to a rectangle (Fig. 8.6). It must be presumed therefore that, for most values of χ, excess x-ray generation predicted by the model in regions A and C is almost exactly compensated for by deficiencies in the regions B and D.

As may also be seen from Fig. 8.6 the profile assumed by the model gives too much x-ray generation in the immediate surface region (A) but this appears to be of little consequence unless absorption is very high. In fact the model works well for oxygen analysis (r.m.s. error 5.6%) on a range of samples with χ values up to 20 000 and also for carbon (Fig. 10.3(e)) where agreement with Monte Carlo data is good for $\chi < 20\,000$ ($f(\chi) > 0.2$).

10.8 CONCLUSIONS AND FUTURE PROSPECTS

This chapter has shown that there is little to choose between the atomic number corrections of Duncumb and Reed (1968), Philibert and Tixier (1968a) and Love *et al.* (1978a) and indeed there would appear little justification in attempting to develop a more refined model.

This is, however, not the case with the absorption correction. Few of those discussed here seem to merit further consideration and it is interesting to note that the simplified Philibert model which was developed as long ago as 1963 is still proving to be one of the most useful methods despite the fact that it is unsuited for light-element ($Z < 11$) analysis. It is to cover the latter area that most alternative correction models have been designed but it would seem that none have been outstandingly successful (Table 10.2). Of those mentioned here only the Love–Scott method appears to be significantly better than the simplified Philibert model in most situations.

Unfortunately, apart from using the Boltzmann transport equation (Brown and Ogilvie, 1966; Brown *et al.*, 1969) which is extremely complex and tedious to solve, there is no theoretical treatment capable of dealing adequately with the x-ray generation caused by a flux of electrons in a specimen. This may explain why most recent work on correction models (Love and Scott, 1978; Parobek and Brown, 1978; Brown *et al.*, 1981; Packwood and Brown, 1981) has been directed at $\phi(\rho z)$ curve fitting which requires no knowledge of electron or x-ray physics. A prerequisite to further progress in this area is, however, a sound data base which must include a large number of experimental results on x-ray depth distributions, supplemented by Monte Carlo calculations to cover conditions which are difficult to obtain experimentally. Much information already exists

(Table 8.1) but this requires careful evaluation to eliminate unreliable data. The x-ray depth distribution is a complex function of incident electron energy, critical excitation energy and atomic number so it may still prove difficult to obtain an equation for $\phi(\rho z)$ which is universally applicable. However, the rather crude rectangular profile used by Love and Scott gives promising results which suggest that the absorption correction is not especially sensitive to the exact shape of the $\phi(\rho z)$ curve. Thus one way of proceeding would be to use a relatively simple shape function, containing the least number of parameters in order to model the x-ray depth distribution, and then ensure that the mean depth of x-ray generation and the surface ionisation terms are incorporated in the expression for $\phi(\rho z)$. Minimising the number of parameters will reduce problems encountered when carrying out least-squares fitting of the data and may enable the formula for $f(\chi)$ to be solved analytically rather than by time-consuming numerical methods.

11

Correction procedures in practice

G. LOVE

After x-ray intensities have been recorded from specimen and standards the usual practice is to feed the data into a computer to convert them into elemental weight concentrations. The alternative of using hand-calculation is very rarely adopted because the ZAF calculations are so lengthy. This may be readily understood if one considers that specimens often contain in excess of six elements, the correction formulae are complex (particularly if continuum fluorescence effects are incorporated) and iterative techniques are employed.

The iterative procedure must be adopted in quantitative electron-probe microanalysis because the equation $k = c\,\mathbf{ZAF}$ cannot be used directly to obtain a value of the weight concentration (c) from the measured x-ray intensity ratio (k); this is due to the fact that **Z**, **A** and **F** are all functions of c rather than k. The method employed for calculating **Z**, **A** and **F** involves making an initial estimate of specimen composition using, commonly, the normalised x-ray intensity ratios. The above equation is then applied to determine c values and these should, of course, be better estimates of the true weight concentrations than the normalised intensity ratios. ZAF factors are next recalculated using the revised information about specimen composition and the whole procedure is repeated to improve further the precision of the corrected data. Several iteration methods have been proposed for use in quantitative electron-probe microanalysis and these are described in section 11.1.

In the following section of the chapter the use of computers and computer software for ZAF calculations is discussed. No attempt is made here to deal with commercial software packages in detail. Rather, the aim is to concentrate upon features which are regarded as essential in any ZAF program and to encourage the microanalyst to adjust input and output in order to suit his own particular requirements. Sometimes the computer system itself may influence the way in which data are processed and hence the merits of mini- and main-frame computer systems are mentioned (section 11.2.1).

Although the ZAF technique is the most widely used method of performing quantititative analyses there are other procedures which may prove advantageous on occasions and these form the subject of the remaining sections in the chapter. One such approach involves the use of empirical alpha factors (section 11.3); these permit the rapid hand-calculation of binary alloy concentrations but have certain limitations when applied to more complex specimens. Other methods, more closely associated with the ZAF concept, include standardless correction procedures (section 11.4) and the peak-to-background method (section 11.5). The former considerably speeds the acquisition of data and also allows quantitative analysis to be performed when suitable standards are not available, whilst the latter offers the prospect of obtaining quantitative data on tilted specimens and rough surfaces.

11.1 ITERATION PROCEDURES

As described earlier, the principle of ZAF iteration depends upon the repeated use of the equation:

$$k = c\,\mathrm{ZAF} \tag{11.1}$$

to compute improved elemental weight concentrations. The process is continued until after j iterations c_j does not differ from c_{j-1} by a significant amount and the two values are said to have converged. It is usual to set $[c_j - c_{j-1}]^2/c_j^2 \leqslant 0.00001$, which is approximately equivalent to 0.3% relative difference between c_j and c_{j-1}. A convergence test is normally applied after each iteration loop and the sequence is halted when the criterion has been satisfied for all elements in the specimen. Reducing the value below ~0.3% relative is found to increase the computation time without any commensurate improvement in the quality of the corrected results. This is because statistical variations of the input data and defects in the correction procedure itself contribute to the magnitude of the error in the final answer.

11.1.1 Simple iteration

Equation (11.1) may be written in the more general form:

$$c_\mathrm{A} = k_\mathrm{A} \,.\, f_\mathrm{A}(c_\mathrm{A}, c_\mathrm{B} \ldots)$$

and by analogy

$$c_\mathrm{B} = k_\mathrm{B} \,.\, f_\mathrm{B}(c_\mathrm{A}, c_\mathrm{B} \ldots) \qquad \text{etc.}$$

These equations indicate merely that the weight concentration (c) of a particular element in the specimen may be expressed as the product of the appropriate x-ray intensity ratio (k) and f, where f is a function of the true composition of the specimen. If k values are available for all elements in the specimen then it is usual practice to normalise them in order to compute f; for example, the first estimate of c_A is taken to be c_{A1}, where

$$c_{A1} = \frac{k_A}{(k_A + k_B + k_C \ldots)}$$

Normalisation is adopted during the iteration cycle because it has been found to aid convergence; however it should be noted that this is not equivalent to normalising the final data, a practice which is to be discouraged (see section 11.2.4).

It is now possible to make a second estimate, c_{A2}, of the weight concentration of element A in the specimen. This is given by:

$$c_{A2} = k_A \,.\, f_A(c_{A1}, c_{B1} \ldots) \;.$$

If it is found that $[c_{A2} - c_{A1}]^2/c_{A2}^2 < 0.00001$ the calculation is complete and c_{A2} should be a good estimate of the weight concentration. Unfortunately, this is an unlikely event and will only occur if the correction factors are small, that is if the compositions of specimen and standard are similar. Hence one is normally left with the problem of computing another value of the weight fraction which is more precise than c_{A2}. Probably the most obvious way of continuing is to normalise the values of c_{A2}, c_{B2} etc., and write:

$$c_{A3} = k_A \,.\, f_A\left[\frac{c_{A2}}{(c_{A2} + c_{B2} + c_{C2} \ldots)}, \frac{c_{B2}}{(c_{A2} + c_{B2} + c_{C2} \ldots)}, \ldots\right]$$

The exercise can be repeated until convergence is achieved, an approach termed simple iteration. However, it is not usually the best way of proceeding and another technique known as hyperbolic iteration offers certain advantages particularly that of rapid convergence.

11.1.2 Hyperbolic iteration

Experimental measurements of Ziebold and Ogilvie (1964) on binary alloy specimens revealed that if c is plotted against k the graph is a hyperbola and can be expressed as

$$\frac{c}{1-c} = \alpha \frac{k}{1-k} \;, \tag{11.2}$$

where α is a constant for a specific element in a binary alloy and for a given set of analysis conditions. The concept of alpha factors can be extended to deal with ternary and higher-order systems and section 11.3 describes its use for converting measured intensity ratios into elemental weight concentrations.

Criss and Birks (1966) recognised that the simple relationship between k and c could be used in an iteration sequence to achieve accelerated convergence.

The method entails using, as a first approximation, the normalised intensity ratios to describe the composition of the specimen, that is, the elemental weight concentrations will be c_{A1}, c_{B1} etc., where

$$c_{A1} = \frac{k_A}{(k_A + k_B + k_C \ldots)}$$

and

$$c_{B1} = \frac{k_B}{(k_A + k_B + k_C \ldots)}$$

Next, a reverse ZAF calculation is carried out to predict the x-ray intensity ratio that would be obtained from a specimen with such a composition, for example

$$k_{A1} = c_{A1}\,\mathbf{ZAF} \ .$$

Now the value of α_{A1} may be estimated by inserting the appropriate data for c_A ($= c_{A1}$) and k_A ($= k_{A1}$) into equation (11.2) as follows:

$$\alpha_{A1} = \frac{c_{A1}(1-k_{A1})}{k_{A1}(1-c_{A1})} \ ;$$

similar formulae will of course apply for other elements in the specimen. The last part of the iteration loop involves incorporating the measured intensity ratios (k_A, k_B, etc.) together with the α values in order to obtain a better estimate of the true elemental weight concentrations. In the case of element A we obtain

$$\frac{c_{A2}}{k_A} = \alpha_{A1} + (1-\alpha_{A1})c_{A2} \ .$$

The second iteration loop is begun by normalising the sum of the weight fractions in the usual way and then the whole of the above procedure is repeated to determine α more accurately and, consequently, to obtain improved values of the weight concentration.

Although the hyperbolic approach makes use of an empirical relationship between c and k in order to accelerate convergence this is not essential and in the method proposed by Wegstein (1958) the problem is treated purely as a mathematical one.

11.1.3 Wegstein iteration

We may write the calculated concentration of element A after the $(j + 1)$th iteration as $c_{A\,j+1}$ and equate this to the concentration determined after the jth iteration (c_{Aj}) plus a small increment h

$$c_{A\,j+1} = c_{Aj} + h \ . \qquad (11.3)$$

It follows that

$$c_{Aj} + h = k_A f_A(c_{Aj} + h) \ .$$

Using a Taylor expansion and ignoring second order effects

$$c_{Aj} + h = k_A(f_A c_{Aj}) + h k_A \delta f_A / \delta c_A) \ .$$

Solving for h

$$h = \frac{k_A f_A(c_{Aj}) - c_{Aj}}{[1 - k_A \,.\, \delta f_A / \delta c_A]} \ .$$

This value may be substituted in (11.3) to give

$$c_{A\,j+1} = c_{Aj} + \frac{k_A f_A(c_{Aj}) - c_{Aj}}{(1 - k_A \,.\, \delta f_A / \delta c_A)} \ . \tag{11.4}$$

Before such a relationship can be employed a value of $\delta f_A / \delta c_A$ is required. If the two previous values of c_A are available (c_{Aj} and $c_{A\,j-1}$) then it is reasonable to write

$$\frac{\delta f_A}{\delta c_A} = \frac{f_A(c_{Aj}) - f_A(c_{A\,j-1})}{c_{Aj} - c_{A\,j-1}} \ .$$

Hence the introduction of $\delta f_A / \delta c_A$ into equation (11.4) permits $c_{A\,j+1}$ to be determined if c_{Aj} and $c_{A\,j-1}$ are known. In the case of the first and second iteration this is not feasible so that simple iteration is conventionally adopted here and the Wegstein formula is applied subsequently.

11.1.4 Comparison of methods

Although convergence is obtained typically within five iterations when using the simple method it is not usually the quickest way of obtaining a solution and indeed there are occasions when convergence is not achieved at all. Divergent oscillations may arise in successively computed values if correction factors are large and a highly absorbing element is calculated by difference. As pointed out by Reed (1975b) this situation is less likely to occur when all elements are analysed or when one element is estimated by stoichiometry because convergence is then aided by normalisation.

Heinrich (1972) reports that hyperbolic iteration has been used extensively in his laboratories and he has never experienced convergence failure. Furthermore, convergence such that $c_j - c_{j-1} < 0.0001$ has always been achieved within four iterations, even when weight fractions have been determined by stoichiometry or by difference. Difficulties can arise with the hyperbolic method of Criss and Birks when compound standards are employed in the analysis (Springer, 1976).

However, this problem is overcome with the modification proposed by Henoc *et al.* (1973) in which the pure standard intensity for the element of interest is calculated using the measured intensity from the compound standard and the appropriate ZAF correction factors.

In most cases the Wegstein method converges quickly, being significantly better than simple iteration when large correction factors are involved (Reed and Mason, 1967; Beaman and Isasi, 1970). Springer (1976) has concluded that it also converges faster than the hyperbolic method in situations where there is a non-linear relationship between c/k and c, and converges as quickly when a linear relationship does apply. However, in the event of $\delta f/\delta c \approx 1/k$ the Wegstein method may give rise to divergence and Springer recommends the adoption of simple iteration in such a case.

Thus there would appear to be little to choose between the hyperbolic and Wegstein iteration techniques and, although the Wegstein method may be slightly faster on occasions, the hyperbolic method is probably the more widely used.

11.1.5 An example

We shall now show how an iteration technique is applied to calculate weight concentrations (c) from measured intensity ratios (k). It is assumed that the specimen contains an iron–aluminium phase which has been analysed at 15 kV in an instrument with an x-ray take-off angle of 40°. Pure standards of aluminium and iron were used to obtain k ratios of 0.385 and 0.418 respectively.

The ZAF correction formula we shall employ embodies the simplified Philibert absorption correction with Heinrich's σ and h values, and the atomic number correction of Duncumb and Reed. Characteristic and continuum fluorescence effects are small in this particular system and therefore will be ignored. Because of its popularity the hyperbolic method has been selected for iteration purposes. To simplify matters the superscripts A, B and AB refer to aluminium, iron and the Fe–Al phase respectively while the subscript A refers to aluminium Kα and B to iron Kα radiations; for example $(\mu/\rho)_A^{AB}$ is the mass absorption coefficient for aluminium Kα in the phase AB.

Data on standards is the first requirement and the following information may be obtained directly from tables:

$$\left.\begin{aligned} R_A^A &= 0.923 \\ R_B^B &= 0.901 \end{aligned}\right\} \text{from Duncumb and Reed (1968)}$$

$$\left.\begin{aligned} (\mu/\rho)_A^A &= 386 \\ (\mu/\rho)_A^B &= 3841 \\ (\mu/\rho)_B^B &= 71 \\ (\mu/\rho)_B^A &= 93 \end{aligned}\right\} \text{from Heinrich (1966a)}$$

$$\left.\begin{aligned}\sigma_A &= 5286\\ \sigma_B &= 7287\end{aligned}\right\}\ \text{from Salter (1970) or by using } \sigma = \frac{4.5 \times 10^5}{E_0^{1.65} - E_c^{1.65}}$$

$$\left.\begin{aligned}h^A &= 0.192\\ h^B &= 0.099\end{aligned}\right\}\ \text{from Salter (1970) or by using } h = 1.2A/Z^2.$$

The stopping power factor (S) is calculated according to Duncumb and Reed (1968)

$$S = \frac{Z}{A}\ln\left(\frac{1.166\bar{E}}{J}\right), \qquad \text{where} \qquad \bar{E} = \frac{2E_0 + E_c}{3}$$

and J is obtained from the expression of Duncumb *et al.* (1969). Thus for aluminium

$$S_A^A = \frac{13}{26.98}\ln\left(\frac{1.166 \times 10.52}{0.142}\right)$$

$$= 2.1481\ ,$$

and for iron

$$S_B^B = 1.7557\ .$$

The absorption correction is determined from

$$f(\chi) = \left[(1 + \chi/\sigma)\left(1 + \frac{h}{1+h} \cdot \chi/\sigma\right)\right]^{-1}\ .$$

Hence for aluminium

$$f(\chi)_A^A = \left[\left(1 + \frac{386 \times \operatorname{cosec} 40°}{5286}\right)\left(1 + \frac{0.192 \times 386 \times \operatorname{cosec} 40°}{1.192 \times 5286}\right)\right]^{-1}$$

$$= 0.8819\ ,$$

and similarly for iron, $f(\chi)_B^B = 0.9838$.

In order to establish values of R, S and $f(\chi)$ for the aluminium–iron phase some initial estimate of its composition has to be made. For this purpose, normalised k ratios are selected, these being based upon the measured values of 0.385 for aluminium and 0.418 for iron. The first estimate of the aluminium concentration, c_{A1}^{AB}, is then

$$c_{A1}^{AB} = \frac{0.385}{0.385 + 0.418} = 0.4795$$

and for iron

$$c_{B1}^{AB} = 0.5205\ .$$

The stopping power factor of a multi-element specimen is calculated according to

$$S = \sum_{i=1}^{i=n} c_i \frac{Z_i}{A_i} \ln\left(\frac{1.166\bar{E}}{J_i}\right) .$$

Hence for aluminium radiation in the phase AB

$$S_{\mathrm{A}}^{\mathrm{AB}} = \left[\frac{0.4795 \times 13}{26.98} \ln\left(\frac{1.166 \times 10.52}{0.142}\right) + \frac{0.5205 \times 26}{55.85} \ln\left(\frac{1.166 \times 10.52}{0.332}\right)\right]$$

$$= 1.0302 + 0.8746 = 1.9048 ,$$

and for iron

$$S_{\mathrm{B}}^{\mathrm{AB}} = 1.9815 .$$

(Note that the value of $\bar{E}$ is different for aluminium K and iron K radiation and therefore $S_{\mathrm{A}}^{\mathrm{AB}} \neq S_{\mathrm{B}}^{\mathrm{AB}}$.)

Values of R for the phase AB are interpolated from the table of Duncumb and Reed (1968), that is

$$R_{\mathrm{A}}^{\mathrm{AB}} = 0.877 ; \qquad R_{\mathrm{B}}^{\mathrm{AB}} = 0.929 .$$

Weight averaging of both h and the mass absorption coefficient allows the absorption correction factor for aluminium radiation in the phase AB to be calculated as follows:

$$f(\chi)_{\mathrm{A}}^{\mathrm{AB}} = \left[\left(1 + \left(\frac{\mu}{\rho}\right)_{\mathrm{A}}^{\mathrm{AB}} \frac{\operatorname{cosec} 40^\circ}{\sigma_{\mathrm{A}}}\right) \times \left(1 + \left(\frac{h^{\mathrm{AB}}}{1 + h^{\mathrm{AB}}}\right) \frac{(\mu/\rho)_{\mathrm{A}}^{\mathrm{AB}} \operatorname{cosec} 40^\circ}{\sigma_{\mathrm{A}}}\right)\right]^{-1}$$

$$= \left[\left(1 + \frac{2184 \times \operatorname{cosec} 40^\circ}{5286}\right)\left(1 + \frac{0.1436 \times 2184 \times \operatorname{cosec} 40^\circ}{1.1436 \times 5286}\right)\right]^{-1}$$

$$= [(1 + 0.6428)(1 + 0.0807)]^{-1} = 0.5633 .$$

Similarly $f(\chi)_{\mathrm{B}}^{\mathrm{AB}} = 0.9805$.

The weight-averaged data based upon the first estimate of composition are summarised in Table 11.1.

Table 11.1.

Parameter	First iteration	Second iteration
R_{A}^{AB}	0.877	0.884
R_{B}^{AB}	0.929	0.934
S_{A}^{AB}	1.9048	1.9402
S_{B}^{AB}	1.9815	2.0173
$f(\chi)_{A}^{AB}$	0.5633	0.5950
$f(\chi)_{B}^{AB}$	0.9805	0.9803

A pair of k ratios can now be established from the equation $k = c\,\mathbf{ZAF}$, that is

$$k_{A1}^{AB} = c_{A1}^{AB} \; \frac{R_{A}^{AB}}{R_{A}^{A}} \; \frac{S_{A}^{A}}{S_{A}^{AB}} \; \frac{f(\chi)_{A}^{AB}}{f(\chi)_{A}^{A}}$$

$$= 0.4795 \times \frac{0.877}{0.923} \times \frac{2.1481}{1.9048} \times \frac{0.5633}{0.9838}$$

$$= 0.2942 \; .$$

Similarly,

$$k_{B1}^{AB} = 0.5205 \times \frac{0.929}{0.901} \times \frac{1.7577}{1.9815} \times \frac{0.9805}{0.9838}$$

$$= 0.4745 \; .$$

A position has now been reached where the α-factors can be calculated:

$$\alpha_{A1}^{AB} = \frac{c_{A1}^{AB}}{k_{A1}^{AB}} \; \frac{(1 - k_{A1}^{AB})}{(1 - c_{A1}^{AB})}$$

$$= \frac{0.4795}{0.2942} \times \frac{(1 - 0.2942)}{(1 - 0.4795)} = 2.2100 \; ,$$

and

$$\alpha_{B1}^{AB} = \frac{0.5205}{0.4745} \times \frac{(1 - 0.4745)}{(1 - 0.5205)} = 1.2022 \; .$$

The second estimate of weight concentration is determined from the α-factor as follows:

$$\frac{c_{A2}^{AB}}{k} = \alpha_{A1}^{AB} + (1 - \alpha_{A1}^{AB}) \, . c_{A2}^{AB} \; ,$$

where in this case k is the measured intensity ratio obtained for aluminium

$$\frac{c_{A2}^{AB}}{0.385} = 2.2100 + (1 - 2.2100)\, c_{A2}^{AB}$$

and

$$c_{A2}^{AB} = 0.5804 \ .$$

Similarly,

$$\frac{c_{B2}^{AB}}{0.418} = 1.2022 + (1 - 1.2022) \,.\, c_{B2}^{AB}$$

and

$$c_{B2}^{AB} = 0.4634 \ .$$

Normalising these two values gives

$$c_{A2}^{AB} = 0.5560 \qquad \text{and} \qquad c_{B2}^{AB} = 0.4439 \ .$$

The normalised values are used to make new estimates of R, S and $f(\chi)$ for the iron–aluminium phase. The data are given in Table 11.1. Using

$$k_{A2}^{AB} = c_{A2}^{AB}\ \mathbf{ZAF}$$

$$= 0.3978$$

and

$$k_{B2}^{AB} = 0.3991 \ .$$

Next revised α values are calculated

$$\alpha_{A2}^{AB} = \frac{c_{A2}^{AB}(1 - k_{A2}^{AB})}{k_{A2}^{AB}(1 - c_{A2}^{AB})}$$

$$= 1.8872$$

and

$$\alpha_{B2}^{AB} = 1.2019 \ .$$

The new weight concentration of aluminium is then given by

$$\frac{c_{A3}^{AB}}{k} = \alpha_{A2}^{AB} + (1 - \alpha_{A2}^{AB}) \,.\, c_{A3}^{AB}$$

that is

$$\frac{c_{A3}^{AB}}{0.385} = 1.8872 + (1 - 1.8872) \,.\, c_{A3}^{AB}$$

and hence

$$c_{A3}^{AB} = 0.542 \ .$$

Similarly,

$$c_{B3}^{AB} = 0.463 \ .$$

After a further iteration (not shown) the value of c_{A4}^{AB} was equal to c_{A3}^{AB} and $c_{B4}^{AB} = c_{B3}^{AB}$. Hence it may be deduced that the composition of the phrase is 46.3% Fe – 54.2% Al, which is consistent with the formula Fe_2Al_5.

11.2 COMPUTING

11.2.1 The computer system

At the time when Beaman and Isasi (1970) examined some forty computer programs which were available for quantitative electron-probe microanalysis, it was evident that most of them were intended to be batch processed on a main-frame computer. Hence, although large sophisticated programs could be used, considerable delays might be experienced between punching the input data on cards or tape and obtaining the corrected results from the computer. If optimum use is to be made of the computer, however, results are needed quickly so that the operator possesses the most up-to-date information on the specimen and is able to plan his future actions accordingly. Consequently, batch processing, being unable to satisfy this requirement, is now used less frequently while multi-access systems or dedicated mini-computers are becoming increasingly popular.

With a multi-access arrangement, a main-frame computer accessed from remote terminals is used on a time-sharing basis and because this is large the software can be permanently stored on hard discs and there is no restriction on program size. Although data retrieval is normally rapid the response may be slowed considerably when the system is heavily loaded. Nevertheless this should not present a major problem in a well designed multi-access configuration.

In the last decade there has been a rapid development in the field of mini-computers and since such instrumentation is now aimed at the mass market they have become very competitively priced. Many laboratories are equipped with dedicated mini-computers, not only because they are relatively cheap but also because of their usefulness in other areas. For example, processing of ED spectra is not easily accomplished with a main-frame computer because of the large amount of data which must be transferred from the analyser, but this problem does not arise if a mini-computer is employed. Again, computer control of the microanalyser is widely recognised as being an efficient method of carrying out repetitive microanalysis measurements and the dedicated mini-computer is ideal for this purpose. A very adequate system would be a 32K mini-computer with dual floppy disc drives. This would permit a lengthy and rigorous program such as COR2 (second version CORrection scheme developed by Henoc *et al.*, 1973) to be accommodated by careful programming and judicious operation of overlay files. Some simplification of the ZAF computer program may be necessary if only a small core store is available but little difficulty should be experienced with random access memory (RAM) in excess of 32K. In recent years the memory size of mini- and micro-computers has increased rapidly (some current micro-computers have 128K of RAM) and this process is likely to continue. Consequently in the future the most complex of ZAF programs should be easily accomodated on mini-computers and the only advantage of the main-frame will be in terms of speed of execution.

11.2.2 The computer program

The majority of ZAF programs surveyed by Beaman and Isasi (1970) incorporated the correction schemes listed in Table 11.2, and most of these have been discussed in Chapter 10. The three atomic number corrections listed differ very little in terms of their ability to correct microanalysis data although the Poole and Thomas method is perhaps slightly less satisfactory than the other two. In the case of the absorption correction the use of Heinrich's constants in the simplified Philibert model is now generally accepted as being preferable to Duncumb and Shields's values, but the latter do not introduce significant errors unless the absorption correction is large. In view of these facts and also that the fluorescence effects are often small it is not surprising to find that the performance of a considerable number of the computer programs examined by Beaman and Isasi is somewhat similar. If a survey of current programs was undertaken one

Table 11.2.

Correction factor	Corrections most commonly employed
Atomic number	Poole and Thomas (1961–62) Duncumb and Reed (1968) Philibert and Tixier (1968a)
Absorption	Simplified Philibert (1963) with Heinrich's (1967) constants for σ and h or Duncumb and Shield's (1966) constants
Characteristic fluorescence	Reed (1965)
Continuum fluorescence	None Springer (1967) Henoc (1968)

might expect the same conclusions to apply because the atomic number, absorption and fluorescence corrections developed in the 1960s are still widely used today. At present the main differences between popular programs lie in the amount of storage space required to run them and in the range of input and output options provided.

Let us now examine in some detail the essential features of a good computer correction program. It should, ideally, cope with an unlimited number of elements per analysis although, in practice, a total of 16 elements is adequate for almost all situations.

Analysis using Kα, Lα and Mα lines ought to be possible where circumstances warrant it, that is, for line energies greater than 0.8 keV in the case of Lα and Mα lines.

Since a wide range of standards is used in microanalysis work, facilities are required which permit the use of compound as well as pure element standards. Where an appropriate standard is not available or the measurement of one particular x-ray line is suspect it should be possible to calculate the concentration by difference or by stoichiometry, that is, assuming certain elements are combined in fixed atomic ratios defined by chemical formulae; for example, in Al_2O_3 the atomic ratio of oxygen to aluminium is 3:2.

There should be no restriction on the kilovoltage used for the analysis provided that all measurements are carried out under identical analysis conditions. Some programs, such as COR2, have an additional facility which allows each element to be measured at a different operating voltage. This increases the complexity of the program but can be extremely useful, especially when light elements are being analysed. However, care must be taken that the position of the probe is not displaced if the voltage is changed.

If several measurements are taken during a single spot analysis some statistical evaluation of the results should be available. This then

(a) permits the rejection of unreliable data,
(b) together with other factors gives an indication of the confidence which can be placed in the corrected analytical results,
(c) provides the operator with information which can be used to assess the performance of the electron-probe microanalyser,
(d) enables small, non-meaningful peak intensities to be eliminated using a significance test which takes into account counting statistics and the precision of background measurement. By carrying out the statistical evaluation at an early stage in the program the apparent concentrations are removed before the iteration sequence and as a result data processing may be speeded up.

Whilst it should not be necessary to include the dead time as input unless the dead time of the various spectrometers is very different, it should be possible to change the stored value easily and to override the correction when, for example, data from energy-dispersive measurements are being used. Although the dead time of modern wavelength-dispersive spectrometers and their associated electronics is generally small ($\sim 1\,\mu s$) it is essential that a correction is performed since it will have a significant effect for count rates of 10 000 and above.

The decision whether or not to carry out a characteristic fluorescence correction should be made within the program because the calculation is lengthy and may be omitted if the correction is small. The decision can be based upon the proximity of the energy of the exciting radiation to the critical energy of the excited line as in FRAME B (Yakowitz *et al.*, 1973).

It is essential in the interest of computation speed to possess the option of short-circuiting the continuum fluorescence correction if one is provided. For example on our version of COR2 running on a 32K Data General Nova computer, the processing time is, on average, 5 minutes per element when a continuum fluorescence correction included but only 2 minutes if it is omitted. In this case the slow response is due not only to the continuum fluorescence correction but also to the extensive use made of overlay files when using a large program on a mini-computer. Nevertheless a doubling of the processing time when using a continuum fluorescence correction is typical.

The iteration procedure ought to incorporate either the hyperbolic or Wegstein method discussed earlier. Convergence tests should take place after each loop and there should also be an upper limit on the number of iterations to allow for cases of non-convergence.

The choice of a computer program depends very much upon the types of analysis being undertaken in the laboratory and the personal preferences of the microprobe analyst. In general the greater the flexibility of the program the more input data are required. It is unlikely that any particular computer program will exactly fit requirements, except those of the person who wrote the program. In this context it is essential that the software is readily accessible to the user so that simple modifications may be easily carried out. However, if software alterations are undertaken it is important that test data are run before and after the changes are made to check that the program is still performing correctly.

11.2.3 Input

Convenience of operation is an important feature of any program and if this is to be achieved the amount of input data needs to be reduced as far as possible. Not only is it time consuming to feed in a large quantity of data but also it increases the possibility of mistakes being introduced by the operator. The amount of input necessary will depend upon the ZAF program being used but in Table 11.3 are listed the minimum requirements.

It is desirable that atomic weights, critical excitation energies, mass absorption coefficients etc. are either calculated in the program or stored as an array.

In addition to the basic requirements listed in Table 11.3 there are other options to consider. The first of these will probably be some form of labelling to define the specimen and area being investigated.

Table 11.3 – Minimum input for ZAF program.

Element identification	Atomic number or chemical symbol
Incident electron energy	
Spectral lines used in the analysis	Kα, etc.
Information on standards	Pure, compound, etc.
Experimental measurements on specimen and standard	Peak and background intensities or *k* ratios

The x-ray take-off angle may or may not be regarded as necessary input depending upon the circumstances. If data are being obtained from a conventional electron-probe microanalyser on which the x-ray take-off angle is fixed, its inclusion as input is obviously unnecessary. However, if measurements are being taken from a scanning electron microscope fitted with spectrometers a variable x-ray take-off angle may be frequently employed. In such cases there is good reason for the operator to be regularly prompted for information regarding its value.

X-ray intensity data from specimen and standard may be introduced either in the form of *k* ratios or as peak and background intensities. The adoption of *k* ratios certainly reduces the amount of input but it does mean that some prior hand-calculation is necessary and also precludes the calculation of dead-time corrections and any proper statistical analysis of the data. For these reasons the separate insertion of peak and background intensities is to be preferred.

Specimen current may be an essential input parameter if the program incorporates a correction for drift. This is rarely worth the extra effort involved because drift does not necessarily vary in a linear manner as assumed in most programs. If the beam current is drifting significantly, the conditons are not suitable for quantitative analysis and the problem should be rectified at source rather than attempting a drift correction.

11.2.4 Output

The output will usually be in the form of elemental weight percentages or atomic percentages. The data should not generally be normalised to 100% as this disregards useful information provided by the program. A total which is substantially different from 100% may be due to one or more of the following reasons:

(a) There is an error in the input – either a wrongly typed value or an erroneous measurement.
(b) There is an unidentified element present.
(c) The correction program is not working satisfactorily – perhaps there is a software error or, alternatively, the ZAF model is not appropriate for the analysis being undertaken.

Although failure to achieve a total close to 100% is a good reason to regard the data with suspicion, it should not be assumed that an analysis is correct if the summation equals 100%. Indeed, there have been instances where over-correction of one elemental concentration has been compensated for by under-correction of another.

As Reed (1975b) points out, although it is desirable to give the estimated error for each concentration this is hard to achieve in practice because it depends not only upon counting statistics but also upon the accuracy of the correction procedure. The latter is difficult to determine and it is probably better to present some statistical information on the accuracy of each measurement and have the option of printing out the individual correction factors so that the operator can judge for himself the reliability of the analysis.

The printing of warnings can be a useful feature. Errors in an analysis may arise from Kronig structure effects (section 8.7.3) when the x-ray line energy of one element is very close to the absorption edge of another. If a warning message is printed the operator may have the opportunity of selecting an alternative x-ray line for analysis. Another useful print-out is a warning if convergence has not been achieved after the specified maximum number of iterations. In such circumstances the weight concentration data are obviously suspect.

a

```
KV=? 30        T=? 15.5
ELEM(Z)            LINE                K-RATIO
? 29               ? 1                 ? .789
? 79               ? 2                 ? .152
? 0
ELEM(Z)            %
 29                 79.3254
 79                 21.0128
```

b

```
KV=? 30        T=? 15.5
ELEM(Z)            LINE                K-RATIO
? 29               ? 1                 ? .789
? 79               ? 0
ELEM(Z)            %
 29                 79.3259
 79                 20.6741
```

Fig. 11.1 – EXCEL computer printout showing data from analysis of copper and gold in a binary alloy; measurements carried out at 30 kV with take-off angle of 15.5°; (a) k ratios for both gold and copper are given; (b) the concentration of gold is found by difference.

a

```
DEFINE
DEFINE PARAMETERS FOR W.D. ZAF CALCULATION
ACCELERATING VOLTAGE (KV) = 30
SPECTROMETER TAKEOFF ANGLE= 15.5
       SPECIMEN TILT ANGLE= 0
NORMALIZE  (NO,ELEM,OXID) ? NO
   LONG OR SHORT PRINTOUT ? LONG
```

b

```
HOW MANY ELEMENTS ? 2

FIRST TELL ME ABOUT YOUR STANDARDS

ELEMENT NUMBER 1: SYMBOL= CU
            SHELL(K,L,M)= K
PURE / COMPOUND STANDARD? PURE
  PURE INTENSITY FOR CU = 10009
  BKG. INTENSITY FOR CU = 9

ELEMENT NUMBER 2: SYMBOL= AU
            SHELL(K,L,M)= L
PURE / COMPOUND STANDARD? PURE
  PURE INTENSITY FOR AU = 10068
  BKG. INTENSITY FOR AU = 68

NOW TELL ME ABOUT THE UNKNOWNS
HOW MANY LOCATIONS ? 1

NAME FOR LOCATION #  1 = COPPER-GOLD ANALYSIS
CU  INTENSITY=8000
CU BACKGROUND=110
AU  INTENSITY=1600
AU BACKGROUND=80
```

c

```
CALCULATED RESULTS FOR UNKNOWN LOCATIONS

LOCATION NUMBER 1  COPP
ELEM          WT%       AT%        OX%       +-REL(%)
 CU        81.100    92.271                    1.518
 AU        21.058     7.729                    2.878
TOTAL :   102.158
```

d

```
CALCULATED RESULTS FOR UNKNOWN LOCATIONS

LOCATION NUMBER 1  COPP
ELEM  SHELL      INTEN        +-REL(%)   K-RATIO
 CU     K      7890.000        1.141     0.78900
 AU     L      1520.000        2.697     0.15200
ELEM  SHELL         Z           A          F
 CU     K       1.02421    0.92864    1.02036
 AU     L       0.86046    0.83755    1.00000
ELEM           WT%        AT%        OX%      +-REL(%)
 CU         81.100     92.271                   1.518
 AU         21.058      7.729                   2.878
TOTAL :    102.158
```

Fig. 11.2 – FRAME B printout of the copper–gold analysis; (a) input information stored on file, (b) input data for specimen and standard, (c) short output, (d) more detailed output.

Figs. 11.1 and 11.2 show examples of input and output obtained using two different programs. Fig. 11.1 is taken from a program developed in our laboratories which incorporates our own atomic number and absorption corrections and the characteristic fluorescence correction of Reed (1965). In the form shown it uses the minimum amount of input – kilovoltage, take-off angle, element identification (expressed as atomic number), x-ray line (Kα = 1, Lα = 2 and Mα = 3) and k-ratios. Typing zero in the element identification column indicates that all k-ratios have been inserted; output is provided in terms of weight percentages. A zero in the x-ray line column as shown in Fig. 11.1(b) indicates that this element is measured by difference. Programs with this type of input/output format are useful for rapidly processing large amounts of data.

In Fig. 11.2 is shown the EDAX version of FRAME. On beginning an analysis information about basic parameters is inserted into a separate file (Fig. 11.2(a)). This is extremely helpful as it reduces the amount of data which has to be introduced for each subsequent analysis, assuming of course that experimental conditions are not altered in the meantime. Fig. 11.2(b) illustrates the data which must be entered for each analysis and is self-explanatory. An (optional) short output is shown in Fig. 11.2(c) and concentrations are given both in terms of weight percent and atomic percent. The values given under 'relative percent' are a measure of the statistical error involved in the experimental measurements and do not indicate errors in the corrected data (which may well be much greater). The long output is illustrated in Fig. 11.2(d) where it may be seen that the individual atomic number, absorption and fluorescence factors are now listed.

11.3 ALPHA FACTORS

Castaing (1951, 1960), in what became known as his second approximation, showed that the difference in x-ray generation efficiencies between specimen and standard could be taken into account by the introduction of empirical alpha coefficients such that

$$\frac{I_A^{sp}}{I_A^{st}} = \frac{\alpha_A c_A}{\sum \alpha_i c_i} , \qquad (11.5)$$

where I_A^{sp} and I_A^{st} are the number of x-rays from element A generated in specimen and standard respectively. In order to carry out quantitative analysis the alpha factors were determined experimentally from measurements on alloys of known composition.

When studying binary alloy systems Ziebold and Ogilvie (1964) noticed that if graphs of c/k versus c were drawn, the plots were approximately linear; (this effect is illustrated in Fig. 11.3 for a silver–gold alloy). Hence one may write

$$\frac{c}{k} = \alpha + \beta c , \qquad (11.6)$$

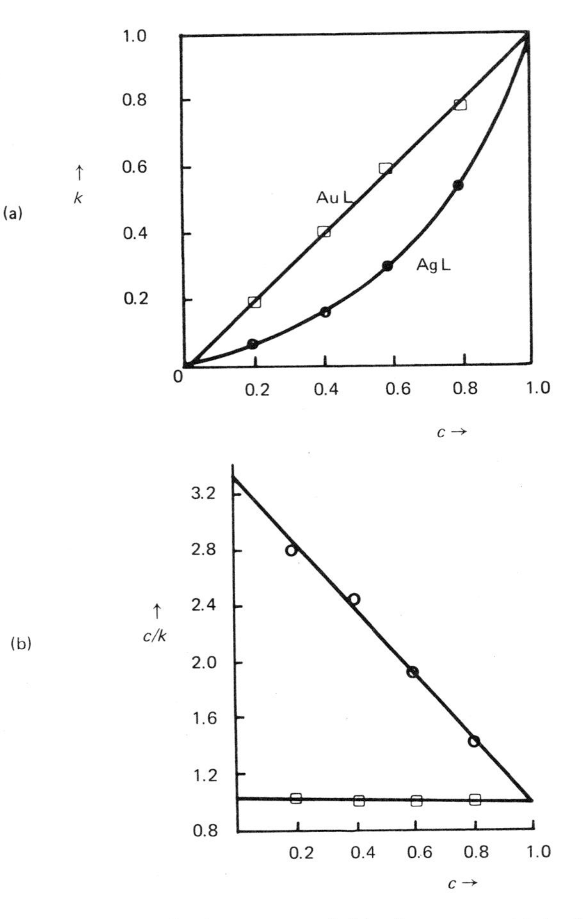

Fig. 11.3 – Plots of (a) k versus c and (b) c/k versus c obtained from a series of silver–gold binary alloys at 30 kV.

where α and β are constants. Now when $c = 1$ the specimen has the same composition as the pure reference standard and $k = 1$. Thus $\beta = 1 - \alpha$ and equation (11.6) may be rewritten as

$$\frac{c}{k} = \alpha + (1-\alpha)c$$

or in a symmetrical form as

$$\frac{1-k}{k} = \alpha\left(\frac{1-c}{c}\right) .$$

The value of α depends both upon the element being analysed and the binary system itself so that when measuring element A in the binary system AB

$$\frac{1-k_A}{k_A} = \alpha_A^{AB}\left(\frac{1-c_A}{c_A}\right) . \tag{11.7}$$

Ziebold and Ogilvie have shown that this is mathematically identical to the formula of Castaing (equation (11.5)). It was somewhat surprising, therefore, to find that the hyperbolic relationship held when absorption effects were present (it will be remembered that Castaing's treatment pertained only to generated x-rays). However, this result was later accounted for by Ziebold and Ogilvie (1966) who showed that equation (11.7) is approximately equivalent to the simplified Philibert absorption correction provided that the correction is not too large.

The relationship between k and c proposed by Ziebold and Ogilvie is extremely useful since only one measurement is needed on a known binary alloy to determine α_A^{AB} and then equation (11.7) may be used directly (avoiding any iteration procedures) to establish any other composition in the binary alloy system. The method therefore lends itself to hand-calculation. Most specimens studied will not be simple binaries but Ziebold and Ogilvie have shown that it may be used on more complex systems, the α factor for element A in a ternary system ABC being given by

$$\alpha_A^{ABC} = \frac{\alpha_A^{AB} c_B + \alpha_A^{AC} c_C}{c_B + c_C} ,$$

that is, the coefficient for the ternary may be determined if the values for the binaries are known. Similarly the treatment may be extended to an n-element system

$$\alpha_A^{AB\ldots n} = \frac{\alpha_A^{AB} c_B + \alpha_A^{AC} c_C \ldots + \alpha_A^{An} c_n}{c_B + c_C \ldots + c_n} .$$

For ternary and higher-order systems a series of simultaneous equations are obtained for $\alpha_A^{AB\ldots n}$ etc. and these must be solved using iterative techniques or alternatively by matrix inversion. Some of the attractive simplicity of the method has therefore been lost but it is still useful for rapid on-line analysis with a small computer (Yakowitz, 1975).

Another drawback to the procedure is the large number of standards required for an analysis. In a conventional ZAF correction method if there are n elements then n standards are needed but with the Ziebold and Ogilvie method $(n-1)n$ measurements of α have to be made and therefore $(n/2)(n-1)$ binary standards are necessary. A way of circumventing this problem is to calculate the matrix of alpha factors using a conventional ZAF program working in reverse, that is, a value for c is assumed and the corresponding k factor determined from the program.

Bence and Albee (1968) showed the usefulness of the alpha coefficient technique in the field of mineralogy. They examined a range of oxides in silicates and assigned each an alpha factor, treating the oxide as an element rather than a compound. The data given by Bence and Albee applied only if a 15 kV probe voltage was used and the x-ray take-off angle was 52.5° (the take-off angle on ARL instruments). Subsequently Albee and Ray (1970) provided results at 15 keV and 20 kV for an x-ray take-off angle of 38.5° and agreed to supply data for other x-ray take-off angles if requested.

Heinrich (1981) has reported that the hyperbolic model works well on the analysis of relatively low atomic number materials such as silicates but is less effective when large atomic number differences exist. It also fails to deal adequately with fluorescence effects because, as shown by Reed (1975b), the fluorescence correction factor is a non-linear function of composition. Although the latter difficulty can be overcome by the introduction of additional alpha coefficients (Claisse and Quintin, 1967) this complicates the calculations and may also increase the number of iterations necessary to achieve convergence (Ogilvie, 1977).

The alpha-factor approach is not merely used as a method for rapid on-line analysis of minerals it is also widely employed as an iterative procedure (see section 11.1.2). Furthermore the relationship between c/k and k discovered by Ziebold and Ogilvie can be employed as a means of rejecting inconsistent experimental data and has been used for this purpose by Heinrich (1968b) and by Love *et al.* (1976). Other areas where the method of alpha factors has proved useful is in establishing minimum detectable levels of elemental concentrations and in assessing the precision of ZAF-corrected results (Ziebold, 1967).

11.4 STANDARDLESS ANALYSIS

11.4.1 Basic principles

It is possible to carry out quantitative analysis without the need for standards provided that the response function of the analysis system can be well characterised. Wavelength-dispersive spectrometers cannot in general meet this criterion owing to such factors as instability of gas-counter gain and alignment problems associated with the crystal analysers. Energy-dispersive systems on the other hand are capable of good reproduction of x-ray data over long periods of time and lend themselves to this technique.

The principle of the method is to calculate the relative intensities of prominent x-ray lines (Kα, Lα etc.) from pure elements using an appropriate theoretical treatment. These intensities are then corrected using a ZAF procedure and further modified to allow for absorption in the detector window, gold coating and silicon dead layer of the energy-dispersive detector. Possessed with this information it is possible to establish the intensity of characteristic x-radiation which the detector will record from any element if x-ray counts have been recorded from a single standard. For example, the number of x-ray counts recorded in the Lα peak from a gold specimen may be predicted using the Kα intensity from a pure copper standard. Hence k ratios can be calculated when such data are used in conjunction with x-ray intensities recorded from the specimen of interest. If no standards are used at all in the analysis a normalisation procedure must be adopted. Initially the sum of the k ratios is set equal to one and re-normalisation is carried out after every ZAF iteration to bring the sum of the weight concentrations to unity. However, in the event of one standard intensity being measured other standard intensities may be obtained directly and consequently normalisation is unnecessary.

11.4.2 Russ

In the method developed by Russ (1974) the generated intensity of K emission (I_g) is determined from

$$I_g = p\omega \frac{N}{A} R \int_{E_0}^{E_c} \frac{Q}{\mathrm{d}E/\mathrm{d}\rho s} \mathrm{d}E \tag{11.8}$$

where p is the transition probability, ω is the fluorescence yield, N is Avogadro's number, R is the backscatter factor, A is the atomic weight and Q is the ionisation cross-section. If the Thomson–Whiddington Law (Whiddington, 1912) is used to describe the rate of electron energy loss and the ionisation cross-section of Bethe (1930), as modified by Green and Cosslett (1961), is used to represent Q, then equation (11.8) may be integrated exactly. The Green and Cosslett expression takes the form

$$Q = \frac{\text{const.}}{E_c^2} \frac{\ln U}{U} \tag{11.9}$$

and the Thomson–Whiddington law may be written as

$$\frac{\mathrm{d}E}{\mathrm{d}\rho s} = \frac{-b}{E} , \tag{11.10}$$

where b depends upon atomic number. Substituting in equation (11.8) and rearranging

$$I_g = \text{const.}\, p\omega \frac{N}{Ab} R \int_{U_0}^{1} (-\ln U)\mathrm{d}U \tag{11.11}$$

$$= \text{const.}\, p\omega \frac{N}{Ab} R\,[U_0 \ln U_0 - U_0 + 1] \; . \tag{11.12}$$

The constant in equation (11.9) will take a fixed value only for a particular shell since the ionisation cross-section will obviously depend upon the number of electrons in each shell. When equation (11.9) is used to represent relative x-ray emission intensities for K, L and M shells an additional term N_{el} is introduced which, if one ignores the small effects of electronic screening, takes the value of 2 for the K shell, 8 for the L shell and 18 for the M shell. We may therefore write

$$I_g = \text{const.}\, p\omega \frac{N_{el}R}{Ab} f(\chi)\, T[U_0 \ln U_0 - U_0 + 1] \tag{11.13}$$

where $T = (1 -$ absorption in window and dead layer of the detector).

In order to calculate the fluorescence yield Russ has used the equation of Wentzel (1927)

$$\omega = \frac{Z^4}{a + Z^4} \quad ,$$

where a takes the values 10^6, 10^8 and 7.5×10^8 for the K, L and M shell respectively. The backscatter factor, R, and the absorption correction, $f(\chi)$, are determined as in the FRAME computer program (Yakowitz *et al.*, 1973). There are two major differences between the early standardless correction of Russ (1974) and the later version (Russ, 1978), both of which are based on equation (11.12). In the earlier method the simplification

$$[U_0 \ln U_0 - U_0 + 1] \simeq (U_0 - 1)^{1.67}$$

was employed, an approximation which can lead to errors of up to 10% in certain instances (Duncumb, 1957). Hence the use of more exact expression was preferred. The second change involved substituting total shell intensities for line intensities. In 1974 Russ used empirical correction factors to describe the relative intensity of a given line. However, the limited resolution of the energy-dispersive detector made it difficult to resolve some of the x-ray lines. For example, the Lα and Lβ lines sometimes overlapped and it was simpler to treat all the L lines as a single peak. The total x-ray intensity of the shell for the standard in question was then obtained from equation (11.13) by setting the transition probability (p) equal to unity. This approach was found to improve significantly the performance of the method (Russ, 1978).

11.4.3 Nasir

A slightly different approach has been adopted by Nasir (1976) in his standardless correction method. He used the formula

$$I_g = \text{const.}\, p\omega \frac{N}{A} c_A \int_{U_0}^{1} \frac{-\ln U \mathrm{d}U}{\sum c_i (Z_i/A_i) \ln (1.166\, \bar{E}/J_i)} \tag{11.14}$$

to describe the intensity of radiation generated by element A in a multi-element specimen. This is basically of the same form as equation (11.11) but the Bethe energy-loss law has been employed instead of the more approximate equation of Thomson and Whiddington.

As pointed out by Bishop (1968) equation (11.14) can be simplified, provided that $E_c/J \gg U_0$, and then integrated to give as a very good approximation

$$I_g = \text{const.}\, p\omega \frac{N}{A} c_A \frac{(U_0 \ln U_0 - U_0 + 1)}{\sum c_i (Z_i/A_i) \ln (1.166\bar{E}/J_i)} \tag{11.15}$$

where $\bar{E}$ is given by $(E_0 + E_c)/2$. Following Nasir we shall consider the case of a binary alloy AB and write

$$\frac{I_g^A}{I_g^B} = \frac{p_A \omega_A c_A A_B}{p_B \omega_B c_B A_A} \frac{(U_0^A \ln U_0^A - U_0^A + 1)}{(U_0^B \ln U_0^B - U_0^B + 1)} \,. \tag{11.16}$$

It will be noticed that a further simplification has been introduced here, namely that the denominator in equation (11.15) is to be the same for both elements so that the ratio of generated intensities (equation (11.16)) contains no atomic number correction. Nasir states that such a correction is not necessary as the backscattering and stopping power corrections are relatively small. This may be correct in many instances and certainly makes the method easier to use but it should be recognised that ignoring the stopping power correction is strictly justifiable only if $E_c^A \simeq E_c^B$ or if $E_0 \gg E_c^A$ and $E_0 \gg E_c^B$. The omission of the backscatter factor R in the development of the correction procedure may also introduce some error because R will be a function of the overvoltage ratio and in general will not be the same for elements A and B. However, if equation (11.16) is accepted as being a reasonable approximation the intensities may be corrected for the effects of absorption ($f(\chi)$ and T) and fluorescence (F) enabling the ratio of the weight concentrations to be determined by

$$\frac{c_A}{c_B} = \frac{I_d^A}{I_d^B} \frac{p_B}{p_A} \frac{\omega_B}{\omega_A} \frac{A_A}{A_B} \frac{(U_0^B \ln U_0^B - U_0^B + 1)}{(U_0^A \ln U_0^A - U_0^A + 1)} \frac{f(\chi)_A}{f(\chi)_B} \frac{T_A}{T_B} \frac{F_A}{F_B} \,,$$

where I_d is the recorded x-ray intensity.

Values of the transition probability for Kα x-ray emission were experimentally determined as 0.874 for elements with $22 \leqslant Z \leqslant 40$ but in all other cases appropriate shell intensities were measured and p was therefore set equal to one. In order to calculate the fluorescence yield, Nasir used the formula of Wentzel described earlier. For K radiation the usual value of $a = 10^6$ has been adopted but for L radiation Nasir proposed the empirical equation $a = 3.527 \times 10^7 - 4.76 \times 10^5 Z$ rather than the value of 10^8 suggested by Reed (1965). Use of these two sets of values give widely differing results (factor of 4 or 5 times) for the L shell fluorescence yield and it is clear that Nasir's values must be incorrect.

The error has arisen because Nasir used the same equation to describe the ionisation cross-section of both the K and L shells. If this is done the term N_{el} should be introduced into equation (11.9). This causes Nasir's values of the L shell fluorescence yield to be reduced by a factor of 4, thereby bringing them in greater accord with the results of Reed. It should be stressed that although Nasir's values of ω_L are wrong this makes no practical difference to the application of the correction method as described by the author because the error in ω_L cancels the error in Q.

In order to apply the standardless procedure the sum of the weight fractions must be set equal to one. The method is readily extended to ternary and higher order systems, and oxides may be dealt with by converting the elemental weight concentrations to oxide fractions before normalisation is carried out.

11.4.4 Comparison of standardless methods

Both the Russ and Nasir methods are based upon the use of equations to represent the physical processes involved in x-ray production. Certain errors will be introduced by the use of these equations and also by other approximations which have been made in developing the standardless techniques. Nevertheless the limited amount of data which have been obtained using the methods of Nasir and Russ appear quite promising. An example of a quantitative analysis on a stainless steel which has been determined using the Russ procedure is shown in Table 11.4. The absolute errors for each element are less than 1% and the relative errors for elemental concentrations which exceed 1 wt% are generally less than 5% although molybdenum is an exception to this. Despite the fact that there are severe

Table 11.4 – Analysis of steel (NBS 348).

Element	NBS certified composition	Calculated; no standards (Russ)	Absolute error (Δc)	Relative error ($\Delta c/c$)
Al	0.23	0.15	−0.08	−34.8%
Si	0.54	0.49	−0.05	−9.2%
Ti	2.24	2.26	0.02	0.9%
V	0.25	0.21	−0.04	−16.0%
Cr	14.54	14.95	+0.41	2.8%
Mn	1.48	1.46	−0.02	1.4%
Fe	53.30	53.66	+0.36	0.7%
Ni	25.80	24.87	−0.93	−3.6%
Cu	0.22	0.35	−0.13	59.1%
Mo	1.30	1.62	0.32	24.6%

overlap problems associated with this particular analysis, results appear to be comparable with those achieved using a conventional ZAF method.

Errors may manifest themselves when the specimen contains elements of widely differing atomic number. This could be more of a problem when using the Russ approach because he treats the term b in equation (11.10) as a constant whereas in fact Cosslett and Thomas (1964c) have shown that its value falls by a factor ~2 from aluminium to gold. However, it is possible that some of this variation may have been partially compensated for in the detection efficiency term, T.

Experimental measurements of Lifshin *et al.* (1977) have shown that

$$I_g \propto (U_0 - 1)^n \quad ,$$

where n is atomic number dependent and varies from 1.35 (for silicon K radiation) to 1.52 (for chromium K radiation). These findings indicate that both standardless correction methods may be prone to error when elements in the spectrum span a wide range of E_c and Z. Some evidence of this effect has been noticed by Russ (1976) when analysing copper–gold alloys (see Table 11.5). Substantial discrepancies were evident especially when the gold Mα line was used and, although normalisation of the data improved matters considerably, the errors were still unacceptably large.

One way of overcoming this problem is to use a purely empirical approach as suggested by Barbi *et al.* (1976). Since the x-ray detection efficiency of K and L x-ray lines varies smoothly with atomic number, a calibration curve may be readily constructed for a particular set of experimental conditions. When a least-squares analysis is carried out on data recorded from a number of pure element standards, empirical equations can be formulated enabling x-ray intensities to be predicted when no suitable standard specimen exists. As pointed out by Barbi, on an individual microanalyser such as approach should be more reliable than the theoretical methods. A disadvantage with the technique is that a large number of calibration curves may be required if analyses are to be carried out at different kilovoltages and different x-ray take-off angles.

The standardless analysis technique is obviously attractive from the point of view of increasing the rate at which data can be acquired. A serious weakness is that a completely standardless analysis requires normalisation of results and this is undesirable for the reasons discussed earlier (section 11.2.4). However, this particular difficulty can be circumvented if a single standard is employed. It remains to be seen whether significant errors are introduced by the use of standardless techniques since a systematic survey on a wide range of samples would be needed to test this. Certainly such methods are more prone to error than if standards are employed; for example small errors in the x-ray take-off angle and kilovoltage settings are partially compensated for if standards are used. The performance of standardless methods will also be strongly influenced by

Table 11.5 – Analysis of copper-gold alloy chemical composition 65% Cu, 35% Au.

(a) ***Comparison of results obtained when using pure standards and no standards***

Radiation	Element	No standards (Russ)	Pure standards
K	Cu	67.4	64.1
L	Au	32.6	33.8
K	Cu	70.3	64.1
M	Au	29.7	34.8

(b) ***Analysis using one standard; element used as standard is bracketed***

Radiation	Element	One standard (Russ)
K	[Cu]	64.1
L	Au	31.1
K	Cu	70.2
L	[Au]	33.9
K	[Cu]	64.1
M	Au	26.4
K	Cu	85.7
M	[Au]	37.0

the spectrometer response function. It is therefore good practice to check this regularly to make sure it has not changed, since factors such as oil contamination of the detector window may significantly modify the value of T (Love *et al.*, 1981; Smith, 1981).

11.5 METHOD OF PEAK-TO-BACKGROUND RATIOS

Peak-to-background ratios (P/B) have been used to determine the composition of a specimen as suggested by Statham and Pawley (1978). If the generally small effects of fluorescence are ignored we may write for element A in a specimen AB

$$\left(\frac{P}{B}\right)_{\mathrm{A}}^{\mathrm{AB}} \propto \frac{c_{\mathrm{A}} R_{\mathrm{A}}^{\mathrm{AB}} \left[\int Q \,\mathrm{d}s\right]^{\mathrm{AB}} f(\chi)_{\mathrm{A}}^{\mathrm{AB}}}{R_{\mathrm{cont}}^{\mathrm{AB}} \left[\int W \,\mathrm{d}s\right]^{\mathrm{AB}} f(\chi)_{\mathrm{cont}}^{\mathrm{AB}}}$$

The terms in the numerator have been discussed previously, and R^{AB}_{cont}, W and $f(\chi)^{AB}_{cont}$ are respectively the backscatter factor, the ionisation cross section and the absorption correction for the continuum radiation. Although data are available on R_{cont} (Rao-Sahib and Wittry, 1974) and $f(\chi)_{cont}$ (Statham, 1976b) the difference between them and the corresponding values for characteristic radiation are small (Statham, 1979) and so one may write

$$\left(\frac{P}{B}\right)^{AB}_{A} \propto c_A \frac{\left[\int Q\mathrm{d}s\right]^{AB}}{\left[\int W\mathrm{d}s\right]^{AB}} \quad . \tag{11.17}$$

The generation term $\int W\mathrm{d}s$ has been examined by Smith *et al.* (1975) and it was suggested that

$$\int W\mathrm{d}s \propto (\bar{Z})^n \tag{11.18}$$

where $\bar{Z}$ is the mean atomic number of the specimen and

$$n = 1.159 + (0.1239 - 0.02857 \ln E_0)(E_\nu - 2.044) \quad ,$$

E_ν being the photon energy. If measurements are now made on a pure element standard we have

$$\left(\frac{P}{B}\right)^{A}_{A} \propto \frac{\left[\int Q\mathrm{d}s\right]^{A}}{\left[\int W\mathrm{d}s\right]^{A}} \quad . \tag{11.19}$$

Combining equations (11.17), (11.18), and (11.19) and taking $\int Q\mathrm{d}s$ as constant for a given element

$$c_A = \left[\left(\frac{P}{B}\right)^{AB}_{A} (\bar{Z}_{AB})^n\right] \Bigg/ \left[\left(\frac{P}{B}\right)^{A}_{A} (Z_A)^n\right] \quad . \tag{11.20}$$

Of course $\bar{Z}_{AB}$ is a function of the composition of the specimen and some initial estimate of this must be made to perform the calculation. The first approximation is to use

$$c_{A1} = \left(\frac{P}{B}\right)^{AB}_{A} \Bigg/ \left(\frac{P}{B}\right)^{A}_{A}$$

and

$$c_{B1} = \left(\frac{P}{B}\right)^{AB}_{B} \Bigg/ \left(\frac{P}{B}\right)^{B}_{B} \quad .$$

Improved concentrations are obtained with equation (11.20) and successively better approximations are achieved by iteration. Although the procedure has been illustrated using a binary specimen and pure element standards this is not a limitation and the method is readily extended to more complex systems.

The importance of proper sample preparation has been dealt with in section 6.1, but in certain instances such practices may not be feasible, for example in the identification of impurity particles present on the fracture surface of a failed component. Polishing the fracture surface would naturally remove evidence of any such effects and one is normally reduced to carrying out semi-quantitative measurements. The peak-to-background method, however, offers the possibility of carrying out quantitative analysis on rough surfaces. This is because the peak-to-background ratio is largely independent of specimen geometry. Tests on a rough sample of ilmenite (Statham, 1979) showed that the variance of the results obtained from different regions of the specimen surface was much less when using the ratio method than when a conventional ZAF approach was employed. There are obviously areas of application for the peak-to-background procedure in other situations where the specimen geometry is not well defined, such as in particle analysis, and these are discussed in section 13.3.

Probably the greatest source of inaccuracy in the method is associated with the practical measurement of the peak-to-background ratio. Small errors in the measurement of background often have little effect in a ZAF analysis because the peak is much larger than the background but the ratio method is much more sensitive to an error of this nature. Great care must therefore be taken to calculate the background as accurately as possible. Statham suggests that his iterative peak-stripping procedure (Statham, 1976c) is useful in this context because it does not require a precise knowledge of experimental parameters such as x-ray take-off angle.

12

The Monte Carlo method

M. G. C. COX

The Monte Carlo method is a mathematical technique, involving random numbers, which can be employed to provide numerical solutions to certain types of problem. The two main areas of application are sampling and simulation.

Sampling is concerned with determining the average behaviour of macro-systems. For example, the centre of gravity of a disc could be found by averaging the coordinates of a few hundred points selected at random within the locus of its periphery. The solution in this case is trivial, but for more complex shapes Monte Carlo methods can prove extremely useful.

In simulation studies, random numbers are used to establish the outcome of a series of events. For instance, a small particle suspended in a fluid undergoes a random walk as a result of bombardment by molecules of the fluid. By using random numbers to determine the direction and velocity of the bombarding molecules the resultant motion of such a particle may be simulated. This simulated movement will not be identical to that of a particle in a real system but a collective property, such as the average displacement per collision, should be similar to that experienced by real particles. It is in this area that the Monte Carlo technique has proved to be so useful in electron-probe microanalysis.

The need for a mathematical method to describe electron behaviour in solids arises because it is not possible to measure experimentally the collective interactions of electrons while within the target. Such information is of vital importance in, for example, calculating the spatial resolution of the electron probe or establishing x-ray distributions with depth in the specimen, both of which are difficult to determine experimentally.

In the present chapter the physical principles involved in the Monte Carlo simulation of electron trajectories are described including the concepts of single and multiple scattering. Factors influencing the accuracy of the method are examined and some applications of the technique are discussed. For a more detailed description of Monte Carlo methods and their applications the reader is referred to the National Bureau of Standards Publication 460 (Heinrich *et al.*, 1976).

12.1 SIMULATION OF AN ELECTRON TRAJECTORY

As we have mentioned earlier (section 2.1) when an electron penetrates into a solid it loses energy by excitation of plasmons, phonons, x-rays, secondary electrons, etc., and it also suffers a series of deflections as a result of close encounters with atomic nuclei. In Monte Carlo programs the usual practice is to treat these processes entirely independently. Hence the former (inelastic) interactions are assumed to cause no deflection of the electron while the latter (elastic) are taken to result in zero energy loss. Although neither assumption is strictly accurate, errors introduced by the approximations are not significant since elastic scattering involves only ~1 eV energy transfer and inelastic scattering causes an electron to be scattered on average, only a fraction of a degree. Between each of the elastic interactions the electron is usually considered to travel in a straight line and to lose energy in a continuous manner.

The essential problem is to calculate the position, energy and direction of travel of the electron after it has suffered an elastic collision given that the values were known prior to the collision. In this way the path of the electron is followed until it either emerges from the sample (for example, is backscattered) or has lost so much energy that (in the case of microanalysis) characteristic x-ray production does not take place.

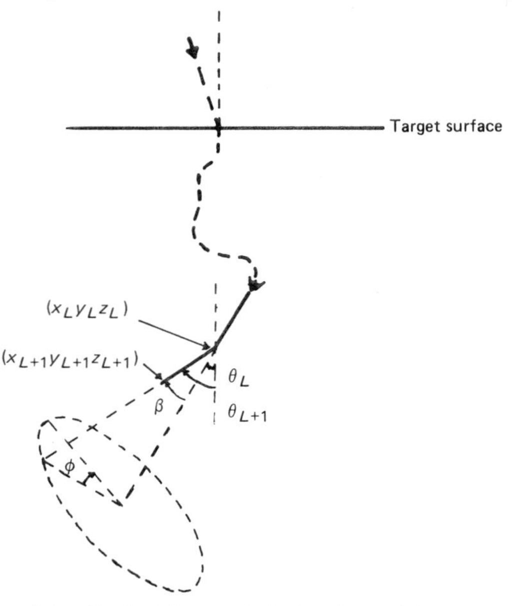

Fig. 12.1 – Typical electron trajectory in a solid.

In Fig. 12.1 is shown a typical electron trajectory in a solid. We shall consider the behaviour of the electron before and after it has undergone a particular elastic interaction, say the $(L + 1)$th collision. The position of the electron is defined with reference to Cartesian coordinates with the origin situated at the point of impact of the electron beam on the specimen. The angles θ and γ are those subtended by the direction of the electron and the x and y axis respectively; δr is the distance between elastic collisions, β is the scattering angle and ϕ is the radial deflection suffered during a collision. From geometrical considerations the following relationships may be obtained:

$$x_{L+1} = x_L + \delta r\,(\sin\theta_L\,.\,\cos\gamma_L)$$

$$y_{L+1} = y_L + \delta r\,(\sin\theta_L\,.\,\sin\gamma_L)$$

$$z_{L+1} = z_L + \delta r\cos\theta_L$$

$$\cos\theta_{L+1} = \cos\theta_L\,.\,\cos\beta - \sin\theta_L\,.\,\sin\beta\,.\,\cos\phi$$

$$\sin(\gamma_{L+1} - \gamma_L) = \frac{\sin\phi\,.\,\sin\beta}{\sin\theta_{L+1}} = A$$

$$\cos(\gamma_{L+1} - \gamma_L) = \frac{\cos\beta - \cos\theta_L\,.\,\cos\theta_{L+1}}{\sin\theta_L\,.\,\sin\theta_{L+1}} = B$$

$$\sin\gamma_{L+1} = B\sin\gamma_L + \mathrm{A}\cos\gamma_L$$

$$\cos\gamma_{L+1} = B\cos\gamma_L + A\sin\gamma_L\ .$$

Calculation of the position and direction of the electron depends upon selecting appropriate values of ϕ, β and δr for each scattering event.

The first of these, the azimuthal angle (ϕ), may be readily obtained using a random number (see section 12.4) between 0 and 1 because any angle of ϕ is equally probable. Thus $\phi = 2\pi R$ where R is the pseudo-random number.

The scattering angle β is also determined using the random number generator, this time in conjunction with an appropriate elastic scattering cross-section (see section 12.2).

If single scattering theory (section 12.2.1) is used, δr is usually taken to be the mean free path between elastic collisions in which case its value will be given by equation (12.4). This is, of course, an approximation since the mean free path is the distance that an *average* electron would travel between elastic collisions and some statistical variation should be incorporated to deal with the behaviour of individual electrons. In practice the more detailed treatment is rarely, if ever, carried out and Bishop (1976) considers the omission is unlikely to introduce any serious error. If multiple scattering theory (section 12.2.2) is employed then it is possible that δr will be fixed and be some fraction (typically one hundredth) of the electron range. In more sophisticated multiple scattering programs δr may be varied such that the amount of scattering which takes place in a given step length remains approximately constant.

Finally the electron energy must be known throughout the whole trajectory because the probability of x-ray production and the values of β and δr are dependent on this parameter. Fortunately, provided that certain simplifying assumptions are made, this may be readily calculated from a knowledge of the rate of energy loss ($-\mathrm{d}E/\mathrm{d}\rho s$) and the distance travelled by the electron in the solid (section 12.3).

Thus monitoring continuously the electron's position, direction of travel and energy allows the x-ray generation at any point to be determined once the appropriate ionisation cross-section and fluorescence yield for the x-ray shell of interest are introduced. Typically some 10 000 of these trajectories may be required to build up a realistic picture of events occurring in the target.

12.2 ELASTIC SCATTERING

The way in which elastic scattering is treated largely distinguishes one Monte Carlo method from another.

12.2.1 Single scattering

The most rigorous approach is to use a single scattering model (for example, Murata *et al.*, 1971, 1972). This involves calculating the mean free path of the electron between elastic collisions and then using a suitable elastic cross-section in conjunction with a random number generator to establish the angle through which it is scattered at each of its interactions with the atomic nuclei.

One of the simplest equations used to describe elastic scattering is that of Rutherford:

$$\sigma'(\beta)\mathrm{d}\Omega = \frac{Z^2 e^4 \mathrm{d}\Omega}{16 \,.\, E^2 \sin^4(\beta/2)} \quad . \tag{12.1}$$

where $\sigma'(\beta)\mathrm{d}\Omega$ represents the probability of an electron being scattered through an angle β in a solid angle $\mathrm{d}\Omega$; e is charge on the electron in esu and E the electron energy in ergs. However, this formula neglects screening of the nucleus by the orbital electrons and results in an infinite cross-section for $\beta = 0$, which implies the mean free path of the electron is zero. A screening factor must therefore be incorporated. Using the Wentzel (1927) model of the atom it is possible to derive a 'screened Rutherford' cross-section ($\sigma'(\beta)$) for scattering into an angle β:

$$\sigma'(\beta) = \frac{Z(Z+1)e^4}{4E^2(1-\cos\beta + 2\alpha)^2} \quad , \tag{12.2}$$

where α is the screening parameter. The $(Z + 1)/Z$ factor is introduced to make some allowance for the contribution to angular scattering by inelastic events.

The total cross section (σ_T) may now be determined by integrating over $d\Omega$, that is,

$$\sigma_T = \int_0^{2\pi}\int_0^{\pi} \frac{Z(Z+1)e^4 \sin\beta \, d\beta d\phi}{4E^4(1-\cos\theta + 2\alpha)^2}$$

$$= \frac{\pi e^4 Z(Z+1)}{4E^2\alpha(\alpha+1)} . \qquad (12.3)$$

We now need an equation which describes the scattering angle (β) at each interaction. The probability (P_β) of an electron being scattered into an angle less than or equal to β is given by

$$P_\beta = \int_0^{2\pi}\int_0^{\beta} \frac{\sigma'(\beta)}{\sigma_T} \sin\beta \, d\beta d\phi .$$

Substituting for $\sigma'(\beta)$ and σ_T from equations (12.2) and (12.3) respectively

$$P_\beta = \int_0^{2\pi}\int_0^{\beta} \frac{\alpha(\alpha+1)\sin\beta \, d\beta \, d\phi}{\pi(1-\cos\beta + 2\alpha)^2}$$

$$= \int_0^{2\pi} \frac{\alpha(\alpha+1)}{\pi} \left[\frac{1}{1-\cos\beta + 2\alpha}\right]_0^{\beta} d\phi$$

$$= \int_0^{2\pi} \frac{(\alpha+1)}{2\pi} \frac{(1-\cos\beta)}{(1-\cos\beta+2\alpha)} d\phi$$

$$= \frac{(\alpha+1)(1-\cos\beta)}{(1-\cos\beta+2\alpha)} .$$

The angle of scattering for a particular elastic collision may be established by setting P_β equal to a random number (R) between 0 and 1. Then re-arranging the above equation:

$$\beta = \cos^{-1}\left[1 - \frac{2\alpha R}{(\alpha+1-R)}\right].$$

If δr is assumed to be equal to the mean free path between elastic collisions

$$\delta r = \frac{1}{\text{No. of atoms per unit vol.} \times \text{total elastic cross-section}}$$

$$= \frac{A}{N\rho\,\sigma_T} , \qquad (12.4)$$

where $N\rho/A$ is the number of atoms per unit volume.

It is useful at this stage to establish the order of magnitude of δr and for this purpose we will consider an electron with an energy of 15 keV travelling in copper. Before beginning the calculation, however, the value for the screening factor (α) must be found. The screening factor is often chosen to be one of the parameters adjusted so that the Monte Carlo model gives a good fit to experimental data. Bishop (1966c) proposed

$$\alpha = 3.4 \times Z^{0.667}/E \ ,$$

where E is the electron energy in electron volts. After substituting the appropriate values for Z and E we obtain $\alpha = 2.1 \times 10^{-3}$. Thus from equations 12.3 and 12.4,

$$\begin{aligned} \delta r &= \frac{A}{N\rho} \frac{4E^2 \alpha(\alpha + 1)}{\pi e^4 Z(Z + 1)} \\ &= \frac{63.55 \times 4 \times (15\,000 \times 1.602 \times 10^{-12})^2 \times 2.1 \times 10^{-3} \times 1.002}{6.02 \times 10^{23} \times 8.96 \times 3.142 \times (4.8 \times 10^{-10})^4\ 29 \times 30} \\ &\cong 40\,\text{Å} \ . \end{aligned}$$

Obviously the value of δr will change with atomic number and electron energy (for a 10 keV electron in copper it is approximately 12 Å) but the important point to note is that a large number of interactions per electron trajectory is involved. For example, if it is assumed that a typical electron range is $\sim 2\mu$m, then at least 1000 elastic scattering events have to be calculated for each electron. Consequently Monte Carlo models based upon single scattering tend to involve a considerable amount of computer time and be correspondingly expensive.

12.2.2 Multiple scattering

Multiple scattering methods require less computation because a number of individual elastic collisions are grouped together and the nett deflection is treated in the same way as a single scattering event. This is achieved by dividing up the electron trajectory into a convenient number of step lengths and assuming that one elastic collision takes place in each step. In some multiple scattering treatments the position where the scattering occurs is taken to be the middle of each step length but it is preferable to allow the scattering position to vary randomly within the step.

One of the earliest Monte Carlo methods developed for electron-probe microanalysis (Green, 1963a) used a multiple scattering approach. Green based his model upon angular scattering distributions obtained experimentally by Thomas (1961) on 0.1μm thick copper foils. From knowledge of the value of the backscatter coefficient for the thin foil (η_i), the probability of electrons being scattered in a forward or backward direction was calculated using a random number with the respective probabilities, $(1 - \eta_i)$ and η_i, varying in steps of 0.01.

The experimental data were then used by Green to establish ten scattering angles (β) of equal probability for electrons scattered in the forward hemisphere and also in the backward hemisphere. The choice of which of the ten angles of equal probability was appropriate for a particular electron-nucleus interaction was again established by a random process. Of course because the angular distribution changes with electron energy, sufficient experimental results must be available to describe the scattering of the electron along its whole trajectory (that is, from E_0 to approximately zero). Therefore although this approach gave encouraging results it was of limited usefulness because of the lack of experimental results on angular-scattering distributions.

A more versatile method was developed by Bishop (1966c) based upon a theoretical study of multiple scattering by Goudsmit and Saunderson (1940). In this treatment the nett effect of the Rutherford-type scattering within a single step length was obtained from a convolution of individual events by expressing the scattering distribution as a series of Legendre polynomials which were numerically evaluated. Using this scheme the angular distribution function for electrons of any energy in any target can be established. Both Bishop (1966c) and Shimizu *et al.* (1966) have employed the method to obtain x-ray depth distributions, Bishop retaining a fixed step length and Shimizu using

$$\Delta \rho r_i = \frac{E_i}{E_0} \rho r_0 \quad ,$$

where the subscript i refers to the number of the step. In the latter case, therefore, the step length is successively reduced as the electron loses energy and scattering increases. Bishop (1976) has suggested that ideally the step length should be equal to the mean free path for single scattering through an angle greater than 10° or 15°, although this criterion is not always easy to satisfy in practice.

Whilst computing time may be significantly reduced by using multiple scattering theory, the complexity of the mathematics is still considerably greater than in the single scattering approach.

A simpler model was proposed by Curgenven and Duncumb (1971) which could be processed by the mini-computers available at the time. An important assumption made was that the multiple scattering which takes place over a given step length can be represented by an equation designed to describe a single scattering act.

They decided to use the simple unscreened Rutherford scattering equation

$$\cot \frac{\beta}{2} = \frac{2p}{b} \quad ,$$

where $\quad b = 1.44 \times 10^{-2} (Z/E) \quad$ (E is in keV) ;

p is the impact parameter, that is, the minimum distance between the scattering nucleus and the original direction of the electron (Fig. 12.2). For each step length in the trajectory the electron deflection was calculated from

$$\cot(\beta/2) = (2p_0\sqrt{R})/b \ ,$$

where R is a random number between 0 and 1. The square root allowed for the greater probability of higher values of p as electrons arrive at random over the whole effective cross-section of radius p_0. The maximum impact parameter (p_0) sets a limit on the minimum scattering angle which can occur within a step length and Curgenven and Duncumb proposed that p_0 should be adjusted empirically to obtain close correspondence between calculated and measured back-scatter coefficients. The Curgenven and Duncumb procedure has been used and refined by a number of workers (Statham 1975; Myklebust *et al.*, 1976; Love *et al.*, 1977a).

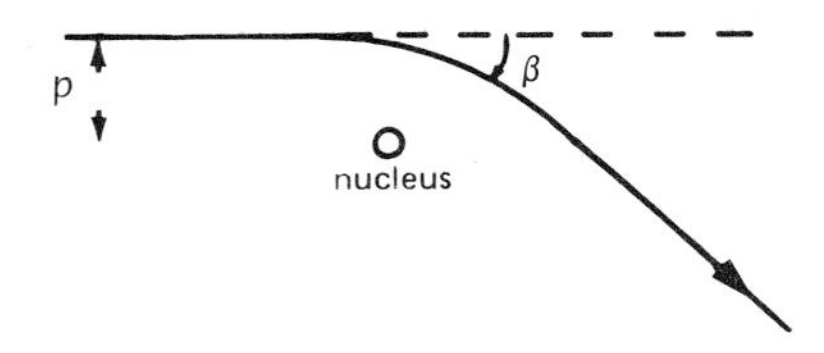

Fig. 12.2 – Deflection of electron in the vicinity of nucleus.

12.3 INELASTIC SCATTERING

Almost all Monte Carlo treatments utilise the Bethe equation (Bethe and Ashkin, 1953) to establish the decrease in energy of the electron as it travels through the target, viz.

$$\frac{\mathrm{d}E}{\mathrm{d}\rho s} = -78\,500 \frac{Z}{AE} \ln\left(\frac{1.166E}{J}\right) \ .$$

The electron energy after each collision or step length may be calculated by numerical integration using for example a Runge–Kutta method (Wilkes, 1966). In the case of multiple scattering, prior knowledge of the electron range (x_r) is required before the Monte Carlo calculation is begun so that the electron trajectory can be divided up into an appropriate number of steps. The electron range can be determined using Bethe's law,

$$x_r = \int_{E_0}^{aJ} \frac{1}{(\mathrm{d}E/\mathrm{d}\rho s)} \mathrm{d}E \ .$$

Notice the integration is not taken over the whole energy range because as mentioned earlier (section 7.1.3) the Bethe law is unrealistic when $E \rightarrow J$. The minimum value of a is 1/1.166 but Curgenven and Duncumb proposed $a = 1.03$. (As an alternative the equation of Love *et al.*, section 7.13, may be used for $\mathrm{d}E/\mathrm{d}\rho s$ in which case the above limitation does not apply.)

The Bethe formula describes the average rate of energy loss experienced by a group of electrons which have traversed a distance ρs. In reality not all electrons will lose energy at the same rate because inelastic scattering involves a discrete loss of energy at each interaction. Hence both the amount of energy lost and the number of interactions will differ from one electron to another. Henoc and Maurice (1976) took this factor into account by adopting an equation of Landau (1944) to describe the energy spread of the electrons as a function of the distance travelled in the target.

This more refined treatment did not lead to significant changes in shape of x-ray depth distributions except for the tail of the curve. On the other hand Shimizu *et al.* (1975) found that neglecting energy straggling resulted in unrealistic energy spectra of electrons transmitted through thin foils. Obviously it is important to consider the purpose for which the Monte Carlo model is being used before deciding whether or not the Bethe energy-loss law is appropriate. It should be borne in mind, however, that if the effects of energy spread are introduced then the unique relationship between path length traversed and residual electron energy is lost and more computer time and storage capacity will be required. Consequently, energy straggling tends to be ignored in most Monte Carlo models.

12.4 RANDOM NUMBER GENERATION

A series of true random numbers possesses certain properties by which it may be distinguished. Firstly, the sequence of numbers never exhibits any long-term tendency to repeat itself. Secondly, short-term sequences of repeating numbers occur at a frequency which may be predicted by probability theory. Finally, the frequency of occurrence of all numbers in a given interval is constant.

Random number sequences can be obtained by measuring a randomly varying parameter, for example by periodically sampling the amplitude of the noise generated by a reversed-biased semiconductor diode. However, in Monte Carlo programs it is more useful to employ mathematical formulae to generate the numbers. In fact, mathematical techniques cannot produce true random numbers because each method is completely deterministic and eventually results in the sequence repeating itself. Fortunately, however, they do provide a pseudo-random series of numbers with very long repeat intervals and these are usually employed in Monte Carlo calculations. For example the series

$$x_{n+1} = 7^9 x_n (\mathrm{mod}\ 10^s) \ ,$$

where 10^s is the random number length and $x_0 = 1$, will generate a long series of pseudo-random numbers.

As we have seen earlier the random numbers are used in Monte Carlo programs to compute scattering and azimuthal angles. The actual number required for the calculation will therefore be twice the number of step lengths (or elastic scattering events) multiplied by the number of trajectories and this might typically be greater than 10^6 (that is, $2 \times 100 \times 5000$).

12.5 DISCUSSION OF MONTE CARLO MODELS

The Monte Carlo method is probably the most accurate theoretical approach presently available for describing electron and x-ray behaviour in solids. However it is not perfect and data from it should be treated with caution especially since uncertainties still exist in many of the formulae incorporated. To quote one example, both Bishop (1966c) and Krefting and Reimer (1973) have indicated that it may be preferable to represent elastic scattering by exact Mott cross-sections rather than the screened Rutherford expression (equation (12.2) used in most methods. Certainly on theoretical grounds the former is more appropriate for low electron energies (< 10 keV) and high atomic number targets. Another problem is posed by the choice of which ionisation cross-section to use (Kyser, 1979). This point has been discussed by Powell (1976b) who suggested that for the K shell the Bethe (1930) equation of the form $QE_c^2 = (\text{const.}/U) \ln (b/U)$ with $b = 0.9$ is preferable to $b = 1$ for $4 < U < 25$. For lower overvoltage ratios empirical formulae of Gryzinski (1965) or Lotz (1970) were recommended.

Hence it follows that it is advisable to check the particular Monte Carlo method by comparing data calculated from the model with experimental values before the program is used for a particular application. Experimental results often used for this purpose include electron backscatter coefficients, backscattered electron energy distributions and $\phi(\rho z)$ curves.

Sometimes empirical adjustments are made to the Monte Carlo model. As mentioned above, Curgenven and Duncumb (1971) chose to optimise the maximum impact parameter such that the correct backscatter coefficient was obtained for each specimen when bombarded with electrons at ninety degrees to the surface. In such a case backscatter coefficients for non-normal incidence may be used to test the Monte Carlo program as illustrated in Table 12.1; clearly agreement here is very good.

A more rigorous test is to examine the energy distributions of backscattered electrons because these are sensitive to the scattering laws incorporated in the Monte Carlo program. Fig. 12.3 shows Monte Carlo data which have been obtained using a multiple scattering and a single scattering model. It may be seen that each model predicts the general shape of the backscattered electron energy distribution reasonably well although both show room for further improvement.

In the study of ZAF correction procedures described in Chapter 11, extensive use has been made of Monte Carlo results, but if this is to be meaningful then it

Table 12.1 – Comparison of Monte Carlo electron backscatter coefficients with the experimental data of Darlington (1975).

Element	U_0	Specimen inclination (deg)	η Monte Carlo	η Darlington
Al	3	30	0.201 ± 0.007	0.197
Al	3	45	0.269 ± 0.007	0.259
Al	3	60	0.376 ± 0.008	0.372
Cu	10	30	0.376 ± 0.008	0.366
Cu	10	45	0.431 ± 0.009	0.431
Cu	10	60	0.526 ± 0.01	0.494

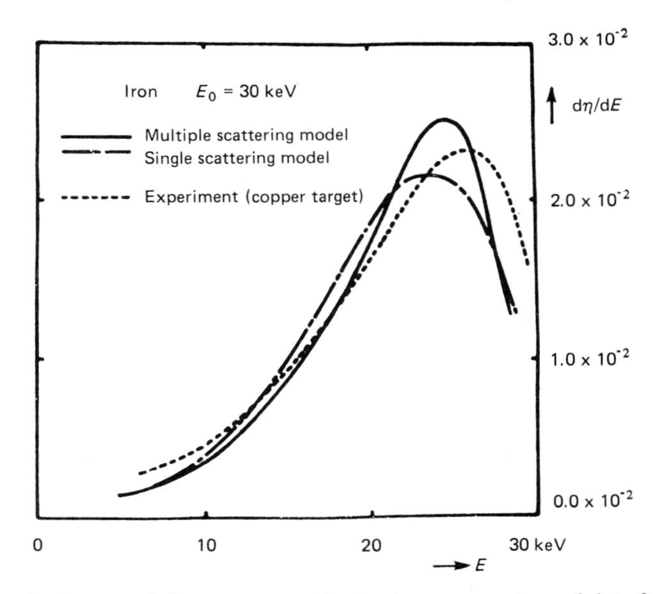

Fig. 12.3 – Backscattered electron energy distributions; comparison of data for single- and multiple-scattering models (Shimizu *et al.*, 1972; courtesy Tokyo U.P.).

is essential that the Monte Carlo model accords well with existing experimental data on x-ray depth distributions. Calculations using the Curgenven-Duncumb model (as modified by Love *et al.*, 1977a) are compared with the experimental $\phi(\rho z)$ data for aluminium (Castaing and Henoc, 1966) in Fig. 12.4 and for copper (Schmitz *et al.*, 1969) in Fig. 12.5; agreement is excellent in each case.

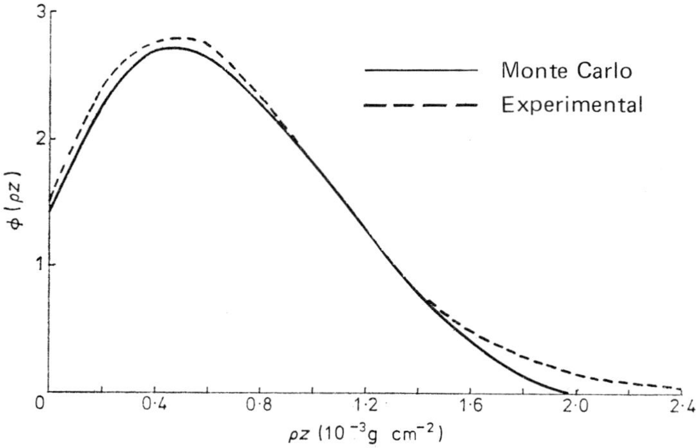

Fig. 12.4 – Comparison of Monte Carlo results with experimental $\phi(\rho z)$ data on aluminium of Castaing and Henoc; 29 kV (Love *et al.*, 1977a; courtesy Inst. of Phys.).

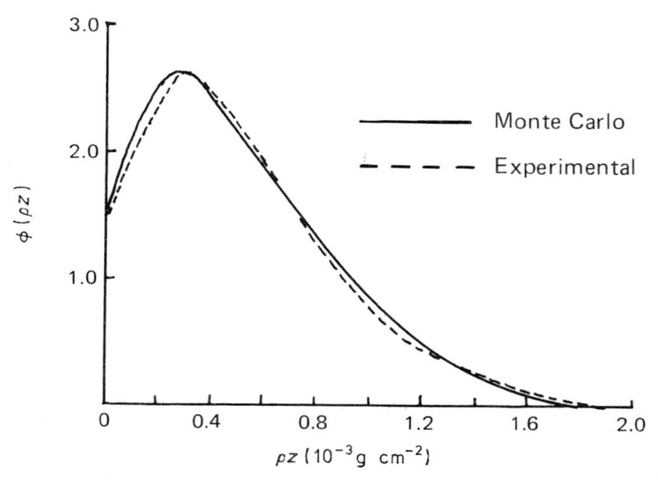

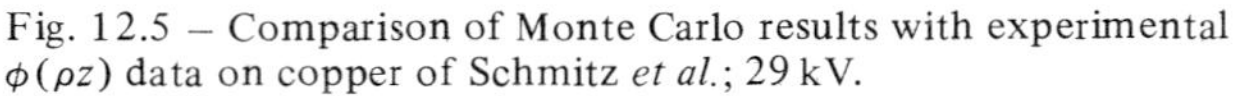

Fig. 12.5 – Comparison of Monte Carlo results with experimental $\phi(\rho z)$ data on copper of Schmitz *et al.*; 29 kV.

The choice of a particular Monte Carlo model is dependent upon the computing power available and the way in which the results of the calculation are going to be applied. For example, Heinrich (1976) has pointed out that where interactions close to the surface are important, as in determining the spatial resolution in thin films, then a single scattering model is to be preferred. However, as a compromise, a mixed model (Myklebust *et al.*, 1976) could be adopted in which single scattering is used in the initial part of the trajectory and multiple scattering is applied subsequently. Generally, the single scattering approach is the more versatile method but this is achieved at the expense of considerably greater computing time. In many instances, such as predicting x-ray distributions in solids, the multiple scattering method appears perfectly satisfactory, and greater complexity does not necessarily result in an improvement of the quality of data.

Since statistical effects will influence the accuracy of Monte Carlo models, it is advisable, before beginning a calculation, to consider the level of precision required. Suppose, for example, the backscatter coefficient for aluminium is required. To achieve a standard deviation of 1% means that 10 000 electrons must be backscattered and, since the backscatter coefficient for aluminium is ~0.15, the number of trajectories simulated would have to be at least 67 000.

12.6 APPLICATIONS OF THE METHOD

Monte Carlo methods have been widely employed in calculating the spatial distribution of the various signals which may be recorded in electron-optical instruments. These calculations have been carried out for secondary electrons, backscattered electrons, Auger electrons, characteristic x-rays, continuum x-rays, etc. The present discussion is confined to those aspects which are of most importance in electron-probe microanalysis.

12.6.1 Electron distributions

The size of the electron interaction volume is difficult to determine experimentally although it is possible to obtain a measure of the electron penetration depth from electron transmission studies on thin foils. A more detailed picture of the size of the interaction volume has been established from a study of etch pits in polymeric materials (Everhart *et al.*, 1972) but this method is restricted to elements of low atomic number. Most estimates of the electron interaction volume are, therefore, made using Monte Carlo calculations.

A number of authors (Curgenven and Duncumb, 1971; Bolon *et al.*, 1975; Newbury and Yakowitz, 1976; Kyser, 1979) have illustrated diagrammatically the behaviour of electrons in solids by plotting electron trajectories projected onto an imaginary plane at right angles to the specimen surface. In this way a direct visual assessment can be obtained of how the electron interaction volume is influenced by parameters such as specimen composition, electron energy and angle of incidence of the electron beam.

The effect of atomic number on the electron distribution has been illustrated in the form of trajectory plots (Figs. 2.13(a) and 2.14(a)). Such results show that little scattering takes place when electrons intially penetrate an aluminium target and only after the electrons have lost a substantial amount of energy do large deflections occur. The nett effect of the relatively weak scattering results in a pear-shaped electron interaction volume being produced. On the other hand gold, with its high nuclear charge, scatters electrons strongly and, since the scattering is centred practically at the surface, about half of the incident electrons manage to escape from the target. This results in an interaction volume which is approximately hemispherical.

The effect on the electron distribution of increasing the electron energy is illustrated in Fig. 12.6 and of altering the angle of electron incidence in Fig. 12.7.

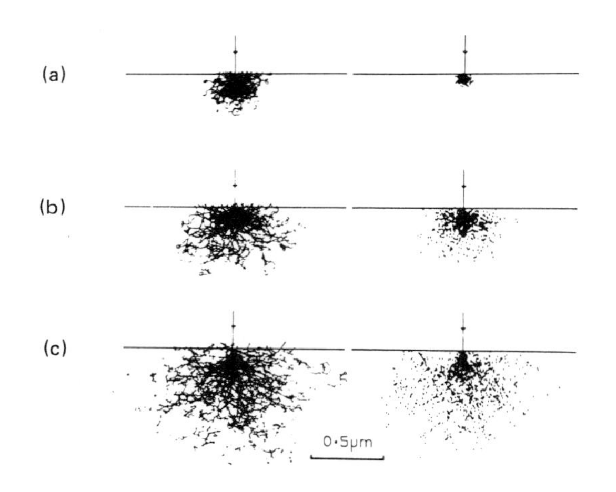

Fig. 12.6 – Effect of changing the electron energy on electron and x-ray distributions in copper (a) 10.5 keV (b) 15 keV and (c) 20 keV (Curgenven and Duncumb, 1971 courtesy Tube Investments Res. Lab.).

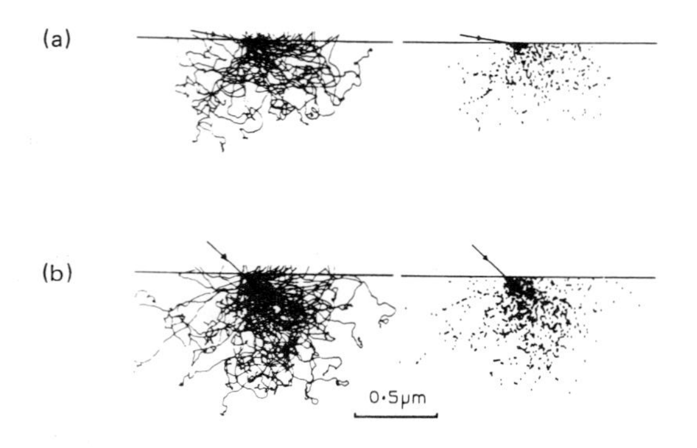

Fig. 12.7 – Effect of changing the angle of incidence of electron beam on electron and x-ray distributions in copper at 20 kV, (a) 10° and (b) 45° to target surface (Curgenven and Duncumb, 1971; courtesy Tube Investments Res. Lab.).

Although a knowledge of the extent of the electron interaction volume in the target is important, it is the volume from which the particular imaging signal is emitted that determines the associated spatial resolution. For instance, Monte Carlo calculations show that electrons are rarely backscattered from depths in the specimen greater than half the total depth of penetration; some results for copper at 30 keV are shown in Fig. 12.8. Similarly when the intensity of backscattered electrons is plotted as a function of escape position (Fig. 12.9),

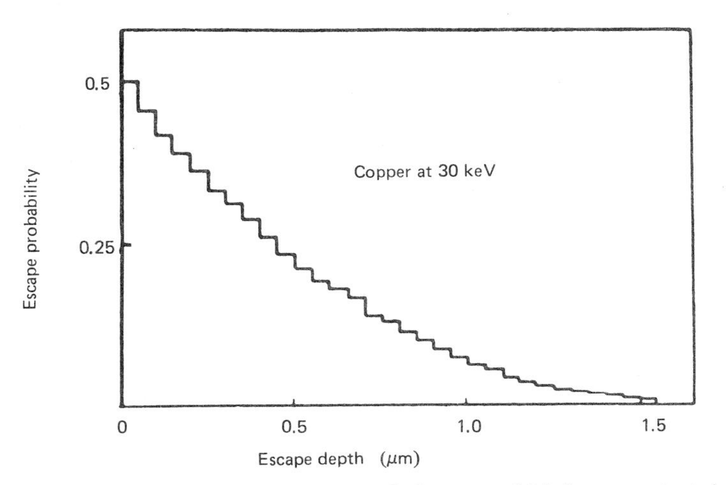

Fig. 12.8 – Probability of escape of electrons which have penetrated various depths into a copper target; maximum depth of penetration is 3 μm, incident electron energy is 30 keV (Shimizu and Murata, 1971; courtesy Dr. K. Murata).

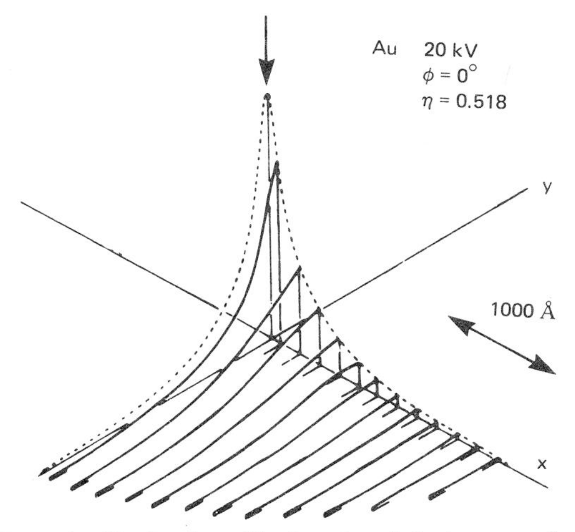

Fig. 12.9 – Intensity distribution of backscattered electrons as a function of position of escape from the surface (Murata, 1974; courtesy Dr. K. Murata).

it is evident that the intensity falls rapidly with increasing distance from the impact point of the probe. These findings explain why the spatial resolution exhibited in scanning electron microscopes is ~100Å whereas the electron interaction volume is much greater, ~1 μm.

12.6.2 Distribution of characteristic x-rays in bulk samples

The spatial resolution achieved in x-ray microanalysis is controlled largely by the incident electron energy and the magnitude of the overvoltage ratio. At high overvoltage ratios the x-ray source size is approximately the same as the electron interaction volume but it will be much smaller when low overvoltage ratios are employed. The depth of x-ray emission is reduced also by decreasing the angle of incidence of the electron beam (Fig. 12.7) and thus both methods can be used as a means of increasing sensitivity in the analysis of surface coatings.

In quantitative electron-probe microanalysis one of the most important uses of Monte Carlo methods is for calculating the distribution of x-rays with depth in the specimen. This has been undertaken by a number of workers (Green, 1963a; Shinoda *et al.*, 1968; Henoc and Maurice, 1972; Love *et al.*, 1977a) and the data used to develop new ZAF correction procedures and to test existing atomic number and absorption corrections (as we have seen in Chapter 10). Agreement between Monte Carlo calculations of $\phi(\rho z)$ curves and experimental tracer measurements is usually good, as shown for example in Fig. 12.4 and 12.5. It has been noted, however, that discrepancies tend to arise when dealing with high atomic number targets and low electron energies. Nevertheless even when agreement is less satisfactory the Monte Carlo results are not necessarily invalid. For example, tracer measurements on heavy elements will be distorted owing to continuum fluorescence, while at low incident electron energies it is difficult to prepare and measure the thin layers needed to obtain accurate $\phi(\rho z)$ curves. Which of the two methods is the more correct will be resolved only when further data are forthcoming.

12.6.3 Distribution of continuum x-rays in bulk samples

Knowledge of the distribution of continuum x-rays is required in certain background subtraction procedures used for ED analysis (section 5.3), such as the method of Reed and Ware (1973). Tracer techniques cannot be used for this purpose because it is not possible to distinguish between continuum x-rays arising in the tracer and the matrix. In fact the only practical route by which such information could be obtained experimentally would be by measuring and analysing the angular distribution of the emitted continuum.

With Monte Carlo methods, on the other hand, the only additional factor required to determine this distribution is the appropriate cross-section for continuum production. Reed (1975a) employed the cross-section of Kirkpatrick and Wiedmann (1945) and scattering results from Monte Carlo calculations of Bishop (1965) to obtain the data. Comparison of $\phi(\rho z)$ curves for characteristic and continuum radiation (see for example Fig. 5.3) then enabled an appropriate expression for the mean depth of continuum x-ray production to be established (via the σ term in the Philibert expression). The treatment was further refined by Statham (1975) who included factors to account for the anisotropic distribution of continuum x-rays.

12.6.4 Particles and thin films

Monte Carlo calculations are easily adapted to simulate electron scattering in small particles and in thin film samples by inserting boundary conditions corresponding to the dimensions of the specimen. In this way they may be used to obtain a better understanding of the effect of x-ray emission from model systems. For example, the effect of increasing the incident electron energy on backscattering from a spherical aluminium particle of diameter $2\,\mu m$ is shown in Fig. 12.10. Here it may be seen that at 5 keV the trajectories are hardly different from those in a flat aluminium target provided that the beam is centred accurately on the particle; hence in this case the traditional ZAF correction will

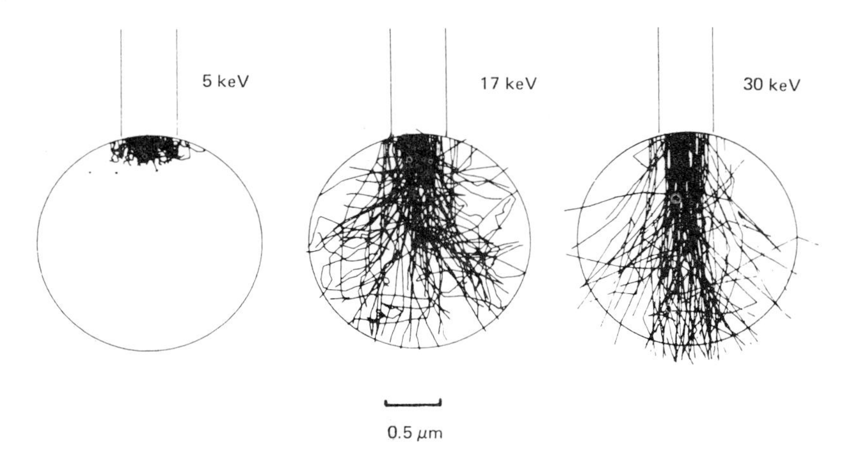

Fig. 12.10 – Interaction volume in aluminium spheres at different incident electron energies (Myklebust *et al.*, 1976; courtesy Nat. Bur. Stands.).

be appropriate. However, increasing the beam energy to 17 keV allows many electrons to escape from the particle and consequently, a further correction is required to take into account the corresponding loss of x-rays. The influence of sphere diameter on x-ray intensity is shown in Fig. 12.11 for aluminium particles irradiated by 17 keV electrons (Bayard, 1973) and results from Monte Carlo calculations are in very good agreement with experimental x-ray data. Close correspondence has also been achieved in similar comparative studies on cylindrical needles, on a copper–gold alloy (Myklebust *et al.*, 1976) and on tantalum carbide

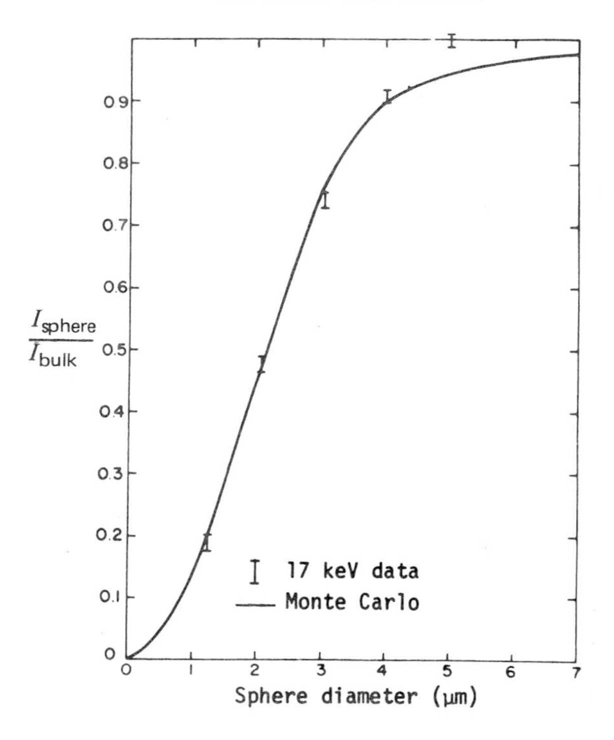

Fig. 12.11 – Effect of diameter of aluminium sphere on x-ray intensity ratio, I_{sphere}/I_{bulk}, 17 keV (after Bayard, 1973).

rods in Ni–Cr alloys (Bolon and Lifshin, 1973). In particle analyses the detector–sample configuration plays an important role in determining the number of x-rays that are detected, as illustrated in Fig. 12.12 which gives results predicted by a Monte Carlo method for two different x-ray take-off angles.

Several workers have used Monte Carlo models to describe the x-ray emission from thin films on substrates (see also section 13.1.3). Kyser and Murata (1974) calculated intensity versus thickness curves for the elements silicon, copper and gold on an alumina substrate and found good agreement with the experimental results. They went on to use the method for deducing the composition and thickness of cobalt–platinum and manganese–bismuth films (see Fig. 13.4(b)) on silica substrates.

Most Monte Carlo work on isolated thin films has been concerned with establishing the spatial resolution in the transmission electron microscope, of a focused beam of electrons (STEM mode) and of the x-ray source. In such studies the finite size of the probe may not be negligible and often the radial distribution of electron density is assumed to be Gaussian. The radial distance of each electron from the centre of the distribution can then be calculated

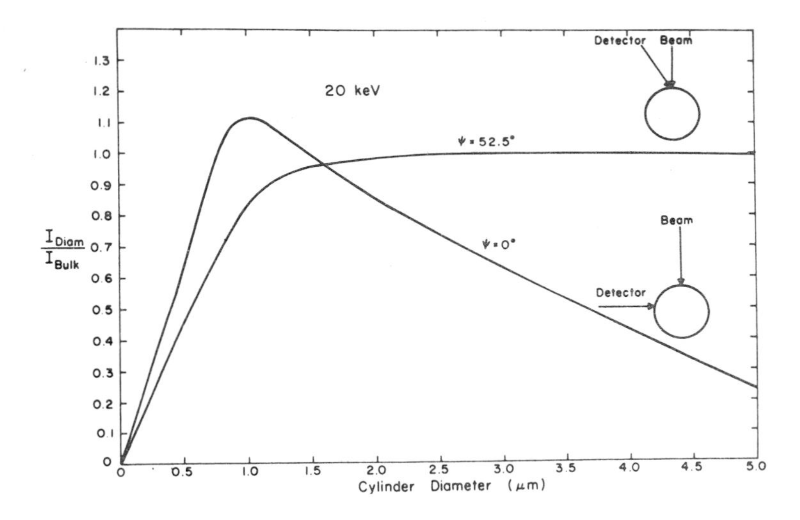

Fig. 12.12 – Silicon x-ray intensity measurements from cylindrical needles (iron–3.22wt% silicon) of different diameters and for two detector geometries (Myklebust *et al.*, 1976; courtesy Nat. Bur. Stands.).

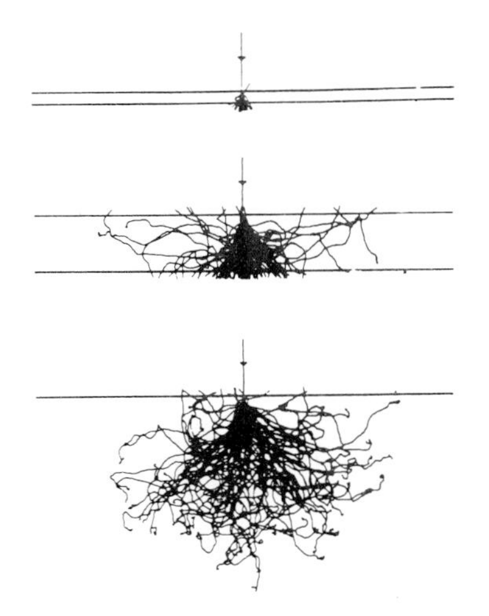

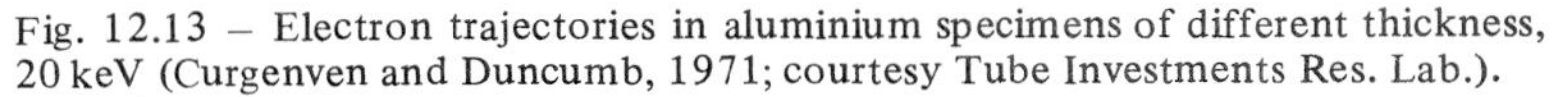

Fig. 12.13 – Electron trajectories in aluminium specimens of different thickness, 20 keV (Curgenven and Duncumb, 1971; courtesy Tube Investments Res. Lab.).

using a Gaussian random number generator (Kyser, 1979). For thin film work single-scattering Monte Carlo models are frequently preferred. This is because (a) it is desirable to represent the scattering as precisely as possible in view of the relatively few collisions experienced by the average electron and (b) computer time is not excessive since the trajectory of each electron is only followed until it leaves the specimen.

Geiss and Kyser have used Monte Carlo methods to study the variation of the x-ray peak-to background as a function of the incident electron energy. They found that the *P/B* increased as the energy was raised, indicating that analytical sensitivity would be greater when using 200 keV rather than 100 keV electrons in the microscope. The variation in the *P/B* ratio with electron energy has also been employed as a test for the presence of stray radiation (Kyser, 1979).

13

Coatings, thin films and particles

V. D. SCOTT

So far it has been assumed that the thickness of the sample is greater than the electron range, but in the present chapter attention is focused on specimen geometries where this is not applicable.

An area of general interest concerns the analysis of surface layers which are different in composition from the underlying material. The measurement can be carried out in a conventional electron-probe microanalyser, although there is some advantage in using inclined electron incidence in order to confine electrons more to the actual film and reduce any substrate excitation. In favourable circumstances sensitivities equivalent to a few atomic layers have been reported, while in films several hundred angstroms thick the detection of differences in thickness of as little as 5 Å has been claimed (Hutchins, 1966). Sensitivities may be increased by reducing the electron accelerating voltage and, in this way, Ranzetta and Scott (1963) were able to show (Fig. 13.1(a)) that the beneficial effect of small additions of calcium on the corrosion resistance of beryllium was associated with a calcium-enriched layer beneath the surface oxide. There may be advantages in using L and M lines with their lower critical excitation potentials and Cox *et al.* (1974) distinguished, by using Lα x-ray intensity measurements on different oxidised alloy steels, between thin (~1000 Å) oxide films of essentially pure Cr_2O_3 and duplex oxide structures containing both chromium and iron, Fig. 13.1(b). Holliday (1973) indicated that it was possible to use the intensity ratio of different L x-ray emissions to estimate the thickness of oxide films on iron and titanium metal. Such approaches have also been widely used to measure the thickness of coatings produced by deposition methods and have special relevance to thin film devices in the semi-conductor field (see, for example, Colby, 1969).

A second area of application involves analysis of thin specimens of dimensions suitable for transmission electron microscopy, a field of study which has rapidly expanded since the advent of transmission microscopes with scanning and x-ray spectrometer attachments. An attraction of such instruments is that chemical

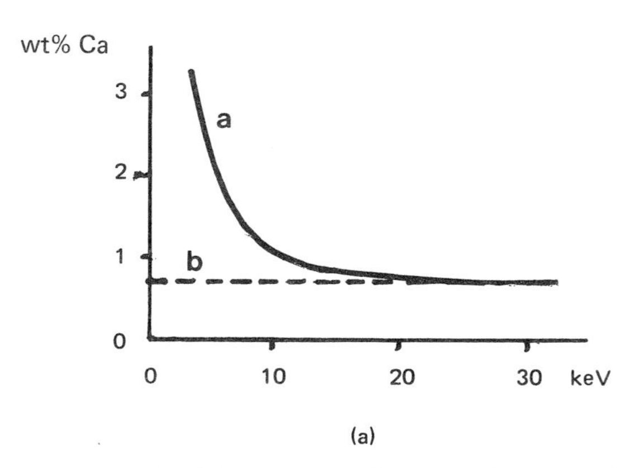

Fig. 13.1(a) – Measured calcium concentration versus electron accelerating voltage showing surface enrichment of calcium in a beryllium alloy; (a) oxidised alloy (b) unoxidised alloy (after Ranzetta and Scott, 1963).

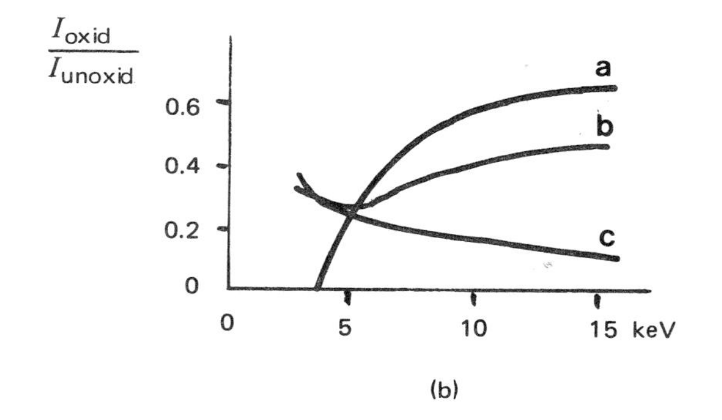

Fig. 13.1(b) – Ratio of x-ray intensities from oxidised and unoxidised iron–chromium alloys as a function of electron accelerating voltage showing surface enrichment of chromium; (a) Fe Lα and (b) Cr Lα radiation. Curve (c) gives the Cr Lα intensity ratio from pure Cr_2O_3 and unoxidised alloy; just below 4 keV this is essentially the same as that from curve (b), indicating that the oxide film on the alloy is Cr_2O_3 rather than a mixed oxide of iron and chromium (after Cox *et al.*, 1974).

composition may be directly correlated with electron diffraction information on selected microscopic features in the specimen, and many examples have been published where possible misinterpretation of structures has been avoided by a combined approach (see, for example, Ranzetta and Scott, 1968). Advantages of the analytical electron microscope are that the need to transfer the specimen from one instrument to another does not arise and the use of electron-transparent specimens results in a reduction in the lateral spread of the electron beam in the sample with an attendant improvement in x-ray spatial resolution (section 13.2.1).

However, thin specimens are usually more difficult to prepare than bulk samples and more often than not, the thickness is non-uniform which means, for quantitative analysis, it has to be known or measured in the localised region of the specimen being chemically analysed.

With coatings on solid substrates or with thin unsupported films, where a large fraction of the incident electrons will be transmitted, the sharp discontinuity in the electron distribution at the interface must be considered and traditional correction methods have to be modified or replaced with an entirely new approach. Investigation of such geometries is ideally suited to Monte Carlo simulation of electron trajectories (see section 12.6.4), where boundary conditions may be set up to correspond to interfacial behaviour. (Of course, it is important to use a Monte Carlo model which gives an accurate representation of events occurring in the immediate surface regions of a target and not one of the more simplified approaches with a limited number of steps in each electron path length). Although, in principle, Monte Carlo methods can be applied to calculate specimen compositions from x-ray intensities they are not convenient for iterative correction procedures and, consequently, are generally used to construct calibration curves. Alternatively, calibration curves may be established directly from x-ray measurements on a series of thin films or coatings of known thickness or composition. These and other techniques for quantitative analysis will be described here. With pure element specimens or samples of known composition, the aim will be to determine thickness. Conversely, if the thickness is known, the chemical composition can be deduced. Some methods may be applied equally to establishing values for both thickness and composition.

The final section of the chapter is devoted to the analysis of small particles and we shall proceed from second phases in solids, arguably the single most successful area of application of electron-probe microanalysis, to problems associated with the analysis of small isolated particles such as airborne dust and precipitates extracted from the bulk.

13.1 THIN COATINGS ON SUBSTRATES

13.1.1 Experimental calibration methods

Thickness measurements on surface films may be carried out simply by first constructing a calibration curve from x-ray measurements made on a series of coated substrate standards whose surface films have an accurately prescribed thickness. Since the characteristic x-ray intensity from the different samples will be related to coating thickness (it will be almost linear up to a certain fraction of the electron range in the sample), the thickness of film on the specimen requiring analysis may be obtained directly by comparing its x-ray emission intensity with the calibration curve. A good example of the method is illustrated in Fig. 13.2, taken from the work of Bolon and Lifshin (1973) on thin gold films evaporated onto various substrate materials. The x-ray intensity measurements from the gold coating (I_C)

have been referred to a bulk sample of gold (intensity I_B) and plotted as the ratio I_C/I_B. The data show (Fig. 13.2(a)) the relative increase in I_C with substrates of higher atomic number due to an increase in the fraction of electrons scattered back into the surface film, and (Fig. 13.2(b)) the reduced intensity ratio with higher incident electron energy due to greater electron penetration in the sample.

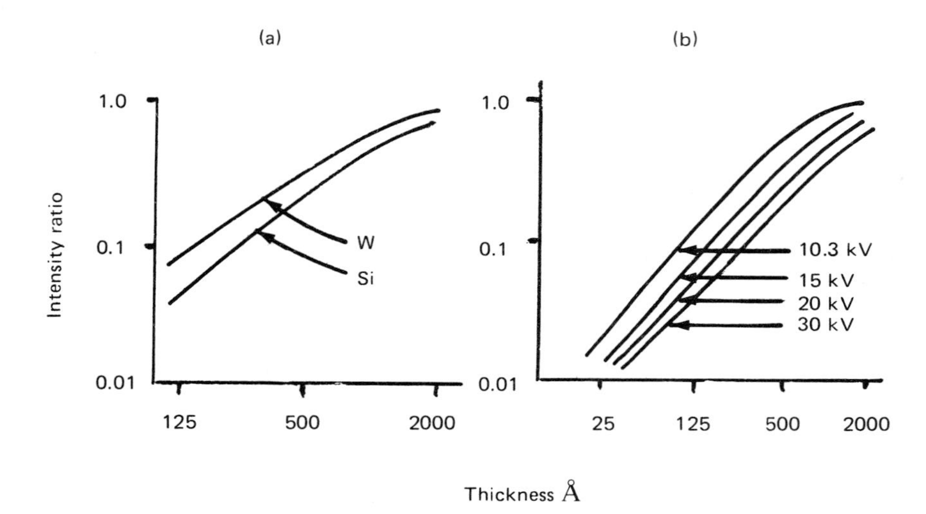

Fig. 13.2 – Measured x-ray intensity ratios from gold film and solid standard, Mα radiation; (a) taken from silicon and tungsten substrates at 20 keV, (b) taken from silicon substrate at different keV; (after Bolon and Lifshin, 1973).

13.1.2 Methods based upon $\phi(\rho z)$ calculations

One of the earliest attempts to calculate the thickness of the surface film was given by Sweeney *et al.* (1960) who expressed the measured intensity ratio as

$$\frac{I_C}{I_B} = \frac{\int_0^{\rho t} \phi'(\rho z) \exp(-\chi \rho z) \, d\rho z}{\int_0^{\infty} \phi(\rho z) \exp(-\chi \rho z) \, d\rho z} \quad . \tag{13.1}$$

This is essentially the same form for expressing x-ray intensities as discussed in section 2.5; ρt is the mass thickness of the film, $\phi'(\rho z)$ and $\phi(\rho z)$ are the x-ray depth distributions in coating and bulk sample respectively which Sweeney assumed to be identical in the surface regions, that is, any difference in electron scattering behaviour in coating and substrate is not taken into account.

The calculation of $\phi(\rho z)$ is, however, difficult and Cockett and Davis (1963) proposed using instead the experimental $\phi(\rho z)$ data of Castaing and Descamps (1955) for this purpose. They obtained values for I_C and I_B from $\phi(\rho z)$ curves and plotted I_C/I_B against ρt; as in the previous work differences between $\phi'(\rho z)$ and $\phi(\rho z)$ were ignored. Similar plots were obtained for the intensity (I_S) of the substrate radiation, I_S/I_B versus ρt, where

$$I_S = \exp[-\chi_C \rho t] \int_{\rho t}^{\infty} \phi(\rho z) \exp(-\chi_S \rho z) \, d\rho z \ .$$

Both expressions could be applied to multi-element coatings by introducing the appropriate mass concentrations into the equations for I_C and I_S. Cockett and Davis suggested that measurements of I_C were preferable for analysing thinner layers (up to one quarter of the electron range), but that measurements of I_S were better for thicker layers. A drawback of their method was that since it relied upon $\phi(\rho z)$ data available at the time, thickness measurements had to be carried out using 29 keV electrons and it could not take advantage of the greater detection sensitivity possible with lower electron accelerating voltages. Furthermore, the assumption that $\phi'(\rho z) = \phi(\rho z)$ meant that it was limited to coatings and substrates close in mean atomic number. Some applications to coatings of chromium, zinc and titanium were described and a minimum measurable film thickness of $0.2\,\mu g\, cm^{-2}$ was claimed, the maximum being ~ 1.5 mg cm^{-2} (thicknesses of a few angstroms to greater than one micrometre.)

Hutchins' (1966) method, which was developed for pure metal films on a substrate, took into account the different electron scattering properties of the two materials. The ratio I_C/I_B (again I_C and I_B were measured x-ray intensities from coating and pure bulk standard respectively) was expressed by three factors:

$$I_C/I_B = I_F \frac{1}{I'_B f(\chi)} f(Z) \ . \qquad (13.2)$$

Here I_F is the x-ray intensity from an isolated thin film of the same mass thickness (ρt) as the coating and is given by

$$I_F = \int_0^{\rho t} \phi''(\rho z) \exp(-\chi \rho z) \, d\rho z \ .$$

For small ρt, $\phi''(\rho z)$ is close to unity and was expressed as $\phi''(\rho z) = 1 + D\rho t$, appropriate values of D being obtained from the initial slope of experimental $\phi(\rho z)$ curves after substrate correction. Writing the exponential as a series and integrating then gave

$$I_F = \{\rho t + (\rho t)^2 [(D/2) - (\chi/2)] - \ldots\} \ .$$

The term $1/[I'_B f(\chi)]$ is a normalisation factor, I'_B referring to the generated x-ray intensity for the bulk material,

$$I'_B = \int_0^\infty \phi(\rho z)\,\mathrm{d}\rho z \;.$$

This factor may be obtained from equation (13.2) by making a single measurement of I_C/I_B on a coating of known thickness.

The term $f(Z)$ in equation (13.2) is the substrate correction factor and, from a study of the manner in which I_C/I_F depended upon atomic number, Hutchins proposed using $1 + G\eta$ where η is the backscatter coefficient for the substrate material. The constant G was found to vary slightly with atomic number, and a value of 1.8 was chosen for $Z \sim 20$ and 2.0 for heavier elements.

The general equation then had the form

$$\frac{I_C}{I_B} = \left\{\rho t + (\rho t)^2\,[(D/2) - (\chi/2)] - \ldots\right\} \frac{1}{\int_0^\infty \phi(\rho z)\,\mathrm{d}\rho z} \, \frac{(1 + G\eta)}{f(\chi)} \,. \tag{13.3}$$

The thickness (t) of the film on the substrate was established by measuring I_C/I_B in the microanalyser and inserting into the expression appropriate values for D, G, η, ρ and the normalisation factor; the procedure would involve several iterations. Duzević and Bonefacić (1978) extended Hutchins' method to alloy thin films on substrates by essentially weight averaging the parameters in equation (13.3). Their modified expression was checked by applying it to NiAl films of known composition deposited on copper and cobalt substrates and satisfactory agreement was claimed.

Colby (1968) took into account the different electron backscattering characteristics of the substrate by treating the electron distribution in the coating as consisting of two parts, one given by the forward scattered electrons and the other by the fraction, η_S, scattered from the coating/substrate interface back into the coating. The method assumed knowledge of coating thickness in order to calculate the composition. The intensity ratio from coating and pure bulk standard was expressed as

$$I_C/I_B = c_A \, \frac{R_C \displaystyle\int_{E_0}^{E_L} Q\Big/\left(\frac{\mathrm{d}E}{\mathrm{d}\rho s}\right)_C \mathrm{d}E + \eta_S \int_{E_L}^{E_{L'}} Q\Big/\left(\frac{\mathrm{d}E}{\mathrm{d}\rho s}\right)_C \mathrm{d}E}{R_B \displaystyle\int_{E_0}^{E_C} Q\Big/\left(\frac{\mathrm{d}E}{\mathrm{d}\rho s}\right)_B \mathrm{d}E} \,.$$

R_C and R_B are the backscatter factors for coating and standard, and $(dE/d\rho s)_C$ and $(dE/d\rho s)_B$ the respective stopping powers; E_L is the mean electron energy at the coating/substrate interface and $E_{L'}$ is the mean energy for those electrons backscattered from the substrate which leave the upper surface of the specimen. The values for E_L and $E_{L'}$ were calculated from an x-ray range equation of the form ρt = const. $(A/Z)(E_0^n - E_c^n)$. Taking $n = 1.5$ (Cosslett and Thomas, 1964c) gave

$$E_L = \left(E_0^{1.5} - \frac{\rho t}{330}\frac{Z}{A}\right)^{0.67} \quad \text{and} \quad E_{L'} = \left(E_0^{1.5} - \frac{\rho t}{165}\frac{Z}{A}\right)^{0.67},$$

where t is in angstrom units, although Warner and Coleman (1973) preferred to use the range equation of Andersen and Hasler (1966),

$$\rho t = 320\,(A/Z)\,(E_0^{1.68} - E_L^{1.68}) \ ,$$

when applying Hutchins's technique to biological materials. Effects due to x-ray absorption were included in an approximated form by Colby and the whole calculation programmed. The method was tested using aluminium coatings on silicon and found to give satisfactory results. Colby pointed out that with thicker films (~2500 Å) the program (MAGIC) could be used for calculating mass thickness as well as chemical composition by carrying out two sets of x-ray measurements at different electron accelerating voltages. The technique was later extended by Oda and Nakajima (1973) to determine coating thickness. Following Duncumb and Reed (1968), they expressed the mean electron energy as $\bar{E} = [E_0 + E_c]/2$ and obtained $I_C/I_B = c_A(R_C/R_B)(S_B/S_C)$, where S_B and S_C are stopping power factors for bulk sample and coating. Rather than adopt Colby's approximate absorption correction, a more accurate form was used. Table 13.1 gives their analysis data for Ag–Cu alloy coatings on Fe–25wt% Ni substrates; coating A consisted of Ag–10wt% Cu alloy 4800 Å thick while coating B was a Ag–50wt%

Table 13.1.

Coating		Ag	Cu	Fe	Ni	Estimated Thickness Å
A	k	0.562	0.061	0.261	0.086	4700
	c	0.908	0.092	0.753	0.247	
B	k	0.023	0.018	0.768	0.244	420
	c	0.510	0.490	0.743	0.257	

Cu alloy 400 Å thick. Considering the large correction factors involved in some of the calculations, the agreement appears remarkably good.

Reuter (1972) calculated $\phi(\rho z)$ data for coating and bulk standard using his own parameters in Philibert's $\phi(\rho z)$ equation. The details of the calculation are given in section 8.3.2. An effective backscatter coefficient (η_{eff}) was introduced to deal with the coated substrate and this varies between the limiting values for coating (η_C) and substrate (η_S) according to

$$\eta_{eff} = \eta_S \frac{n_{\rho t}}{n_0} + \eta_C \left(1 - \frac{n_{\rho t}}{n_0}\right) ,$$

where $n_{\rho t}$ is given by $n_0 \exp(-\sigma \rho t)$ or $n_0 [1 - 4.10^{-4} Z^{0.5} (\rho t / E_0^{1.7})]$, whichever is the larger.

For very thin coatings $n_{\rho t} \approx n_0$, and then the backscatter coefficient for the substrate material should be used in calculating the surface ionisation function, $\phi(0)$, see equation (8.11). The coating thickness is obtained from equation (13.1) by inserting the appropriate $\phi(\rho z)$ data.

With regard to possible x-ray fluorescence effects, it should be noted that none of the above methods take this factor into account. Such effects may arise when elements in the coating are fluoresced by radiation in the substrate, and could constitute a major source of error when the substrate radiation has an energy slightly greater than the critical ionisation energy of the element in the coating. As pointed out by Cox et al. (1979), fluorescence enhancement can amount to ~20% and they proceeded to develop a correction which could be used to calculate satisfactorily fluorescence effects in pure element and in multi-element coatings.

13.1.3 Methods incorporating Monte Carlo calculations

Bolon and Lifshin (1973) modified the simplified Monte Carlo model of Curgenven and Duncumb (1971) (see section 12.2.2) to calculate electron and x-ray distributions in thin films on substrates. Values for I_C/I_B were derived from the program and compared with x-ray intensity ratios on gold-coated samples of silicon (Fig. 13.2). Good agreement between the two sets of data was obtained, although Bolon considered some refinement of the Monte Carlo model would be beneficial.

The method for measuring coating thickness proposed by Bishop and Poole (1973) utilised the Monte Carlo calculations of Bishop (1967, 1968) to establish electron scattering distributions. From a study of these distributions as a function of target composition, incident electron energy E_0 and overvoltage ratio U_0 ($=E_0/E_c$), it was noted that their shape varied little with incident beam energy. Hence curves were plotted using the overvoltage ratio and a parameter z/r, the fraction of the Bethe range r, and these were shown to be virtually invariant with incident energy for a given element. A series of curves of $1/U_0$ versus z/r was then constructed (Fig. 13.3(a)) for a range of values of I_C/I_B. To establish the thickness of coating on a specimen, the fraction z/r of the Bethe range was determined from the curves using the measured intensity ratio I_C/I_B and the appropriate value of U_0, interpolating between the curves if necessary. The Bethe range (r) was then estimated using curves of E_0 versus r provided for a range of pure element targets (Fig. 13.3(b)) and z calculated. In systems where the

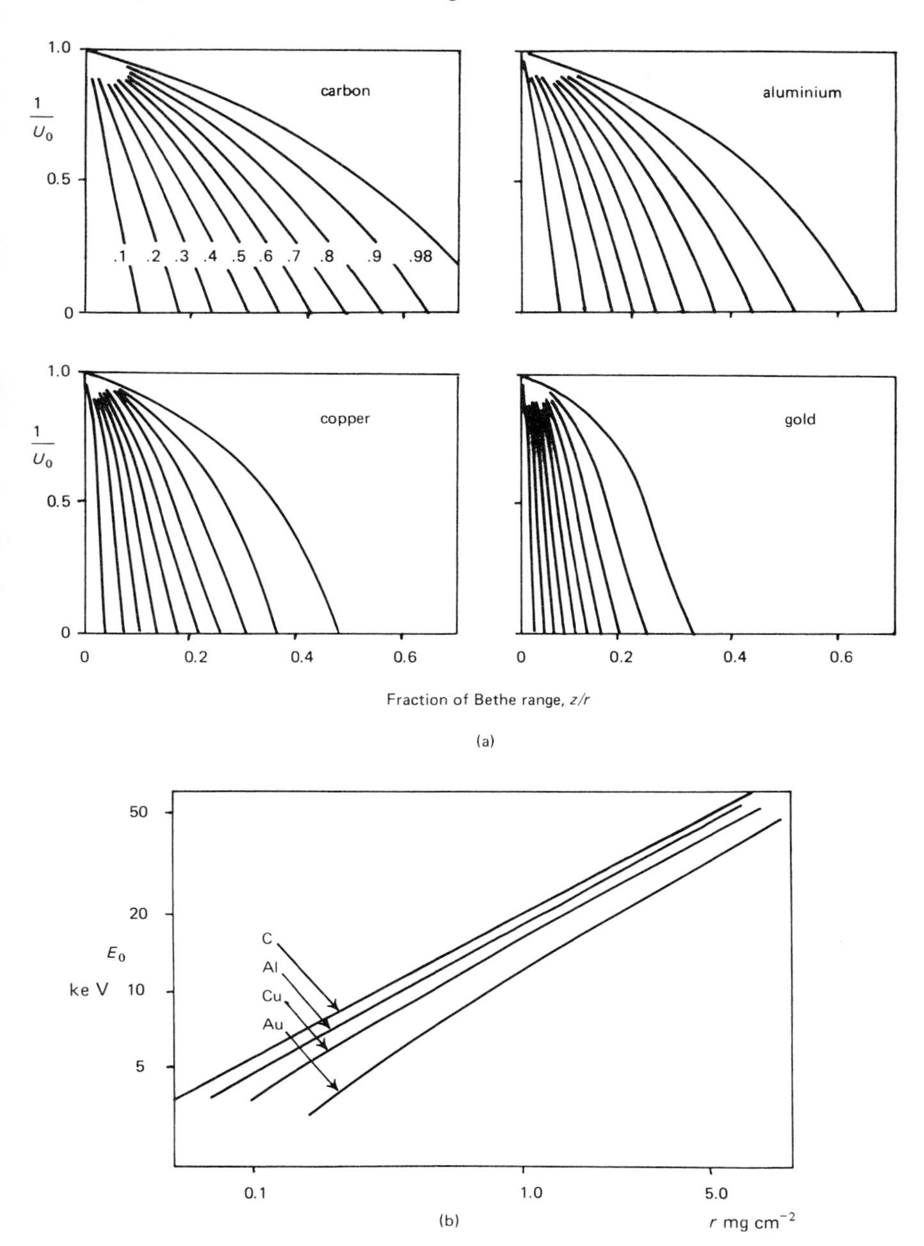

Fig. 13.3 – (a) Curves of generated x-ray intensity ratio from surface film (I_C) and solid standard (I_B) plotted for I_C/I_B values of 0.1, 0.2, . . . , 0.98 as a function of the fraction of the Bethe range (z/r) and overvoltage ratio (U_0). (b) Bethe range (r) as a function of electron energy (E_0) for different elements (after Bishop and Poole, 1973).

surface film and the substrate do not differ greatly in atomic number and in x-ray absorption, the measured x-ray intensity ratio from surface and bulk sample can, of course, be related directly with I_C/I_B. Where absorption is important, Bishop and Poole proposed that absorption in the thin film may be calculated by assuming a simple point source (Green, 1964) whereby x-ray generation takes place at a depth equal to one half the film thickness; x-ray absorption in the bulk sample would be estimated using either the simplified approach of Philibert (1963) (see section 8.2) or the Bishop (1974) model (section 8.4.3). In connection with atomic number differences between surface film and substrate, it was suggested that for very thin films, where the substrate dominates electron scattering behaviour, the curves for the substrate element should be used to obtain z/r. For thicker surface films, say greater than 30% of the electron range (see Cosslett and Thomas, 1965), curves for the atomic number of the film should be employed. Intermediate cases would require interpolation. Bishop and Poole showed also how their approach may be applied using measurements of substrate radiation. Furthermore, the case of multi-element systems was considered, from which it was demonstrated that compositional analysis could be carried out on surface films using their curves provided that the film thickness was known. Accuracies of ~10% were claimed for a range of systems with films of mass thickness between 3 and 200 μg cm^{-2}.

A more rigorous Monte Carlo model (Murata *et al.*, 1971) was used by Kyser and Murata (1974) in studies of thin films on substrates. Fig. 13.4(a) shows the calculated distributions of gold Mα x-rays when 20 keV electrons are incident normally on a specimen consisting of a (0.18 mg cm^{-2}) layer of gold on alumina and on a gold target. The Monte Carlo model was used to construct calibration curves of $(I_C/I_B)_A$ versus c_A, where c_A is the mass concentration of element A in the coating, as a function of its mass thickness ρt. Fig. 13.4(b) gives calculated calibration curves for the system $Mn_x Bi_y$ on a silica substrate. The non-linearity of the curves reflects the difference between the electron scattering properties of manganese and bismuth. By measuring manganese and bismuth radiations from the coatings, it was possible to calculate both the chemical composition and the mass thickness by checking that the results were self-consistent.

13.2 THIN UNSUPPORTED FILMS

13.2.1 Spatial resolution

The much more restricted spread of the electron beam in thin specimens compared with that in bulk samples has been referred to in section 12.6.4, see Fig. 12.13. The degree of spreading will be a function of specimen composition and thickness as well as accelerating voltage but, for specimens of dimensions commensurate with transmission work, resolutions achieved may be close to probe sizes.

Some Monte Carlo calculations of beam spreading taken from Newbury

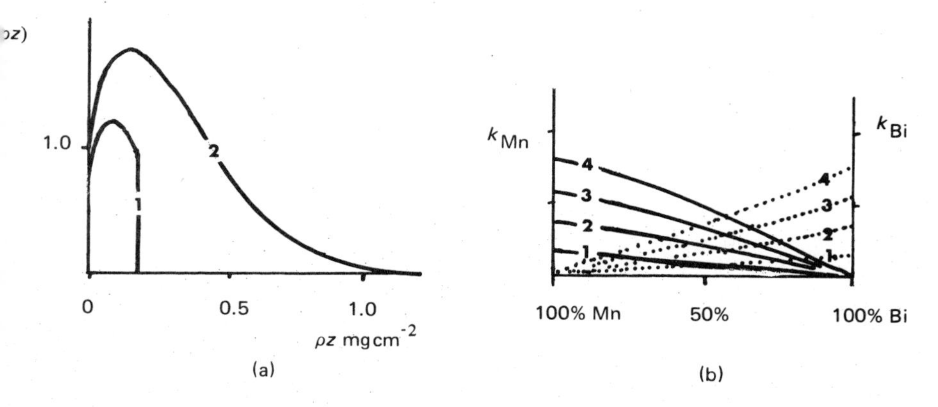

Fig. 13.4 – (a) Calculated x-ray distributions with depth of gold Mα radiation in (1) 95 nm thick gold film on alumina substrate and (2) solid gold; 20 keV. (b) Calibration curves (k versus c) for $Mn_x Bi_y$ films of mass thickness 25 (1), 50 (2), 75 (3), and 100 (4) μg cm^{-2} on silica substrate; Mn Kα unbroken lines, Bi Mα broken lines; 20 keV, 52.5° take-off angle (after Kyser and Murata, 1974).

and Myklebust (1979) are given in Table 13.2; Geiss and Kyser (1979) have published similar values. Such data indicate that with a 100 keV electron beam, spreading is ~200 Å to 300 Å for thin films with mass thicknesses of ≈1 μg mm^{-2}, approximately equivalent to copper films of 1000 Å thickness or aluminium films of 3000 Å thickness. To these values must be added in quadrature (Reed, 1982) the electron beam diameter (d_0) in order to obtain the size of the x-ray source (d_1):

$$d_1^2 = d_0^2 + b^2,$$

where b is the beam broadening.

Table 13.2 – Monte Carlo calculations of beam spreading (Å) at 100keV in thin foil specimens (after Newbury and Myklebust, 1979).

Element	Film thickness (Å)				
	100	500	1000	3000	5000
Carbon	2.2	19	41	160	330
Aluminium	4.1	30	76	300	664
Copper	7.8	58	175	970	2440
Gold	17.1	150	522	5990	17250

Goldstein *et al.* (1977) used a simple model based on Rutherford's classical treatment of scattering by the atomic nucleus to derive an expression for the broadening of an infinitely narrow electron beam passing through a foil of thickness t:

$$b = k\frac{Z}{E_0}\left(\frac{\rho}{A}\right)^{1/2} t^{3/2} \quad , \tag{13.4}$$

where k is a constant. It was originally assumed that scattering occurs at the centre of the foil, giving a value of 6.25×10^5 for k (assuming b and t are in centimetres); integration with respect to depth results, however, in k becoming 7.22×10^5, although the formula remains otherwise unchanged (Jones and Loretto, 1981). The beam breadth at the lower surface of the foil, as given by equation (13.4), contains 90% of the electron trajectories. Since one need consider only relatively large scattering angles (for example, several degrees) the neglect of nuclear screening is a justifiable approximation (Brown, 1981). Furthermore, the probability that a single large-angle deflection exceeds that of two or more lower-angle scattering events combining to produce a comparable total deflection is a valid assumption up to the thickness where multiple scattering becomes significant (Reed, 1982). The model gave rather similar values to those in Table 13.2 for films below $\sim 1\,\mu\text{m mm}^{-2}$ mass thickness but higher values were obtained for thicker films.

As Goldstein (1979) has argued, there would appear to be little justification in working with electron beam diameters less than ~50 Å to improve spatial resolution unless very much thinner films (for example, 200 Å for copper or 500 Å for aluminium) are used. This requirement may, however, present preparation difficulties and an alternative way of achieving better spatial resolution would be to operate the electron microscope at a higher voltage. Equation (13.4) suggests that beam broadening may be reduced by a factor of 2 by using 200 kV rather than a 100 kV beam.

There have been a number of experiments carried out with the aim of measuring x-ray spatial resolution. One method is to record the x-ray emission from some known microstructural feature in the thin foil and to relate the x-ray measurements with predicted data. However, attempts using precipitates present in the specimen have usually failed to give the required information because of x-ray generation in the portion of matrix traversed by the electron beam, that is, even if the precipitates are of similar size to the electron beam diameter, the foil thickness will generally be much larger. Faulkner *et al.* (1977) gave a technique for determining x-ray spatial resolution which was based upon a simple model of a spherical particle embedded in a cylindrical column of foil, both of which have known geometrical dimensions and composition. Measured x-ray intensities from an element unique to the precipitate and to matrix respectively were then corrected to give the weight fraction and hence the volume fraction of the precipitates. From studies on titanium carbides, 450–700 Å in diameter, in an austenitic steel foil 1500–2000 Å thick, Faulkner deduced x-ray source diameters of ~1000 Å at 100 kV and ~500 Å at 200 kV. These values are more than twice those calculated above, which suggests that the assumption of a uniformly cylindrical x-ray source requires modification. Measurements made on features which extend throughout the thickness of the foil might, therefore, be more appropriate for estimating beam spreading and Doig and Flewitt (1977) have described measurements of grain-boundary segregation of tin and phosphorus in Fe–3% Si alloys. However, these workers chose to estimate the minimum width of grain-boundary phase which could be detected, assuming the likely composition of the phase and the shape of the x-ray source, rather than make a direct measurement of spatial resolution. Romig and Goldstein (1979) used an interface with a sharp composition discontinuity and plotted a concentration profile. The effect of beam spreading was then deduced from the distortion of the profile in the vicinity of the interface. An x-ray source size of ~500 Å was quoted from measurements made at 100 keV with a beam diameter of 200 Å on a heat-treated Fe–14.7% Ni foil of ~1500 Å thickness. Other studies (Rao and Lifshin, 1977; Lyman *et al.*, 1978) have also given estimates of source size between 300 and 500 Å.

13.2.2 X-ray intensities

With such small mass volumes of material being excited, x-ray yields will be correspondingly less than in conventional microanalysis of bulk specimens. The average number of ionisations (dn) for each incident electron is given by

$$dn = \frac{N\rho}{A} Q ds \quad \text{(cf. equation (7.1))}$$

and with a specimen of thickness, t, the intensity (I) of characteristic x-rays which are generated by the beam is

$$I = \omega \frac{N\rho}{A} Q t$$

photons per incident electron.

The equation for x-ray intensity in this form assumes:

(i) the incident electrons experience a negligible loss of energy in traversing the specimen,
(ii) specimen thickness (t) can be directly equated with the average electron path length, and
(iii) no electrons are backscattered.

Such approximations are reasonable for specimen thicknesses involved in transmission electron microscopy. Furthermore, if the film is very thin, the effects of x-ray absorption and fluorescence may be neglected when deriving concentrations from emitted x-ray intensities. These conditions have been termed the 'thin film criterion' and methods for quantitative analysis based upon this are described in the following sections. Later (section 13.2.8), the limiting conditions are discussed and an indication given when correction factors may have to be considered.

As an example, for a copper specimen 100 nm thick, and supposing the fluorescence yield $\omega = 0.42$, $\rho = 9\,\mathrm{g\,cm^{-3}}$ and $A = 63.5$, the x-ray intensity, I, is $3.6 \times 10^{15} Q$ photons/electron, Now Q, the ionisation cross-section, is energy dependent but at the high incident energies used in electron microscopy it varies only slowly with excitation voltage E (as $(1/U) \ln U$, where $U = E/E_c$ and E_c is the critical excitation energy, see equation (7.6)). Assuming then a value for Q of $2 \times 10^{-20}\,\mathrm{mm}^2$, $I \approx 6 \times 10^{-5}$ photons/electron; hence for a typical electron beam current of 0.3 nA ($\sim 10^9$ electrons/s) falling on the specimen, I is approximately 6×10^4 photons/s. Measurement of these intensities may be satisfactorily carried out with an ED system (Chapter 4) where the x-ray detector can be placed close (say $\sim$20 mm) to the specimen. With a detector area of some $30\,\mathrm{mm}^2$, approximately 0.6% of the x-rays (that is, 360 photons/s) may be collected, many more of course than would be recorded with a WD system fitted to the electron microscope.

Examples of ED spectra obtained from a 150 nm thick specimen of Al–5% Cu–0.5% Ag alloy are reproduced in Fig. 13.5. The material was in a slightly over-aged condition to give well developed matrix precipitates, Fig. 13.5(a),

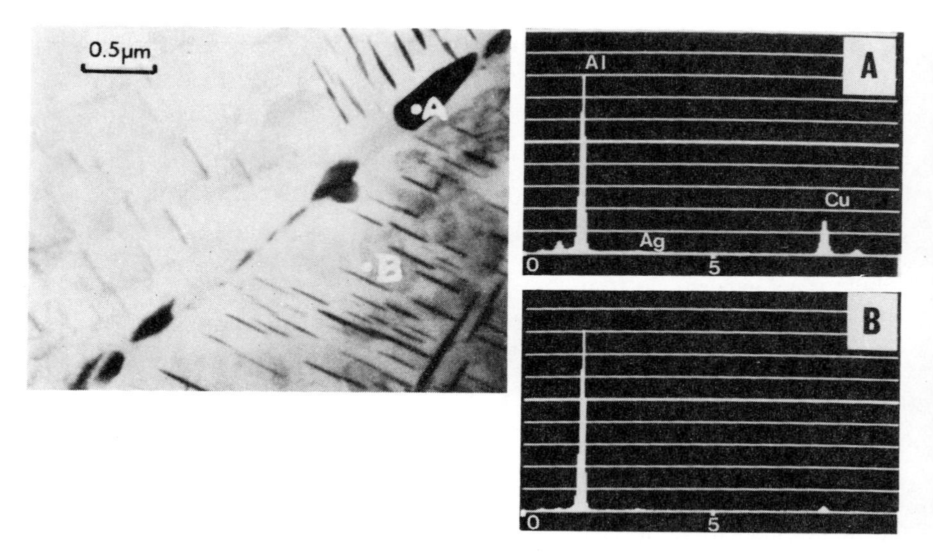

Fig. 13.5 – Analysis of grain-boundary precipitate (A) in aluminium–copper alloy.

together with rather larger precipitates at the grain boundaries. Spectra from a grain-boundary precipitate and from the matrix, accumulated using a counting period of ~200 s, are illustrated in Figs. 13.5(b) and 13.5(c) respectively. Not only is the ~5% copper level within the metal grain clearly revealed but the spectra show that the much smaller level of silver (~0.5%) is also detectable.

13.2.3 Analysis using a solid standard

One of the earliest methods proposed for composition analysis of thin films utilised a solid standard (Philibert and Tixier, 1968b). Now the number of ionisations ($\mathrm{d}n$) in path length, $\mathrm{d}s$, is given by

$$\mathrm{d}n = cQ \frac{N\rho}{A} \mathrm{d}s$$

and the number of x-rays generated by

$$\mathrm{d}I = cQ \frac{N}{A} \omega \, \mathrm{d}\rho s \ .$$

In a thin film specimen Philibert and Tixier assumed that electron backscattering and x-ray absorption could be ignored. Hence the x-ray emission from a thin film of thickness ρt is

$$I_\mathrm{F} = c \frac{N}{A} \omega \int_0^{\rho t} Q \, \mathrm{d}\rho s \ .$$

Further the ionisation cross-section, Q, may be taken as constant in the specimen, that is,

$$I_\mathrm{F} = c \frac{N}{A} \omega Q \int_0^{\rho t} \mathrm{d}\rho s$$

$$= c \frac{N}{A} \omega Q \rho t \ . \qquad (13.5)$$

Substituting for Q using the Bethe cross-section

$$Q = \frac{\text{const.}}{E_\mathrm{c}^2} \frac{\ln U_0}{U_0} \ ,$$

where U_0 is the overvoltage ratio (E_0/E_c),

$$I_\mathrm{F} = \frac{\text{const.}}{E_\mathrm{c}^2} c \frac{\ln U_0}{U_0} \rho t = \text{const.} \, c \frac{\ln U_0}{E_0 E_\mathrm{c}} \rho t \ .$$

For a pure bulk standard of element A, taking into account atomic number and absorption effects, the x-ray emission is

$$I_B = \text{const.}\frac{R}{S}f(\chi) \; .$$

Hence

$$\frac{I_F}{I_B} = \text{const.}\, c\, \frac{\ln U_0}{E_c}\,\frac{S}{R}\,\frac{1}{f(\chi)}\,\rho t \; .$$

While the constant may be readily evaluated, errors may arise due to the problem of estimating the mass thickness (ρt) of the specimen at the point of analysis. Consequently, the method is more effective for establishing concentration ratios for two elements (A and B) present in the specimen since, provided that measurements are taken at the same point, the mass thickness drops out of the equation, that is,

$$\left(\frac{I_F}{I_B}\right)_A \Bigg/ \left(\frac{I_F}{I_B}\right)_B = \frac{c_A}{c_B}\left(\frac{\ln U_0}{E_c}\,\frac{S}{R}\,\frac{1}{f(\chi)}\right)_A \Bigg/ \left(\frac{\ln U_0}{E_c}\,\frac{S}{R}\,\frac{1}{f(\chi)}\right)_B \; .$$

Philibert and Tixier applied their method to the analysis of nickel and aluminium in precipitates extracted from stainless steel and, after carrying out the correction, obtained a nickel : aluminium ratio close to (±10%) the composition Ni Al. Without the correction they considered the phase Ni_2Al or Ni_3Al might have been incorrectly deduced.

A problem with the Philibert and Tixier method is that ZAF correction procedures have been developed for low kV and are not appropriate for electron accelerating voltages used in transmission electron microscopy. In particular the deep electron penetration into a solid standard will produce a very large absorption correction in many cases and the greatest source of error may well lie with the standard rather than the specimen itself. Duncumb (1968) suggested, therefore, measuring x-ray intensities at lower kV (15 to 35 kV) and claimed good results, while Jacobs and Baborovska (1972) reported apparently satisfactory data when using a high kV (100 kV) on the foil and a low kV (40 kV) on the solid standard. However, methods such as these which utilise solid standards are inconvenient for practical analysis.

13.2.4 Analysis using a thin film standard

If a thin film standard of known composition is available, the ratio of the x-ray intensities from specimen AB and standard A may be simply reduced to

$$\frac{I_A^{AB}}{I_A^{A}} = k_A^{AB} = \frac{c_A^{AB}}{c_A^{A}}\,\frac{(\rho t)^{AB}}{(\rho t)^{A}} \; .$$

To apply this equation it is necessary to know the mass thickness of the standard as well as that of the specimen. Alternatively, if intensity ratios are obtained using pure element standards of known thickness (ρt^{A} and ρt^{B} respectively) then, since $c_{\mathrm{A}}^{\mathrm{A}} = c_{\mathrm{B}}^{\mathrm{B}} = 1$,

$$\frac{k_{\mathrm{A}}^{\mathrm{AB}}}{k_{\mathrm{B}}^{\mathrm{AB}}} = \frac{c_{\mathrm{A}}^{\mathrm{AB}}}{c_{\mathrm{B}}^{\mathrm{AB}}} \frac{(\rho t)^{\mathrm{B}}}{(\rho t)^{\mathrm{A}}} \; . \tag{13.6}$$

Now knowledge of the mass thickness of the specimen is not required. Philibert *et al.* (1970) have employed the method using thin film standards prepared by thermal evaporation but there are few other reports of its use, presumably because of the difficulty of manufacturing suitable thin film standards and measuring their thickness.

13.2.5 Double ratio method using characteristic x-rays

Here the standard is a compound material in the form of a thin film which contains both elements of interest, say A and B, in known concentration. Thus the mass thickness term in equation (13.6) is eliminated, that is

$$\frac{k_{\mathrm{A}}^{\mathrm{AB}}}{k_{\mathrm{B}}^{\mathrm{AB}}} = \left(\frac{c_{\mathrm{A}}^{\mathrm{AB}}}{c_{\mathrm{A}}^{0}} \frac{(\rho t)^{\mathrm{AB}}}{(\rho t)^{0}}\right) \Bigg/ \left(\frac{c_{\mathrm{B}}^{\mathrm{AB}}}{c_{\mathrm{B}}^{0}} \frac{(\rho t)^{\mathrm{AB}}}{(\rho t)^{0}}\right) \; ,$$

where the superscript 0 refers to the standard, and hence

$$\left(\frac{k_{\mathrm{A}}}{k_{\mathrm{B}}}\right)^{\mathrm{AB}} = \left(\frac{c_{\mathrm{A}}}{c_{\mathrm{B}}}\right)^{\mathrm{AB}} \Bigg/ \left(\frac{c_{\mathrm{A}}}{c_{\mathrm{B}}}\right)^{0} \; .$$

This principle can be used for estimating the copper concentration of the grain-boundary precipitate in the Al–Cu foil illustrated in Fig. 13.5. Comparison of the Cu : Al x-ray intensity ratio shows that it is seven times higher in the precipitate (Fig. 13.5(c)) than in the matrix (Fig. 13.5(b)). Since the matrix contains 5 wt% copper (balance almost all aluminium) we may write

$$7 = \left(\frac{c_{\mathrm{Cu}}}{c_{\mathrm{Al}}}\right)^{\mathrm{ppt}} \Bigg/ \left(\frac{c_{\mathrm{Cu}}}{c_{\mathrm{Al}}}\right)^{\mathrm{matrix}}$$

$$= \left(\frac{c_{\mathrm{Cu}}}{c_{\mathrm{Al}}}\right)^{\mathrm{ppt}} \Bigg/ \left(\frac{0.05}{0.95}\right)$$

that is

$$\left(\frac{c_{\mathrm{Al}}}{c_{\mathrm{Cu}}}\right)^{\mathrm{ppt}} \approx 2.7 \; .$$

Hence the precipitate contains 27 wt% copper and 73% aluminium. (The copper level is lower than the expected 54 wt% for $\mathrm{CuAl_2}$, presumably due to some of the surrounding matrix being excited as well.)

The sample could contain more than two elements and the method would still work, provided that all elements of interest are present in the standard and that their mass concentrations are known. The idea of using emission from part of the specimen as a 'standard' is not new to microanalysis and has frequently been adopted in studies on bulk material. It has the added advantage that the respective correction factors tend to be similar in both x-ray intensity measurements, thus producing a compensation effect.

13.2.6 Standardless methods

In section 11.4 of this book methods of carrying out quantitative analysis without using standards were discussed. The treatment was concerned essentially with bulk specimens but the same principles and practice may be applied to thin film analysis when an ED system, with its good long-term stability, is fitted to the electron microscope.

The method was proposed by Cliff and Lorimer (1975) and involves calibration of the instrument to produce a series of values which give an 'instrument response' for different pure elements. The 'instrument response' term for a particular element includes x-ray generation factors, such as the fluorescence yield for the x-ray of interest, the ionisation cross-section appropriate for the electron beam energy being used, and the efficiency of the x-ray detector. Provided that the thin film criterion (see section 13.2.8) is not exceeded then the recorded x-ray intensity ratio from elements A and B in a thin specimen AB is given by

$$\frac{I_A}{I_B} = \frac{c_A^{AB}}{c_B^{AB}} \frac{g_A}{g_B} ,$$

where c refers to elemental weight concentrations and g is the 'instrument response' factor. Measurements of g were carried out using carefully selected and prepared samples and standardised conditions in the electron microscope. Cliff chose to use silicon-containing samples and to refer all g values to silicon Kα measurements. Thus the above equation may be written as

$$\frac{c_A^{AB}}{c_B^{AB}} = \frac{I_A}{I_B} \frac{g_{Si}}{g_A} \frac{g_B}{g_{Si}} .$$

The 'instrumental response' factors, that is the ratios g_{Si}/g_A, etc., were given the designation $k^\dagger$ by Cliff ($k = 1$ for pure silicon). A typical master curve for a range of pure elements using Kα lines is shown in Fig. 13.6. However, it should be noted that such a curve is appropriate only for the instrumental configuration and accelerating voltage chosen for its construction. Experimental values for Lα

† This should not be confused with the use of k for x-ray intensity ratios as customarily adopted in electron-probe microanalysis.

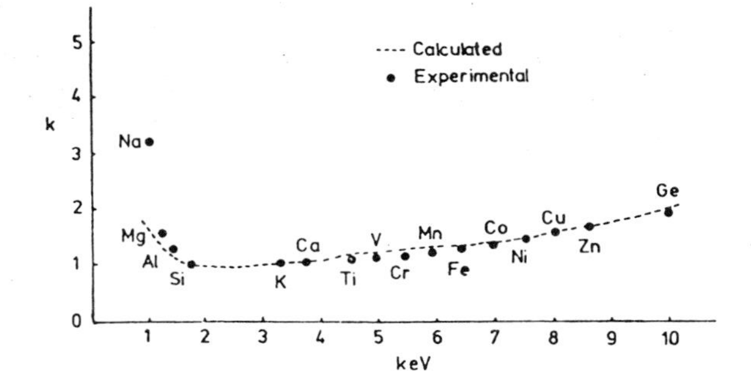

Fig. 13.6 – Comparison of experimental and calculated k values, 100 keV electrons (Cliff and Lorimer, 1975; courtesy J. of Microscopy).

lines using the EMMA-4 instrument operating at 100 kV are reported by Goldstein *et al.* (1977), while k values for the total L shell are given by Sprys and Short (1976) using a Philips EM300 electron microscope.

It is, of course, possible to calculate k values theoretically by taking into account the efficiency of x-ray production and of x-ray collection. The principle of the technique as applied to bulk specimens has been described in section 11.4 and extended to thin film analysis by Goldstein *et al.* (1977) to produce k values for any element and any accelerating voltage.

Let us begin by considering a thin foil containing the elements A and Si in weight concentrations c_A and c_{Si} respectively. By employing equation (13.5) it follows that the generated K intensity ratio of A to Si is

$$\left(c_A \frac{N}{A_A} \omega_A Q_A \rho t\right) \Bigg/ \left(c_{Si} \frac{N}{A_{Si}} \omega_{Si} Q_{Si} \rho t\right) = \left(\frac{c\omega Q}{A}\right)_A \Bigg/ \left(\frac{c\omega Q}{A}\right)_{Si} .$$

To obtain the corresponding ratio for just Kα intensities it is necessary to introduce p, the relative transition probability ($p = I_{K\alpha}/(I_{K\alpha} + I_{K\beta})$ where I refers to the intensity of the respective x-ray lines). Now the Kα x-ray intensity ratio recorded by the x-ray detector (I_A/I_{Si}) will not be equal to the generated intensity ratio because the x-rays from element A and Si will be absorbed by different amounts in the detector itself. Thus we may write

$$\frac{I_A}{I_{Si}} = \frac{g_A}{g_{Si}} = \frac{1}{k_A} = \left(\frac{\omega p Q T}{A}\right)_A \Bigg/ \left(\frac{\omega p Q T}{A}\right)_{Si} ,$$

where T refers to detector efficiency as in section 11.4.2. The value of T is calculated by summing all the different components (beryllium window, gold layer and dead layer) giving rise to absorption as described in section 5.3.1, that is

$$T_A = \exp\left[-\left(\frac{\mu}{\rho}\right)_{Be}^{A} (\rho t)_{Be} - \left(\frac{\mu}{\rho}\right)_{Au}^{A} (\rho t)_{Au} - \left(\frac{\mu}{\rho}\right)_{Si}^{A} (\rho t)_{Si}\right] .$$

Comparisons of measured and calculated k values have been made by Goldstein (1979) and a more recent comparison (Wood *et al.*, 1981) is included as Table 13.3. It is apparent that for elements above $Z = 14$ agreement between calculated and experimental k ratios is reasonable in most cases although the differences often exceed the estimated experimental error. The result for silver is anomalous and it is probable that the calculated value is in error here since some silver radiation may pass straight through the silicon detector without absorption; consequently unless the thickness of the silicon is well characterised the exact detector response will be impossible to calculate. The data on elements

Table 13.3 – Comparison of experimental and calculated k values (from Wood *et al.*, 1981).

Element	I	II	III	IV
Na	5.77	3.2	3.57 ± 0.21	1.83
Mg	2.07 ± 0.1	1.6	1.49 ± 0.007	1.31
Al	1.42 ± 0.1	1.2	1.12 ± 0.03	1.14
P	–	–	0.99 ± 0.016	1.04
S	–	–	1.08 ± 0.05	1.07
K	–	1.03	1.12 ± 0.027	1.05
Ca	1.0 ± 0.07	1.06	1.15 ± 0.02	1.04
Ti	1.08 ± 0.07	1.12	1.12 ± 0.046	1.17
Cr	1.17 ± 0.07	1.18	1.46 ± 0.03	1.24
Mn	1.22 ± 0.07	1.24	1.34 ± 0.04	1.32
Fe	1.27 ± 0.07	1.3	1.30 ± 1.28†	1.36
Ni	1.47 ± 0.07	1.48	1.67 ± 0.06	1.48
Cu	1.58 ± 0.07	1.60	1.59 ± 0.05	1.65
Mo	4.3	–	4.95 ± 0.17	4.68
Ag	8.49	–	12.4 ± 0.63	7.08

I Cliff and Lorimer, 1975 (100 kV experimental).
II Lorimer *et al.*, 1977 (100 kV experimental).
III Wood *et al.*, 1981 (120 kV experimental).
IV Wood *et al.*, 1981 (120 kV calculated).

† Presumably this error band as quoted in the original paper is a printing mistake.

below silicon are far less satisfactory, particularly the value for sodium. These lighter elements are very sensitive to the value of T_A and significant errors may be introduced by uncertainties in the exact value of $(\rho t)_{Si}$, $(\rho t)_{Au}$ and particularly $(\rho t)_{Be}$. Goldstein and Williams (1981) have indicated the degree to which k is influenced by variations in these parameters. Although errors associated with the thickness of the beryllium window may be removed by using windowless detectors to measure k ratios (Thomas, 1980), this approach is of little practical value for microanalysts using standard detectors. An additional factor affecting measurements of elements sodium, magnesium and aluminium is the build-up of carbon contamination on the specimen and/or deposition of hydrocarbons on the detector window itself (Love *et al.*, 1981). As a consequence the value of k will be altered but it is difficult to make allowance for this using the calculation method.

A further problem associated with calculating k values arises from uncertainties in the most appropriate expressions to use for Q and ω. Hence the majority of workers (see, for example Wood *et al.*, 1981) prefer to use the experimental method especially for the lighter elements. The Cliff–Lorimer technique has proved to be very popular because it is easy to apply once the appropriate calibration curve has been constructed.

13.2.7 Double ratio method using continuum x-rays

Quantitative analysis, by referring the measurements of characteristic x-ray intensities to that of the continuum radiation, is based upon the proposition that the intensity of continuum x-rays is a function of atomic number and mass thickness of the specimen (section 2.2.4). Hall (1968) used the method initially on biological samples where the mean atomic number of the material could be regarded as substantially constant. Hence the continuum intensity would then be directly proportional to the mass thickness and Hall proposed measuring a selected band of wavelengths in a little-absorbed part of the continuum in order to monitor variations of mass thickness in the specimen.

The continuum intensity then provided a correction factor which could be applied to the characteristic x-ray measurements made at the same point on the specimen. Thus the ratio of the concentrations, c_1 and c_2, of a given element at two different points on the specimen is

$$\left(\frac{I_1}{I_2}\right)_{\text{char}} = \frac{c_1}{c_2}\left(\frac{I_1}{I_2}\right)_{\text{cont.}}$$

However, since continuum intensity is a function of atomic number, the continuum method in this form is applicable only when the mean atomic number of the material analysed is nearly constant, as in biological specimens where the tissue analysed is of approximately constant composition with respect to major elements (for example, carbon, oxygen and hydrogen).

A different version of the continuum approach has been developed by Marshall and Hall (1968) to overcome this limitation. They expressed the ratio of characteristic intensity (I_A) to continuum intensity (I_{cont}) for a thin film as

$$\frac{I_A}{I_{cont}} = \text{const.} \frac{c_A}{Z^2} \frac{1}{\ln[1.166(E_0/J)]}$$

which, since the logarithmic term is relatively unimportant, gives

$$\frac{I_A}{I_{cont}} = \text{const.} \frac{c_A}{Z^2} \ .$$

Hall and Werba (1971) showed that this expression could be satisfactorily applied to the analysis of sodium and potassium in a range of minerals and further that the choice of continuum energy band was not too critical. However, the continuum method has not been used extensively outside the field of biology. The reader should be warned moreover of the problems of continuum measurement (section 13.2.10).

13.2.8 Atomic number, absorption and fluorescence effects

Clearly there is a limiting thickness of specimen above which the simple methods of quantitative analysis described in the foregoing will not give satisfactory results. The questions to be decided are what is the thickness, in what way is it affected by specimen composition and probe voltage, and what corrections may be applied in order to produce accurate data? We shall begin by considering the magnitude of atomic number, absorption and fluorescence effects in thin films.

The question of introducing an atomic number correction has been discussed by Tixier and Philibert (1969) and Tixier (1979). Tixier pointed out that since the backscatter coefficient for thin film specimens is very small (Cosslett and Thomas, 1965), the backscatter correction factor (R) is essentially unity; the published electron backscattering data of Badde *et al.* (1971) would support the conclusion. Badde gave also transmission coefficients for electrons as a function of film thickness and with 100 keV electrons this is close to unity. Moreover, the fraction of energy lost by an electron is small as may be seen from the Bethe energy-loss equation (Bethe and Ashkin, 1953),

$$\frac{dE}{d\rho s} = -78\,500 \frac{Z}{AE} \ln\left(1.166 \frac{E}{J}\right) .$$

Thus, for example, the energy loss (dE) for 100 keV electrons incident upon a 2000 Å thick specimen of copper is

$$78\,500 \times \frac{29}{63.5} \times \frac{1}{100} \ln\left(\frac{1.166 \times 10^2}{0.39}\right) 1.78 \times 10^{-4}\,\text{keV} = 0.36\,\text{keV} \ .$$

Hence the stopping power factor (S) may also be ignored in thin film analysis.

Philibert and Tixier (1975) stated that characteristic fluorescence corrections are negligible when $(\mu/\rho)_{B}^{AB}\rho t < 0.1$, where B refers to the element that excites element A. This suggests that even in a case of strong fluorescence excitation, such as iron exciting chromium in an alloy of 90% Fe and 10% Cr, fluorescence should be negligible up to a foil thickness of about 12 000 Å. However, Nockolds *et al.* (1980) have disputed the basis for this calculation and furthermore obtained experimental evidence on Fe–Cr alloys which agrees with equation (9.9) in predicting significant fluorescence at much lower foil thicknesses (see Fig. 13.8). Thus characteristic fluorescence is not always negligible, though it is reasonable to ignore the less important effect of continuum fluorescence in thin films.

Most workers have, however, concentrated upon absorption effects when deducing a critical film thickness below which corrections may be safely ignored.

Absorption in a thin foil of thickness t may be treated in a similar manner to absorption in bulk samples (see section 2.5), the integration limits now being 0 and the mass thickness ρt (see Fig. 13.7(a)).

Tixier and Philibert (1969) assumed that ionisation through the layer (mass thickness ρt) is constant. Thus from equation (2.1),

$$f(\chi) = \frac{\int_0^{\rho t} \phi(\rho z) \exp(-\chi \rho z) \mathrm{d}\rho z}{\int_0^{\rho t} \phi(\rho z) \mathrm{d}\rho z}$$

$$= \frac{\text{const.} \int_0^{\rho t} \exp(-\chi \rho z) \mathrm{d}\rho z}{\text{const.} \int_0^{\rho t} \mathrm{d}\rho z}$$

$$= \frac{\left[-\frac{1}{\chi} \exp(-\chi \rho z)\right]_0^{\rho t}}{[\rho z]_0^{\rho t}}$$

$$= \frac{1 - \exp(-\chi \rho t)}{\chi \rho t} \quad . \tag{13.7}$$

The correction is simple to apply, once the mass thickness of specimen at the point of analysis is known (see section 13.2.9), although several iterations may be needed to achieve the final result. However, Lorimer *et al.* (1977) decided to derive an alternative correction and, following Green (1964) and Bishop and Poole (1973), assumed all x-rays were produced in the centre of the foil. This gave an absorption factor of

$$\exp(-\chi \, . \, \rho t/2) \quad .$$

So far it has been assumed that the foil has uniform thickness but this is seldom the case in practice. Metal foils produced by electropolishing or ion-beam thinning are usually wedge-shaped in cross-section and more often than not the surface topography is irregular. Consequently the x-ray absorption path length in the specimen will depend not only upon foil thickness but also upon the orientation of the specimen relative to the position of the detector. Fig. 13.7(b) shows the difference in path lengths when the thin part of the wedge is pointed firstly towards the detector (position A) and then away from the detector (position B). Glitz *et al.* (1981) in their studies on Ni Al foils have shown that the x-ray intensity ratios of Ni Kα to Al Kα radiation can change by a factor of two when the foil is rotated 180°, from position A to position B. The effect will, of course, be most serious when x-radiation from one element is strongly absorbed

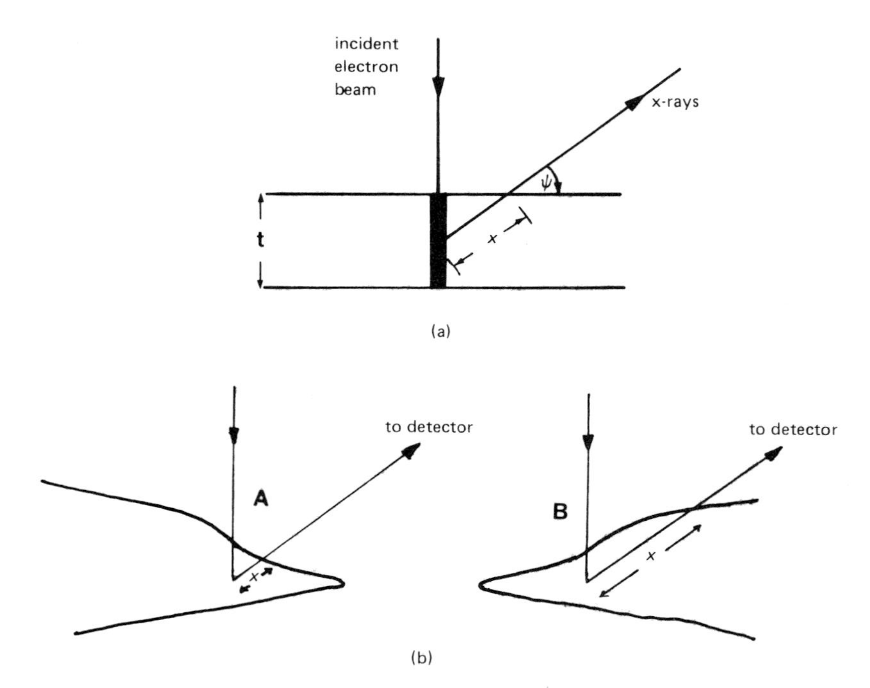

Fig. 13.7 – Absorption path length in thin films of (a) uniform thickness and (b) wedge-shaped cross-section.

in the other element and when the wedge angle is large. Unfortunately, unless a large number of thickness measurements are carried out the exact wedge profile cannot be characterised and all that can be done is to minimise any errors incurred. This is best achieved by ensuring that the thin part of the wedge points towards the detector since the absorption path length is reduced (position A in Fig. 13.7(b)).

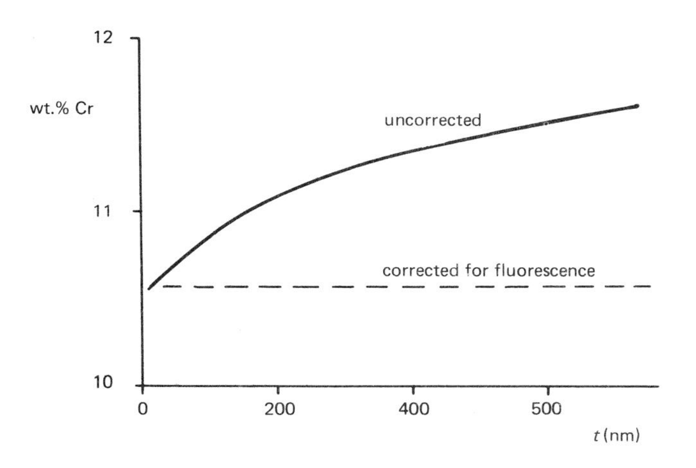

Fig. 13.8 – Variation in apparent concentration of chromium in 10.5% Cr–89.5% Fe as a function of foil thickness (t) before and after fluorescence correction (after Nockolds *et al.*, 1980).

The thin film criterion proposed by Tixier and Philibert is based upon the proposition that x-ray absorption in the foil should be less than 5%, that is, $f(\chi)$ should be greater than 0.95. This figure was related to the precision with which x-ray intensity measurements could be carried out on thin film specimens. Rewriting $f(\chi)$ as the series

$$f(\chi) = 1 - \chi\frac{\rho t}{2!} + \chi^2\frac{(\rho t)^2}{3!} - \ldots .$$

then leads to the criterion that absorption corrections may be ignored if

$$\chi\rho t < 0.1 \; . \tag{13.8}$$

The critical specimen thickness can hence be established from χ and ρ values calculated for a given specimen composition. Using this criterion, a number of critical film thicknesses (t_c) have been calculated (Table 13.4) for a range of specimen compositions. The values assume an x-ray take-off angle of 30° and normal incidence of the electron beam; other instrumental geometries would

Table 13.4 – Critical specimen thicknesses (t_c) for several systems; A and B refer to radiations being measured.

Specimen AB	μ/ρ ($cm^2 g^{-1}$)		ρ ($g\,cm^{-3}$)	t_c (Å)		
				Eqn (13.8)		Eqn (13.9)
	A	B		A	B	
Al–4w/o Cu	586	50	2.77	3090	36200	3370
Al_2Cu	3056	52	4.3	380	22300	390
AlNi	3435	60	5.8	250	14300	255
NiFe	218	81	7.9	2890	7810	4620
CuAu (Au L line)	171	157	15.0	1930	2100	23810

give different values. The data indicate that severe absorption problems would be experienced when measuring aluminium in either copper-rich or nickel-rich matrices. With a typical transmission specimen thickness of 2000 Å, $f(\chi)$ values (equation (13.7)) for Al Kα emission are 0.78 and 0.76 in $CuAl_2$ and AlNi respectively, that is, more than 20% of the generated x-rays would be absorbed in the specimens. When the other element in these binary systems is measured then absorption is negligible. The thin film criterion should, therefore, be applied to the more heavily absorbed radiation and inspection of the respective mass absorption coefficients for the different radiations in the compound will establish which one this is.

It has been argued that the actual difference in mass absorption coefficients is the more appropriate factor for establishing the thin film criterion. Again taking a value of 5% as the maximum absorbed x-ray intensity we may, for a specimen AB, express this condition as

$$\frac{f(\chi)_A}{f(\chi)_B} > 0.95 \ ,$$

that is

$$\left[1 - \frac{(\chi\rho t)_A}{2!}\right]\left[1 - \frac{(\chi\rho t)_B}{2!}\right]^{-1} > 0.95$$

$$\left[1 - \frac{(\chi\rho t)_A}{2!}\right]\left[1 + \frac{(\chi\rho t)_B}{2!}\right] > 0.95$$

$$1 - \frac{(\chi\rho t)_A}{2!} + \frac{(\chi\rho t)_B}{2!} > 0.95$$

and we have

$$(\chi\rho t)_A - (\chi\rho t)_B < 0.1 \quad .$$

Where elements A and B are measured at the same point on the specimen (as in the ratio method) this reduces to

$$(\chi_A - \chi_B)\rho t < 0.1 \quad . \tag{13.9}$$

Values for the critical thickness of film calculated using this criterion are included also in Table 13.4 and results for the specimens quoted are generally close to figures derived for the more heavily absorbed radiation using the Tixier and Philibert (1969) criterion. Goldstein (1979) provided data on x-ray absorption in thin films at an x-ray take-off angle of 45°; his thin film criterion is given by $(\chi_B - \chi_A)(\rho t)/2 < 0.1$, a less conservative 10% absorption factor being adopted. He concluded that absorption corrections were necessary in most of the systems he cites apart from Ni–Fe alloy, even though the foils were transparent in transmission electron microscopy.

The thin film criterion may be determined experimentally from measurements of the intensity of a particular characteristic x-ray line and a band of continuum x-rays at various points on a specimen of uniform composition where the thickness is known. A plot of the ratio of x-ray intensities against specimen thickness will then indicate where the thin film criterion fails by giving a departure from linearity. The principle has been used for nickel-chromium alloys (Lorimer *et al.*, 1972) and iron–manganese alloys (Cliff and Lorimer, 1972) where, although no actual thickness measurements were quoted, it was at that time deduced that if the specimen was transparent to 100 keV electrons corrections were not required. As mentioned above, however, electron transparency at the operating voltage does not necessarily satisfy the thin film criterion, x-ray absorption effects having been reported in many thin film studies (Hutchins, 1966; Jacobs and Baborovska, 1972; Lorimer *et al.*, 1976; Zaluzec and Fraser, 1976; Goldstein *et al.*, 1976).

13.2.9 Measurement of foil thickness

We have seen in the preceding sections of this chapter how an accurate measurement of the foil thickness at the point of analysis may be needed to establish chemical compositions. Knowledge of foil thickness is required also when determining the population of fine structures in a specimen, such as defects produced by irradiation or precipitates formed in heat-treated alloys (Love *et al.*, 1980).

Early methods of foil thickness measurement were based upon the analysis of either extinction contours or known crystallographic features such as slip plane traces (Hirsch *et al.*, 1965). However, the analysis of extinction contours cannot be generally applied since the contours are measurable only on those edges of foil which have a sharply tapering section; moreover, imaging conditions must be adjusted such that the deviation parameter is zero otherwise large errors may arise. Techniques of thickness measurement based upon measuring crystallographic features in the specimen are, in principle, simple to carry out. The idea is to measure the projected width of a feature which intersects the top and bottom surfaces of the foil, before and after tilting through a known angle; usually the parallax position, where the fault lies parallel to the electron beam, is chosen as one of the tilt orientations. Slip plane traces, stacking faults and grain boundaries have all been used for this purpose. Unfortunately, however, such crystallographic features are not always present in specimens and even when they do occur they may not be located in the particular region of foil requiring analysis. In an attempt to overcome this limitation but still exploit the same principle Heimendahl (1973) proposed applying latex spheres to the foil. Nevertheless, all these techniques are prone to errors (Kelly, 1975; Hall and Vander Sande, 1975; Clareborough and Head, 1976).

The incorporation of scanning transmission (STEM) and x-ray analysis (ED) facilities on the transmission electron microscope has now provided alternative methods for carrying out foil thickness determinations. These may be divided into 'absolute' methods such as contamination spot measurements Lorimer *et al.*, 1976) and convergent beam diffraction (Kelly *et al.*, 1975), and those requiring calibration in which electron backscattering (Niedrig, 1978) or x-ray emission from the foil is used (Miller and Scott, 1978; Nockolds *et al.*, 1980; Cliff and Lorimer, 1980). Although the former do not strictly come within the scope of this book, a brief description of these methods will be given for comparison purposes.

The contamination spot method requires the electron beam to be focused onto the foil specimen in order to form small contamination spots upon its upper and lower surfaces. The foil is then tilted through a known angle and the separation of the spots measured. From the tilt angle φ and the separation distance x the thickness of foil, t, in the beam direction may be found from $x/\sin\varphi$, Fig. 13.9. Only, however, if the foil is perpendicular to the beam while the spots are being deposited will the actual foil thickness be given. Also, as mentioned by Love *et al.* (1977b) the contamination spot method measures the total thickness of specimen, which includes any layers of contamination and/or oxide present on the foil surfaces. Rae *et al.* (1981) have discussed such errors and pointed out that these may be as much as 750Å on stainless steel foils, a major factor being the difficulty in locating precisely the points on the image from which to make measurements. It must be concluded, therefore, that although the method is convenient to apply in many respects it can give misleading results by overestimating the true foil thickness.

The convergent beam technique also requires a focused beam of electrons, convergence of the beam being controlled such that divergent cones of diffracted electrons are imaged. A two-beam condition is required which usually necessitates

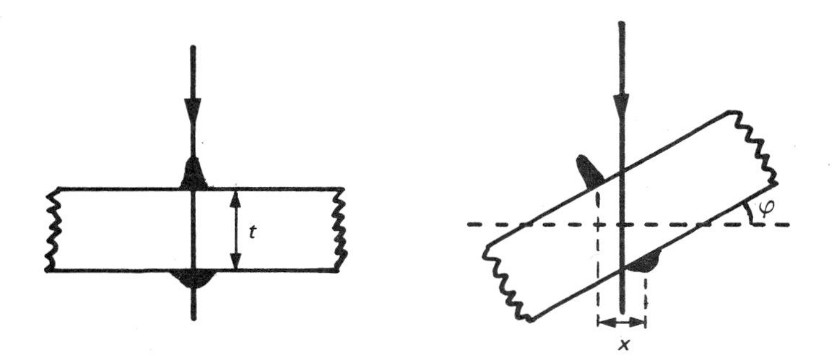

Fig. 13.9 – Principle of contamination spot method for determining foil thickness (t); after deposition of the spots the foil is tilted through a known angle φ and the separation (x) measured.

some few degrees tilt adjustment of the foil. The diffraction spots then appear as discs, each disc containing a series of fringes and we may write (MacGillavry, 1940)

$$\left(s_i^2 + \frac{1}{\xi_g^2}\right)t^2 = n_i^2 \tag{13.10}$$

where s_i is the deviation of the ith minimum from the exact Bragg position, ξ_g is the extinction distance, and n_i is an integer. The value of s_i can be found from measurements of the fringe spacing (x_i) and the distance between central fringe and the centre spot (X), that is

$$s_i = \frac{x_i}{X}\frac{\lambda}{d^2} ,$$

where d is the interplanar spacing for the operating diffraction and λ is the electron wavelength. Hence, provided ξ_g is known accurately, the thickness of foil can be calculated from equation (13.10).

Kelly *et al.* (1975) have shown that if equation (13.10) is rearranged as

$$\left(\frac{s_i}{n_i}\right)^2 + \frac{1}{n_i^2}\frac{1}{\xi_g^2} = \frac{1}{t^2} ,$$

then a plot of $(s_i/n_i)^2$ against $1/n_i^2$ using different values of n_i should give a straight line of slope $-1/\xi_g^2$ with an intercept on the $(s_i/n_i)^2$ axis of $1/t^2$. Hence the value for ξ_g is not required in thickness determinations, although it is useful to know the approximate value in order to check that the slope of the line is correct since this can confirm that the proper sequence of integers for n_i has been chosen. Kelly describes convergent beam diffraction measurements on thin films of copper and stainless steel and, from comparisons with foil thickness determinations using slip plane traces, claimed an accuracy of ±2% for thicknesses in the range 500 Å to 3000 Å, with aluminium films the maximum thickness that could be reliably measured was 5000 Å. Allen (1980) considered possible errors in the method and recommends using higher order diffractions than 200, 220 and 311 in order to reduce multiple beam interactions which can lead to errors when measuring very thin films; the upper thickness limit is determined by anomalous absorption and for 100 keV electrons in aluminium, copper and gold was estimated to be 6000 Å, 1300 Å and 600 Å respectively.

The measurement of x-ray intensities to establish film thickness can involve either characteristic or continuum x-ray emission. The important point is to choose radiation which is little absorbed in the specimen so that the thin film criterion (section 13.2.8) is applicable over a sufficiently wide range of thickness. The x-ray intensity from the point of interest is usually compared with a calibration curve constructed from x-ray intensity measurements on foils of similar composition whose thickness has been established by an 'absolute' method such as convergent beam diffraction. Nockolds *et al.* (1980) have proposed an equation for calculating specimen thickness directly from measured characteristic x-ray intensities. Their thin film measurements were compared with x-ray intensities obtained from a solid standard under identical beam conditions and absorption correction terms were included in the expression. The corrected x-ray intensity ratios showed a linear dependence upon film thickness, although some doubt has been expressed (Cliff and Lorimer, 1980) concerning the validity of the equation when radiation from lighter ($Z < 20$) elements is used.

As an alternative to constructing calibration curves, 'absolute' methods of using x-ray intensity measurements have been proposed (Morris *et al.*, 1980) which rely upon knowing the mass absorption coefficient for the specimen and then deducing its thickness from measurements of x-ray attenuation. Two approaches were described. The first involved measuring the ratio of K:L x-ray intensities, which meant that the method could be applied only to specimens containing an element whose atomic number was above $Z = 27$; using copper x-radiations it appeared to give reasonable results on Al–Cu foils. The second idea was to carry out x-ray measurements with the specimen tilted through known angles and, from the differences in x-ray attenuation, to calculate the mass thickness of specimen; this method is not restricted to samples containing elements of $Z > 27$, although it is more difficult to apply in practice and will be subject to errors if the specimen has a wedge-shaped profile for the reasons discussed in section 13.2.8. Both methods require iterative procedures.

In discussing methods for measuring foil thickness, Love *et al.* (1977b) have pointed out some of their advantages and disadvantages. Apparent inconsisitencies were explained by the fact that each technique measures something different. The contamination spot method determines the total foil thickness including any surface layers of oxide and contamination; convergent beam diffraction gives the metal thickness, while x-ray methods will overestimate if radiation generated in any surface layer of oxide, etc., is included in the measurement of x-ray intensity. An advantage of using x-ray intensity measurements is that the method is fast, virtually instantaneous once the calibration curve has been obtained. On the other hand, the contamination spot and convergent beam diffraction techniques require measurements to be made on film, the latter method being particularly slow because a graphical construction is subsequently needed before a thickness value can be established. With all methods, however, it should be noted that the values derived refer to the foil thickness in the beam direction and if the specimen is not accurately aligned perpendicular to the beam, a further error may be introduced when estimating the population density of fine structures. It may be concluded, therefore, that determining the thickness of the specimen remains still one of the more difficult measurements to carry out accurately in thin film analysis.

13.2.10 Spurious x-ray signals and other artefacts

In the foregoing we have treated the recorded x-ray signal as if it originated entirely from a selected region of the specimen and assumed that the region is representative of the material requiring analysis. However, this may not be the case and spurious effects may be caused by:

(a) instrumental factors such as stray radiation;
(b) specimen surface artefacts caused by the method of preparation or by contamination before and during examination in the electron microscope.

Sources of stray radiation have been reported by many workers (see Goldstein and Williams, 1977, 1978; Zaluzec, 1979) and the occurrence may differ from one design of instrument to another. Spurious x-rays may be generated by electrons, scattered within the specimen chamber, striking various components and specimen areas remote from the region of analysis; in addition, the detected x-ray signal may include fluorescence x-rays produced by x-rays generated from apertures (Zaluzec and Fraser, 1976), etc. in the system. The unwanted x-rays may give rise to additional peaks in the spectrum as well as a general increase in background intensity. A measure of the amount of spurious radiation may be obtained by recording the residual x-ray spectrum when the electron beam is positioned off the specimen; Goldstein and Williams (1978) describe the use of a silver disc containing a hole for this purpose. Although subtraction of the 'hole count' from the specimen has been proposed as a means of obtaining the true x-ray intensities, a better solution is to eliminate the unwanted x-rays altogether. This may be achieved by carrying out modifications to electron microscope and specimen holder, and manufacturers now supply a kit which reduces spurious x-rays to a tolerable level. Figures 13.10(a) and 13.10(b) illustrate the beneficial effect of fitting such a kit† to the JEOL 100CX instrument; in addition to removing x-ray peaks from copper, niobium and tin which were unrelated to the specimen, the reduction in background intensity improved significantly the overall detection sensitivity.

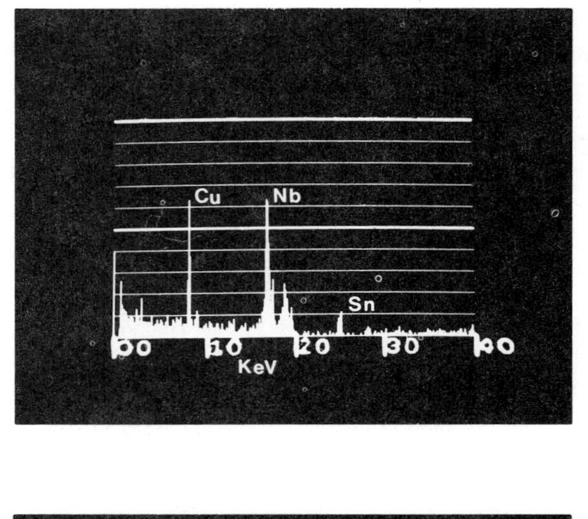

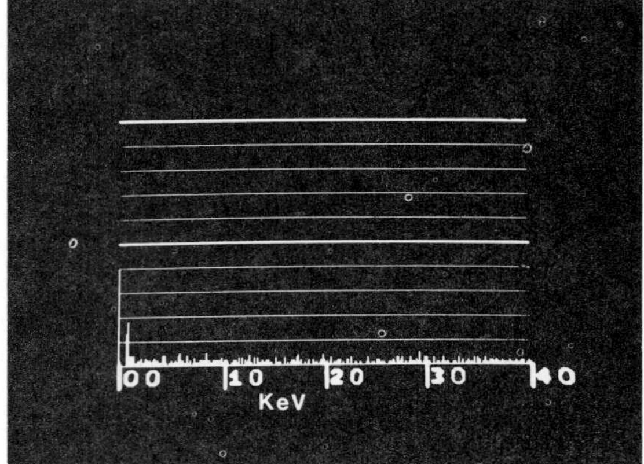

Fig. 13.10 – Reduction of spurious x-ray counts after fitting special kit to transmission electron microscope.

† Supplied by Hexland Ltd., East Challow, Oxford.

The extreme care that is needed when preparing specimens for analysis is continually referenced in the literature and has been mentioned in section 6.2. If the surface layers are not representative of the bulk material then problems will arise, the more specific the analytical technique being used the greater the problems. In the case of thin film specimens any non-representative surface layers may constitute a significant proportion of the total amount of material being analysed and introduce substantial errors. An example of this is the copper-rich oxide formed on aluminium–copper alloys during electropolishing (Thompson *et al.*, 1977). The effect is visible as an increase in the intensity ratio of CuK and AlK radiations with decrease in thickness of the thin film specimen (Fig. 13.11). Morris *et al.* (1977) described similar results on aluminium–silver alloys and showed that such oxide films were not present on specimens which were thinned using an ion beam provided that the operation was carried out in a low residual partial pressure.

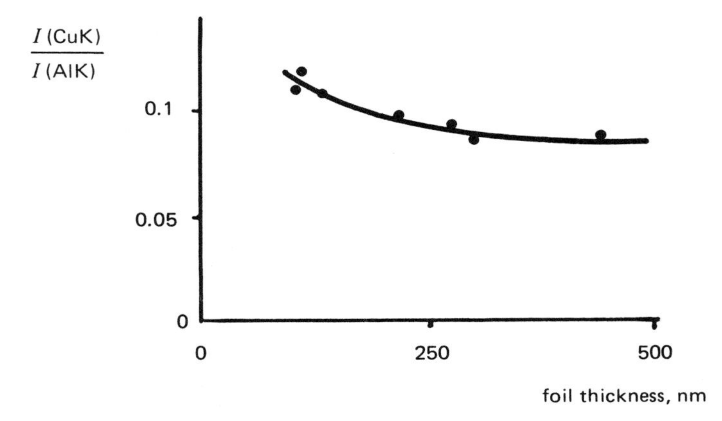

Fig. 13.11 – Measured x-ray intensity ratio (CuK/AlK) from foil of aluminium–copper alloy as a function of foil thickness showing surface enrichment of copper, 100 keV.

The question of specimen contamination with carbon has been mentioned before, (sections 6.3.4 and 13.2.9) and the subject has been excellently reviewed by Hren (1979). In the transmission electron microscope the deposition of carbonaceous materials during specimen examination will take the form of a layer covering both surfaces where electrons impinge upon and emerge from the thin film. As illustrated in Fig. 6.5(a), when a focused beam is used the deposit piles up around the points of impact and emergence. The rate at which the carbon builds up on a specimen depends upon the cleanliness of the vacuum system, but it may not take long before it produces a measurable effect on the emitted x-ray intensity. For example, Zaluzec (1978) observed an increase with time in the Ni Kα: Al Kα intensity ratio from a Ni Al specimen and attributed this to the progressively increased absorption of Al Kα by the growing accumulation of carbon (Ni Kα radiation is little absorbed by carbon). Goldstein (1979) points out that lack of agreement between published k (instrument response) values for elements such as aluminium, magnesium and sodium may be due in part to specimen contamination. To this occurrence should be added the likelihood that contamination in liquid form may be deposited directly on the detector window and cause attenuation of the softer x-radiations (Love *et al.*, 1981).

In conclusion, it has been shown how some of the spurious effects which may arise in thin film analysis in the transmission electron microscope can be overcome or at least made tolerable. The important thing is to be aware of all possible problem sources and, when presenting analytical data, to take account of those which cannot be removed.

13.3 PARTICLES

Many of the difficulties experienced with quantitative electron-probe microanalysis of coatings and thin films are accentuated when dealing with particles. This applies whether the particles are present as second phases in a solid matrix or whether they are isolated.

Let us begin by considering the case familiar to many microanalysts where a specimen containing particles has been carefully prepared by appropriate metallographic/petrographic techniques, and assume that no preparation artefacts (see section 6.1) are present. Possible problems arising in analysing the particles may be conveniently discussed using the following series of diagrams which depict a few selected particle matrix geometries. Figure 13.12(a) illustrates a particle AB, sufficiently large to contain all the incident electrons, which is embedded in a matrix A. Here the generated x-ray distribution would be identical to that in bulk sample of AB and can be deduced using the formulae for the atomic number correction described in Chapter 7. The absorption correction is, however, more difficult to apply since those x-rays generated in the particle which travel through matrix material on their way to the detector will experience differing degrees of attenuation, particularly where $(\mu/\rho)_{\mathrm{A}}^{\mathrm{AB}} \gg (\mu/\rho)_{\mathrm{A}}^{\mathrm{A}}$. The magnitude of the effect will be dependent upon x-ray take-off angle and size of particle and although such geometrical factors could, in principle, be taken into account in ZAF calculations they are complex in practice to introduce. In addition, the possibility of fluorescence effects across the particle/matrix interface may have to be considered, as discussed in section 9.7, whereby fluorescence may emanate from regions many micrometres distant from the particle (Henoc *et al.*, 1969). The particle size problem was well illustrated in some early work of Birks *et al.*, (1960) aimed at differentiating chi and sigma phases, notorious embrittling agents in stainless steels. The data showed that, although the phases may be distinguished by comparing the intensity ratio of molybdenum and chromium radiations, ratios corresponding to bulk analyses were not achieved unless the particles were several micrometres in size. A further complication, which occurs with opaque specimens, is that it may be impossible to know for certain that the particle *is* sufficiently thick to stop all the electrons.

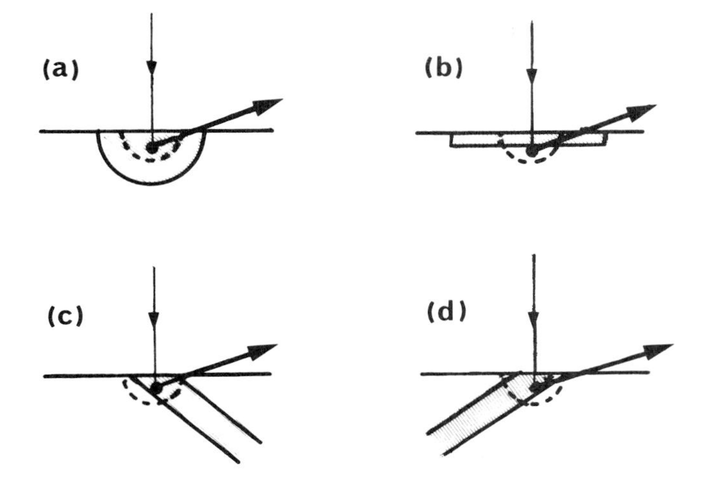

Fig. 13.12 – Some examples of phase analysis where geometrical effects can complicate any attempt to apply conventional ZAF corrections. Dotted lines indicate region of x-ray generation.

Figure 13.12(b) depicts a particle which is thinner than the depth of electron penetration but, provided that its lateral dimensions exceed that of the x-ray source, it may be treated by one of the methods applicable to coating analysis. If its diameter is smaller, a situation arises where the correction procedures developed for a coating are further complicated by the addition of factors relating to the geometry shown in Fig. 13.12(a). Even greater analytical uncertainties are introduced with the particle geometry illustrated in Fig. 13.12(c), and again with that in Fig. 13.12(d). These are by no means the most difficult cases facing operators, as many readers will be aware, and it is not surprising that methods based upon ratio techniques are preferred, wherever possible, for quantitative work.

These considerations apply equally to isolated particles, which may be treated along the lines discussed above as more difficult cases of thin film analysis. However, here additional factors which further complicate quantitative analysis are that the incident electrons will not strike the surface at the same angle at all points and that the x-ray take-off angle and absorption path length will vary with position of the generated x-ray. Even simple geometries such as a sphere

(Fig. 13.13) are difficult to treat rigorously and most particles have a more complex shape than this. The specimen illustrated in an earlier chapter (Fig. 6.1) is a good example which demonstrates the variation of recorded x-ray emission across the surface even though the particle (sulphur) is homogeneous in composition.

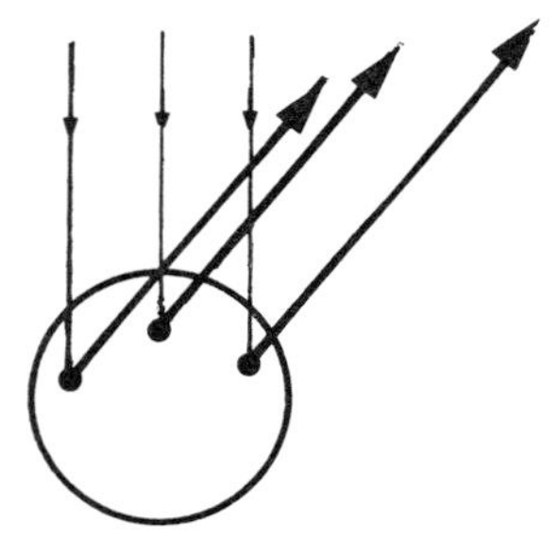

Fig. 13.13 – Analysis of particles (or rough surfaces) may show a variation in the recorded x-ray intensity even though the region being sampled is homogeneous in chemical composition.

Some early work on particle analysis was described by Duncumb (1966) using his combined electron microscope and microanalyser (EMMA). He showed that analysis of two different sizes of titanium sulphide particle extracted from a steel gave similar Ti Kα:S Kα x-ray intensity ratios, which indicated that a ratio technique may be applied down to particle sizes approaching the beam diameter of 1000 Å. The limiting size of particle, below which the x-ray intensity ratio shows a departure from linearity, will be a function of its chemical composition and the electron accelerating voltage used. In general, if factors associated with absorption and atomic number effects are likely to be significant for the system then changes in measured x-ray intensity ratios will tend to be appreciable. To some extent the non-linearity between intensity ratio and particle size could be taken into account by first constructing a calibration curve. Armstrong and Buseck (1977) plotted intensity ratios of manganese and zinc K x-rays measured from a range of mineral particles based upon the willemite–tephroite system and showed a good correlation with the actual composition ratios for sizes down to 0.5 μm; their results on Fe Kα and Si Kα x-ray intensity measurements in the fayalite–forsterite system gave poorer correlation due, it was presumed, to the larger difference in atomic number between the elements of interest. Hence there are good arguments for carrying out analyses at high probe voltages (say ~100 kV) in order to minimise any atomic number effects (as in thin film analysis, section 13.2).

An interesting ratio technique (Small *et al.*, 1980), which was based upon Hall's approach to thin film analysis (section 13.2.7), utilised a portion of the continuum radiation in order to take into account particle size effects. The authors argued that, to a first approximation, the continuum radiation and characteristic emission are influenced by particle effects to a similar extent, that is the depth distribution of continuum and characteristic x-rays is approximately the same. Hence the intensities of the characteristic peak (P) and background (B) from particle (p) and from bulk material (b) of the same composition may be equated, $(P/B)_p = (P/B)_b$. Thus the problem was essentially to deduce P_b, which could then be converted into mass concentration using a conventional ZAF correction routine. Rearranging the above equation gave $P_b = P_p(B_b/B_p)$, and B_b/B_p may be regarded as a 'particle size correction factor'. The procedure utilised an ED system for making x-ray measurements, the method adopted for estimating background (see section 5.3) being based upon the formula developed by Lifshin *et al.* (1975) which is itself an extension of the Kramers equation (Kramers, 1923). Although B_b could not be estimated directly, since it is presumed that no such specimen is available, it could be calculated form $\Sigma_i c_i B_i$ when its composition was known. Consequently, x-ray measurements were made on the appropriate pure standard(s) and the recorded intensities corrected for atomic number, absorption and fluorescence effects within the iterative ZAF program; as a first approximation, the peak and background values obtained on the particle were used. In terms of the k factor familiar to microanalysts, the procedure may be summarised as

$$k = \frac{P_b}{P_s} = \frac{P_b}{B_b}\frac{B_b}{P_s}, \quad \text{and since} \quad \frac{P_b}{B_b} = \frac{P_p}{B_p},$$

$$k = \frac{P_p}{B_p}\frac{B_b}{P_s}.$$

The method was tested (Small *et al.*, 1978) on glass microspheres and found to give satisfactory results and was later (Small *et al.*, 1979, 1980) used to analyse particles of talc, FeS_2 and ZnS. The authors have pointed out the dangers of ignoring such particle-size corrections, particularly when concentrations deduced are normalised to 100% since errors may be increased rather than reduced by this practice. Statham and Pawley (1978) have also proposed using peak and background measurements but in their method the final result was expressed as a concentration ratio.

Monte Carlo treatments as an aid to developing formulae for quantitative analysis of particles have been discussed in section 12.6. The reader is referred especially to the article by Newbury *et al.* (1980) in NBS Special Publication No. 533 (Heinrich, 1980), a document which contains other useful papers on the subject of particle analysis.

References

Albee, A. L., and Ray, L. A. (1970), *Anal. Chem.*, **42**, 1408.

Allen, S. M. (1980), *Phil. Mag.*, **A43**, 325.

Andersen, C. A., and Hasler, M. F. (1966), *X-ray Optics and Microanalysis*, eds. R. Castaing, P. Deschamps, and J. Philibert, (Paris: Hermann), p. 310.

Andersen, C. A., and Wittry, D. B. (1968), *J. Phys. D: Appl. Phys.*, **1**, 529.

Archard, G. D. (1961), *J. Appl. Phys.*, **32**, 1505.

Archard, G. D. and Mulvey, T. (1963), *Brit. J. Appl. Phys.*, **14**, 626.

Armstrong, J. T., and Buseck, P. R. (1977), Proc. 12th Ann. Conf. Microbeam Analysis Soc., Boston, Mass., Paper 41.

Badde, H. G., Drescher, H., Krefting, E. R., Reimer, L., Seidel, H., and Buhring, W. (1971), *Electron Microscopy and Analysis*, ed. W. C. Nixon, (London: Inst. Physics), p. 74.

Baker, M. A., Holland, L., and Laurenson, L. (1971), *Vacuum*, **21**, 479.

Bambynek, W., Craesman, B., Fink, R. W., Freund, H. V., Mark, H., Swift, C. D., Price, R. E., and Venugopala Rao, P. (1972), *Rev. Mod. Phys.*, **44**, 716.

Barbi, N. C., and Lister, D. B. (1981), *Energy-Dispersive X-ray Spectrometry*, eds. K. F. J. Heinrich, D. E. Newbury, R. L. Myklebust, and C. E. Fiori, Nat. Bur. Stands. Spec. Publ. 604, (Washington: US Dept. Commerce), p. 35.

Barbi, N. C., Skinner, D. P., and Blinder, S. (1976), Proc. 11th Ann. Conf. of the Microbeam Analysis Soc., Miami, Florida, Paper 8.

Barkla, C. G., and Sadler, C. A. (1909), *Phil. Mag.*, **17**, 739.

Bartosek, J., Masek, J., Adams, F., and Hoste, J. (1972), *Nucl. Instr. Meth.*, **104**, 221.

Baun, W. L. (1969), *Advances in Electronics and Electron Phys. Suppl. 6*, eds. A. J. Tousimis, and L. Marton, (New York: Academic Press), p. 155.

Bayard, M. (1973), *Microprobe Analysis*, ed. C. A. Andersen, (New York: Wiley), p. 323.

Beaman, D. R. (1969), *Mikrochim. Acta*, 117.

Beaman, D. R., and Isasi, J. A. (1970), *Anal. Chem.*, **42**, 1540.

Beaman, D. R., and Isasi, J. A. (1972), *Electron Beam Microanalysis*, Amer. Soc. for Testing and Materials Spec. Tech. Publ. 506.

Beatty, R. T. (1912), *Proc. Roy. Soc.*, **A87**, 511.

Beatty, R. T. (1913), *Proc. Roy. Soc.*, **A89**, 314.

Bence, A. E., and Albee, A. L. (1968), *J. Geol.*, **76**, 382.

Berger, M. J., and Seltzer, S. M. (1964), Studies of Penetration of Charged Particles in Matter, Nat. Res. Council. Publ. 1133, (Washington: Nat. Acad. Sciences), p. 205.

Bethe, H. A., (1930), *Ann. Phys. Leipz*, **5**, 325.

Bethe, H. A., and Ashkin, J. (1953), *Experimental Nuclear Physics*, (New York: Wiley), **1**, 252.

Birks, L. S. (1963), *Electron Probe Microanalysis*, (New York, Interscience), p.78.

Birks, L. S., Siomkajlo, J. M., and Koh, P. K. (1960), *Trans. Amer. Inst. Min. Engrs.*, **218**. 806.

Bishop, H. E. (1965), *Proc. Phys. Soc.*, **85**, 855.

Bishop, H. E. (1966a), *X-ray Optics and Microanalysis*, eds. R. Castaing, P. Deschamps, and J. Philibert, (Paris: Hermann), p. 112.

Bishop, H. E. (1966b), *X-ray Optics and Microanalysis*, eds. R. Castaing, P. Deschamps, and J. Philibert, (Paris: Hermann), p. 153.

Bishop, H. E. (1966c), Ph.D. Thesis, University of Cambridge.

Bishop, H. E. (1967), *Brit. J. Appl. Phys.*, **18**, 703.

Bishop, H. E. (1968), *J. Phys. D: Appl. Phys.*, **1**, 673.

Bishop, H. E. (1974), *J. Phys. D: Appl. Phys.*, 7, 2009.

Bishop, H. E. (1976), *Use of Monte Carlo Calculations in Electron-Probe Microanalysis and Scanning Electron Microscopy*, eds. K. F. J. Heinrich, D. E. Newbury, and H. Yakowitz, Nat. Bur. Stands. Spec. Publ. 460, (Washington: US Dept. Commerce), p. 5.

Bishop, H. E., and Poole, D. M. (1973), *J. Phys. D: Appl. Phys.*, **6**, 1142.

Bloch, F. (1933), *Zeit. Phys.*, **81**, 363.

Blodgett, K. B., and Langmuir, I. (1937), *Phys. Rev.*, **51**, 964.

Blokhin, M. A. (1965), *Methods of X-ray Spectroscopic Research*, (Oxford: Pergamon).

Bloomfield, D. J., Love, G., and Scott, V. D. (1981), *Electron Microscopy and Analysis 1981*, ed. M. J. Goringe, (Bristol: Inst. Physics), p. 182.

Boersch, H. (1939), *Zeit. Tech. Phys.*, **20**, 346.

Bolon, R. B., and Lifshin, E. (1973), *SEM/1973*, ed. O. Johari, (Chicago: IITRI), p. 281.

Bolon, R. B., Lifshin, E., and Ciccarelli, M. F. (1975), *Practical Scanning Electron Microscopy*, eds. J. I. Goldstein, and H. Yakowitz, (New York: Plenum Press), p. 299.

Bothe, W. (1929), *Zeit. Phys.*, **54**, 161.

Bracewell, B. L., and Veigele, W. J. (1971), *Developments in Applied Spectroscopy*, **9**, (New York: Plenum Press), p. 357.

Bragg, W. H., and Bragg, W. L. (1913), *Proc. Roy. Soc.*, **A88**, 428.
Brown, D. B., and Ogilvie, R. E. (1966), *J. Appl. Phys.*, **37**, 4429.
Brown, D. B., Wittry, D. B., and Kyser, D. F. (1969), *J. Appl. Phys.*, **40**, 1627.
Brown, J. D. (1966), Ph.D. Thesis, University of Maryland.
Brown, J. D. (1969), *Advances in Electronics and Electron Physics Suppl. 6*, eds. A. J. Tousimis, and L. Marton, (New York: Academic Press), p. 45.
Brown, J. D. (1981), *Energy-Dispersive X-ray Spectrometry*, eds. K. F. J. Heinrich, D. E. Newbury, R. L. Myklebust, and C. E. Fiori, Nat. Bur. Stands. Spec. Publ. 604, (Washington: US Dept. Commerce), p. 381.
Brown, J. D., and Parobek, L. (1972), *X-ray Optics and Microanalysis*, eds. G. Shinoda, K. Kohra, and T. Ichinokawa, (Tokyo: Tokyo Univ. Press), p. 163.
Brown, J. D., and Parobek, L. (1973), *Advances in X-ray Analysis*, **16**, eds. L. S. Birks, C. S. Barrett, J. B. Newkirk, and C. O. Rudd, (New York: Plenum Press), p. 198.
Brown, J. D., and Parobek, L. (1974), Proc. 9th Ann. Conf. Microbeam Analysis Soc., Ottawa, Ontario, Paper 39.
Brown, J. D., and Parobek, L. (1976), *X-ray Spectrom.*, **5**, 36.
Brown, J. D., and Robinson, W. H. (1979), *Microbeam Analysis 1979*, ed. D. E. Newbury, (San Francisco: San Francisco Press), p. 238.
Brown, J. D., von Rosenstiel, A. P., and Krisch, T. (1979), *Microbeam Analysis 1979*, ed. D. E. Newbury, (San Francisco: San Francisco Press), p. 241.
Brown, J. D., Packwood, R. H., and Milliken, K. (1981), *Microbeam Analaysis 1981*, ed. R. H. Geiss, (San Francisco: San Francisco Press), p. 174.
Brown, L. M. (1981), *J. Phys. F: Metal Phys.*, **11**, 1.
Buchner, A. R., and Pitsch, W. (1971), *Zeit. Metall.*, **62**, 392.
Buchner, A. R., and Wepner, W. (1971), *Archiv. Eisenh.*, **42**, 565.
Burhop, E. H. S. (1955), *J. Phys. Radium*, **16**, 625.
Busch, H. (1926), *Ann. Phys.*, **4**, Series 81, 974.
Busch, H. (1927), *Ann. Phys.*, **4**, Series 83, 849.
Castaing, R. (1951), Ph.D. Thesis, University of Paris.
Castaing, R. (1960), *Advances in Electronics and Electron Physics*, **13**, ed. L. Marton, (New York: Academic Press), p. 317.
Castaing, R., and Derian, J. C. (1966), *X-ray Optics and Microanalysis*, eds. R. Castaing, P. Deschamps, and J. Philibert, (Paris: Herman), p. 193.
Castaing, R., and Descamps, J. (1954), *C. R. Acad. Sci. Paris*, **238**, 1506.
Castaing, R., and Descamps, J. (1955), *J. Phys. Radium*, **16**, 304.
Castaing, R., and Guinier, A. (1949), Proc. 1st Internat. Conf. on Electron Microsc., ed. A. L. Houwink, (The Hague: Martinus Nijhoff), p. 60.
Castaing, R., and Guinier, A. (1950), *Congr. Internat. de Microsc. Electronique, Paris*, p. 319.
Castaing, R., and Henoc, J. (1966), *X-ray Optics and Microanalysis*, eds. R. Castaing, P. Deschamps, and J. Philibert, (Paris: Hermann), p. 120.
Cauchois, Y. (1932), *J. de Phys.*, **3**, 320.

Claisse, F., and Quintin, M. (1967), *Canad. Spectrosc.*, **12**, 129.
Clareborough, L. M., and Head, A. K. (1976), *Phil. Mag.*, **33**, 557.
Clark, J. C. (1935), *Phys. Rev.*, **48**, 30.
Cliff, G., and Lorimer, G. W. (1972), Proc. Fifth Europ. Congr. Elec. Microsc., (London: Inst. Physics), p. 140.
Cliff, G., and Lorimer, G. W. (1975), *J. Microsc.*, **103**, 203.
Cliff, G., and Lorimer, G. W. (1980), *Electron Microscopy 1980*, Vol. 3, Analysis, eds. P. Brederoo, and V. E. Cosslett, (Leiden: Electron Microscopy Foundation), p. 182.
Coates, D. G. (1980), *Electron Microscopy 1980*, Vol. 3, Analysis, eds. P. Brederoo, and V. E. Cosslett, (Leiden: Electron Microscopy Foundation), p. 26.
Cockett, G. H., and Davis, C. D. (1963), *Brit. J. Appl. Phys.*, **14**, 813.
Colby, J. W. (1968), *Advances in X-ray Analysis*, **11**, eds. J. B. Newkirk, G. R. Mallet, and H. G. Pfeiffer, (New York: Plenum Press), p. 287.
Colby, J. W. (1969), *Thin Film Dielectrics*, ed. F. Vratny, (New York: Electrochem. Soc.).
Conru, H. W., and Laberge, P. C. (1975), *J. Phys. E: Sci. Instr.*, **8**, 136.
Cooke, C. J., and Duncumb, P. (1969), *X-ray Optics and Microanalysis*, eds. G. Mollenstedt, and K. H. Gaukler, (Berlin: Springer), p. 245.
Cosslett, V. E., and Thomas, R. N. (1964a), *Brit. J. Appl. Phys.*, **15**, 235.
Cosslett, V. E., and Thomas, R. N. (1964b), *Brit. J. Appl. Phys.*, **15**, 883.
Cosslett, V. E., and Thomas, R. N. (1964c), *Brit. J. Appl. Phys.*, **15**, 1283.
Cosslett, V. E., and Thomas, R. N. (1965), *Brit. J. Appl. Phys.*, **16**, 779.
Covell, D. F., Sandomire, M. M., and Eichen, M. S. (1960), *Anal. Chem.*, **32**, 1086.
Cox, M. G. C., McEnaney, B., and Scott, V. D. (1974), *Phil. Mag.*, **29**, 585.
Cox, M. G. C., Love, G., and Scott, V. D., (1979), *J. Phys. D: Appl. Phys.*, **12**, 1441.
Criss, J., and Birks, L. S. (1966), *The Electron Microprobe*, eds. T. D. McKinley, K. F. J. Heinrich, and D. B. Wittry, (New York: Wiley), p. 217.
Curgenven, L., and Duncumb, P. (1971), Tube Investments Res. Lab. Rep. 303.
Darbord, R. (1922), *J. Phys. et Radium*, **3**, 218.
Darlington, E. H. (1971), Ph.D. Thesis, University of Cambridge.
Darlington, E. H. (1975), *J. Phys. D: Appl. Phys.*, **8**, 85.
de Broglie, M. (1914), *J. Phys. et Radium*, **4**, 265.
Doig, P., and Flewitt, P. E. J. (1977), *J. Microsc.*, **110**, 107.
Dolby, R. M. (1959), *Proc. Phys. Soc.*, **73**, 81.
Dolby, R. M. (1963), *J. Sci. Instr.*, **40**, 345.
Duerr, J. S., and Ogilvie, R. E. (1972), *Anal. Chem.*, **44**, 2361.
Duncumb, P. (1957), Ph.D. Thesis, University of Cambridge.
Duncumb, P. (1966), *The Electron Microprobe*, eds. T. D. McKinley, K. F. J. Heinrich, and D. B. Wittry, (New York: Wiley), p. 490.
Duncumb, P. (1968), *J. de Microscopie*, **7**, 581.

Duncumb, P., and Cosslett, V. E. (1956), Proc. Symp. X-ray Microsc. and Microradiog., Camb., p. 374.

Duncumb, P., and Melford, D. A. (1966a), Paper presented at 1st Nat. Conf. Electron-Probe Analysis Soc., College Park, Maryland.

Duncumb, P., and Melford, D. A. (1966b), *X-ray Optics and Microanalysis*, eds. R. Castaing, P. Deschamps, and J. Philibert, (Paris: Hermann), p. 240.

Duncumb, P., and Reed, S. J. B. (1968), *Quantitative Electron-Probe Microanalysis*, ed. K. F. J. Heinrich, Nat. Bur. Stand. Spec. Publ. 298, (Washington: US Dept. Commerce), p. 133.

Duncumb, P., and Shields, P. K. (1966), *The Electron Microprobe*, eds. T. D. McKinley, K. F. J. Heinrich, and D. B. Wittry, (New York: Wiley), p. 284.

Duncumb, P., Shields-Mason, P. K., and Da Casa, C. (1969), *X-ray Optics and Microanalysis*, eds. G. Mollenstedt and K. H. Gaukler, (Berlin: Springer), p. 146.

Durr, G., Hofer, W. O., Schulz, F., and Wittmaack, K. (1971), *Zeit. Phys.*, **246**, 312.

Duzević, D., and Bonefacić, A. (1978), *X-ray Spectrom.*, 7, 152.

Ehlert, R. C., and Mattson, R. A. (1967), *Advances in X-ray Analysis*, **10**, eds. G. R. Mallett, M. Fay, and W. M. Mueller, (New York: Plenum Press), p. 389.

Erasmus, D. A. (1982), *Electron-Probe Microanalysis in Biology*, (London: Chapman and Hall).

Ershov, O. A. (1967), *Optics and Spectrosc. USA*, **22**, 252.

Everhart, T. E., Herzog, R. F., Chung, M. S., and Devore, W. J. (1972), *X-ray Optics and Microanalysis*, eds. G. Shinoda, K. Kohra, and T. Ichinokawa, (Tokyo: Tokyo Univ. Press), p. 81.

Fabian, D. F., Watson, L. M., and Marshall, C. A. W. (1971), *Rep. Prog. Phys.*, **34**, 601.

Fairstein, E. (1975), *IEEE Trans. Nucl. Sci.*, **22**, 463.

Fano, U. (1947), *Phys. Rev.*, **72**, 26.

Faulkner, R. G., Hopkins, T. C., and Norrgard, K. (1977), *X-ray Spectrom.*, **6**, 73.

Fink, R. W. (1981), *Energy-dispersive X-ray Spectrometery*, eds. K. F. J. Heinrich, D. E. Newbury, R. L. Myklebust, and C. E. Fiori, Nat. Bur. Stands. Spec. Publ. 604, (Washington: US Dept. Commerce), p. 5.

Fink, R. W., Jopson, R. C., Mark, H., and Swift, C. D. (1966), *Rev. Mod, Phys.*, **38**, 513.

Fioratti, M. P., and Piermattei, S. R. (1971), *Nucl. Inst. Meth.*, **96**, 605.

Fiori, C. E., Myklebust, R. L., Heinrich, K. F. J., and Yakowitz, H. (1976), *Anal. Chem.*, **48**, 172.

Fiori, C. E., Myklebust, R. L., and Gorlen, K. (1981), *Energy-Dispersive X-ray Spectrometry*, eds. K. F. J. Heinrich, D. E. Newbury, R. L. Myklebust, and C. E. Fiori, Nat. Bur. Stands. Spec. Publ. 604, (Washington: US Dept. Commerce), p. 233.

Fischer, D. W. (1970), *Advances in X-ray Analysis*, **13**, eds. B. L. Henke, J. B. Newkirk, and G. R. Mallett, (New York: Plenum Press), p. 159.

Fisher, G. L., and Farningham, C. D. (1972), ASM Mat. Eng. Congr., Cleveland, Ohio.

Fitzgerald, R., Keil, K., and Heinrich, K. F. J. (1968), *Science*, **159**, 528.

Franks, A. (1972), *X-ray Optics and Microanalysis*, eds. G. Shinoda, K. Kohra, and T. Ichinokawa, (Tokyo: Tokyo Univ. Press), p. 57.

Franks, A., and Lindsey, K. (1966), *The Electron Microprobe*, eds. T. D. McKinley, K. F. J. Heinrich, and D. B. Wittry, (New York: Wiley), p. 88.

Furuno, S., and Izui, K. (1971), *Jap. J. Appl. Phys.*, **10**, 1077.

Gehrke, R. J., and Davies, R. C. (1975), *Anal. Chem.*, **47**, 1537.

Geiss, R. H., and Kyser, D. F. (1979), *Ultramicrosc.*, **3**, 379.

Glitz, R. W., Notis, M. R., Williams, D. B., and Goldstein, J. I. (1981), *Microbeam Analysis 1981*, ed. R. H. Geiss, (San Fransico: San Francisco Press), p. 309.

Goldstein, J. I., (1979), *Introduction to Analytical Electron Microscopy*, eds. J. J. Hren, J. I. Goldstein, and D. C. Joy, (New York and London: Plenum Press), p. 83.

Goldstein, J. I., and Williams, D. B. (1977), *SEM/1977*, ed. O. Johari, (Chicago: IITRI), p. 651.

Goldstein, J. I., and Williams, D. B. (1978), *Scanning Electron Microscopy/1978*, ed. O. Johari, (Chicago: SEM Inc.), p. 427.

Goldstein, J. I., and Williams, D. B. (1981), *Quantitative Microanalysis With High Spatial Resolution*, eds. G. W. Lorimer, M. H. Jacobs, and P. Doig, (London: Metals Society), p. 5.

Goldstein, J. I., Lorimer, G. W., and Cliff, G. (1976), Proc. Sixth Europ. Congr. Elec. Microsc. TAL Internat., Israel, p. 56.

Goldstein, J. I., Costley, J. L., Lorimer, G. W., and Reed, S. J. B. (1977), *SEM/1977*, ed. O. Johari, (Chicago: IITRI), **1**, p. 315.

Goldstein, J. I., Newbury, D. E., Echlin, P. Joy, D. C., Fiori, C. E., and Lifshin, E. (1981), *Scanning Electron Microscopy and X-ray Microanalysis*, (New York and London: Plenum Press).

Goudsmit, S., and Saunderson, J. L. (1940), *Phys. Rev.*, **57**, 24.

Gouy, M. (1916), *Ann. Physique*., **5**, 241.

Green, M. (1962), Ph.D. Thesis, University of Cambridge.

Green, M. (1963a), *Proc. Phys. Soc.*, **82**, 204.

Green, M. (1963b), *X-ray Optics and X-ray Microanalysis*, eds. H. H. Pattee, V. E. Cosslett, and A. Engstrom, (New York: Academic Press), p. 361.

Green, M. (1964), *Proc. Phys. Soc.*, **83**, 435.

Green, M., and Cosslett, V. E. (1961), *Proc. Phys. Soc.*, **78**, 1206.

Green, M., and Cosslett, V. E. (1968), *J. Phys. D: Appl. Phys.*, **1**, 425.

Gryzinski, M. (1965), *Phys. Rev.*, **138A**, 336.

Hall, E. L., and Vander Sande, J. B. (1975), *Phil. Mag.*, **32**, 1289.

Hall, T. A. (1968), *Quantitative Electron-Probe Microanalysis*, ed. K. F. J. Heinrich, Nat. Bur. Stands. Spec. Publ. 298, (Washington: US Dept. Commerce), p. 269.

Hall, T. A., and Werba, P. R. (1971), *Electron Microscopy and Analysis*, ed. W. C. Nixon, (London: Inst. Physics), p. 146.

Hallerman, G., and Picklesimer, M. L. (1969), *Advances in Electronics and Electron Physics Suppl. 6*, eds. A. J. Tousimis, and L. Marton, (New York: Academic Press), p. 197.

Heimendahl, M. von (1973), *Micron.*, **4**, 111.

Heinrich, K. F. J. (1966a), *The Electron Microprobe*, eds. T. D. McKinley, K. F. J. Heinrich, and D. B. Wittry, (New York: Wiley), p. 296.

Heinrich, K. F. J. (1966b), *X-ray Optics and Microanalysis*, eds. R. Castaing, P. Deschamps, and J. Philibert, (Paris: Hermann), p. 159.

Heinrich, K. F. J. (1967), Proc. 2nd Nat. Conf. Electron-Probe Analysis Soc., Boston, Mass., Paper 7.

Heinrich, K. F. J. (1968a), *Quantitative Electron-Probe Microanalysis*, ed. K. F. J. Heinrich, Nat. Bur. Stands. Spec. Publ. 298, (Washington: US Dept. Commerce), p. 5.

Heinrich, K. F. J. (1968b), *Advances in X-ray Analysis*, **11**, eds. J. B. Newkirk, G. R. Mallett, and H. G. Pfeiffer, (New York: Plenum Press), p. 40.

Heinrich, K. F. J. (1972), *Anal. Chem.*, **44**, 350.

Heinrich, K. F. J. (1976), *Use of Monte Carlo Calculations in Electron-Probe Microanalysis and Scanning Electron Microscopy*, eds. K. F. J. Heinrich, D. E. Newbury, and H. Yakowitz, Nat. Bur. Stands. Spec. Publ. 460, (Washington: US Dept. Commerce), p. 1.

Heinrich, K. F. J. (1980), *Characterization of Particles*, Nat. Bur. Stands. Spec. Publ. 533, (Washington: US Dept. Commerce).

Heinrich, K. F. J. (1981), *Electron Beam X-ray Microanalysis*, (New York: Van Nostrand Reinhold).

Heinrich, K. F. J., Myklebust, R. L., Rasberry, S. D., and Michaelis, R. E. (1971), *Preparation and Evaluation of SRM's 481 Gold-Silver and Gold-Copper Alloys for Microanalysis*, Nat. Bur. Stands. Spec. Publ. 260-28, (Washington: US Dept. Commerce).

Heinrich. K. F. J., Yakowitz, H., and Vieth, D. L. (1972), Proc. 7th Nat. Conf. Electron Probe Analysis Soc., San Francisco, California, Paper 3.

Heinrich, K. F. J., Newbury, D. E., and Yakowitz, H. (1976), *Use of Monte Carlo Calculations in Electron-Probe Microanalysis and Scanning Electron Microscopy*, Nat. Bur. Stands. Spec. Publ. 460, (Washington: US Dept. Commerce), p. 12.

Henke, B. L. (1963), *X-ray Optics and Microanalysis*, eds. H. H. Pattee, V. E. Cosslett, and A. Engstrom, (New York: Academic Press), p. 157.

Henke, B. L., and Ebisu, E. S. (1974), *Advances in X-ray Analysis*, **17**, eds. C. L. Grant, C. S. Barrett, J. B. Newkirk, and C. O. Rudd, (New York: Plenum Press), p. 150.

Henke, B. L., Elgin, R. L., Lent, R. E., and Ledingham, R. B. (1967), *Norelco Reporter*, **14**, 112.

Henoc, J. (1962), Ph.D. Thesis, University of Paris.

Henoc, J. (1968), *Quantitative Electron-Probe Microanalysis*, ed. K. F. J. Heinrich, Nat. Bur. Stands. Spec. Publ. 298, (Washington: US Dept. Commerce), p. 197.

Henoc, J., and Maurice, F. (1972), *X-ray Optics and Microanalysis*, eds. G. Shinoda, K. Kohra, and T. Ichinokawa, (Tokyo: Tokyo Univ. Press), p. 113.

Henoc, J., and Maurice, F. (1976), *Use of Monte Carlo Calculations in Electron-Probe Microanalysis and Scanning Electron Microscopy*, eds. K. F. J. Heinrich, D. E. Newbury, and H. Yakowitz, Nat. Bur. Stands. Spec. Publ. 460, (Washington: US Dept. Commerce), p. 61.

Henoc, J., Maurice, F., and Zemstoff, A. (1969), *X-ray Optics and Microanalysis*, eds. G. Mollenstedt, and K. H. Gaukler, (Berlin: Springer), p. 187.

Henoc, J., Heinrich, K. F. J., and Myklebust, R. L. (1973), *A Rigorous Correction Procedure for Quantitative Electron-Probe Microanalysis (COR2)*, Nat. Bur. Stands. Tech. Note 769, (Washington: US Dept. Commerce).

Hillier, J. (1943), *Phys. Rev.*, **64**, 318.

Hillier, J. (1947), US Patent 2418 029, (Applied for in 1943, published in 1947).

Hillier, J., and Baker, R. F. (1944), *J. Appl. Phys.*, **15**, 663.

Hirsch, P. B., Howie, A., Nicholson, R. B., Pashley, D. W., and Whelan, M. J. (1965), *Electron Microscopy of Thin Crystals*, (London: Butterworths).

Holliday, J. E. (1966), *The Electron Microprobe*, eds. T. D. McKinley, K. F. J. Heinrich, and D. B. Wittry, (New York: Wiley), p. 3.

Holliday, J. E. (1973), *Advances in X-ray Analysis*, **16**, eds. L. S. Birks, C. S. Barrett, J. B. Newkirk, and C. O. Rudd, (New York: Plenum Press), p. 53.

Hren, J. J., (1979), *Introduction to Analytical Electron Microscopy*, eds. J. J. Hren, J. I. Goldstein, and D. C. Joy, (New York and London: Plenum Press), p. 481.

Hubbell, J. H., and Veigele, W. J. (1976), *Comparison of Theoretical and Experimental Photoeffect Data 0.1keV to 1.5MeV*, Nat. Bur. Stands. Tech. Note 901, (Washington: US Dept. Commerce).

Hutchins, G. A. (1966), *The Electron-Microprobe*, eds. T. D. McKinley, K. F. J. Heinrich, and D. B. Wittry, (New York: Wiley), p. 390.

Jacobs, M. H., and Baborovska, J. (1972), Proc. Fifth Europ. Congr. Elec. Microsc., (London: Inst. Physics), p. 136.

Jaklevic, J. M., and Goulding, F. S. (1971), *IEEE Trans. Nucl. Sci.*, **18**, 187.

Jaklevic, J. M., Goulding, F. S., and Landis, D. A. (1972), *IEEE Trans. Nucl. Sci.*, **19**, 392.

Johann, H. H. (1931), *Zeit. Phys.*, **69**, 185.

Johansson, T. (1932), *Naturwiss*, **20**, 758.

Johansson, T. (1933), *Zeit. Phys.*, **82**, 507.

Jones, I. P., and Loretto, M. H. (1981), *J. Microsc.*, **124**, 3.

Kandiah, K. (1975), *Physical Aspects of Electron Microscopy and Microprobe Analysis*, eds. B. Siegel, and D. R. Beaman, (New York: Wiley), p. 395.

Kandiah, K., Smith, A. J., and White, G. (1975), *IEEE Trans. Nucl. Sci.*, **22**, 2058.

Kanter, H. (1957), *Ann. Phys. Leipz.*, **20**, 144.

Karlovac, N. (1975), *IEEE Trans. Nucl. Sci.*, **22**, 457.

Kaye, G. W. C. (1909), *Phil. Trans. Roy. Soc.*, **A209**, 123.

Kehl, G. L. (1949), *Principles of Metallographic Laboratory Practice*, 3rd Edit., (New York: McGraw-Hill).

Keith, H. D., and Loomis, T. C. (1976), *X-ray Spectrom.*, **5**, 93.

Kelly, P. M. (1975), *Phys. Stat. Sol.*, **32**, 529.

Kelly, P. M., Jostsons, A., Blake, R. G., and Napier, J. G. (1975), *Phys. Stat. Sol.*, **31**, 771.

Khan, M. R., and Karimi, M. (1980), *X-ray Spectrom.*, **9**, 32.

Kirianenko, A., Maurice, F., Calais, D., and Adda, Y. (1963), *X-ray Optics and X-ray Microanalysis*, eds. H. H. Pattee, V. E. Cosslett, and A. Engstrom, (New York: Academic Press), p. 559.

Kirkpatrick, P., and Wiedman, L. (1945), *Phys. Rev.*, **67**, 321.

Knoll, M. (1935), *Zeit. Tech. Phys.*, **16**, 467.

Knoll, M., and Ruska, E. (1932a), *Ann. Phys.*, **12**, 607.

Knoll, M., and Ruska, E. (1932b), *Ann. Phys.*, **12**, 641.

Kohlhass, E., and Scheiding, F. (1970), *Archiv. Eisenh.*, **41**, 97.

Kozlenkov, A. I., Belov, Yu. I., Bogdanov, V. G., and Shulgin, A. I. (1981), Proc. Fifth Conf. Mikrosonde, (Leipzig: Phys. Soc. DDR), p. 47.

Kramers, M. A. (1923), *Phil. Mag.*, **46**, 836.

Krefting, E. R., and Reimer, L. (1973), *Quantitative Analysis with Electron-Microprobe and Secondary Ion Mass Spectroscopy*, ed. E. Preuss, (Julich: Zentralbibliothek der KFA), p. 114.

Kulenkampff, H., and Spyra, W. (1954), *Zeit. Phys.*, **137**, 416.

Kyser, D. F. (1972), *X-ray Optics and Microanalysis*, eds. G. Shinoda, K. Kohra, and T. Ichinokawa, (Tokyo: Tokyo Univ. Press), p. 147.

Kyser, D. F. (1979), *Introduction to Analytical Electron Microscopy*, eds. J. J. Hren, J. I. Goldstein, and D. C. Joy, (New York and London: Plenum Press), p. 199.

Kyser, D. F., and Murata, K. (1974), *IBM J. Res. and Develop.*, **18**, 352.

Landau, L. (1944), *J. Phys. USSR*, **8**, 201.

Landis. D. A., Goulding, F. S., Pehl, R. H., and Walton, J. T. (1971), *IEEE Trans. Nucl. Sci.*, **18**, 115.

Lenard, P., and Becker, A. (1927), *Handbuch der Experimental Physik*, **14**: Kathodenstrahlen, (Leipzig: Akad. Verlagsyes), p. 178.

le Poole, J. B. (1964), Proc. 3rd Europ. Conf. Elec. Microsc., (Prague: Czech Acad. Sci.), p. 439.

Leroux, J. (1961), *Advances in X-ray Analysis*, **5**, ed. W. M. Mueller, (London: Pitman), p. 153.

Leroux, J., and Thinh, T. P. (1977), Revised Tables of Mass Attenuation Coefficients, Available from Corp. Scientifique Claisse Inc., 7-1301 Place de Merici, Quebec, Canada.

Lifshin, E., Ciccarelli, M. F., and Bolon, R. B. (1975), *Practical Scanning Electron Microscopy*, eds. J. I. Goldstein, and H. Yakowitz, (New York: Plenum Press), p. 291.

Lifshin, E., Ciccarelli, M. F., and Bolon, R. B. (1977), *X-ray Optics and Microanalysis*, eds. R. E. Ogilvie, and D. B. Wittry, Boston, Mass., Paper 104.

Lineweaver, J. L. (1963), *J. Appl. Phys.*, **34**, 1786.

Long, J. V. P., and Cosslett, V. E. (1957), *X-ray Microscopy and Microradiography*, eds. V. E. Cosslett, A. Engstrom, and H. H. Pattee, (New York: Academic Press), p. 215.

Lorimer, G. W., Al-Salman, S. A., and Cliff, G. (1977), *Developments in Electron Microscopy and Analysis 1977*, ed. D. L. Misell, (Bristol: Inst. Physics), p. 369.

Lorimer, G. W., Nasir, M. J., Nicholson, R. B., Nuttall, K. Ward, D. E., and Webb, J. R. (1972), *The Strucutre and Properties of Materials – Techniques and Applications of Electron Microscopy*, ed. G. Thomas, (Berkeley: Univ. California Press), p. 222.

Lorimer, G. W., Cliff, G., and Clark, J. N. (1976), *Developments in Electron Microscopy and Analysis 1975*, ed. J. A. Venables, (New York: Academic Press), p. 153.

Lotz, W. (1970), *Zeit. Phys.*, **232**, 101.

Love, G., and Scott, V. D. (1978), *J. Phys. D: Appl. Phys.*, **11**, 1369.

Love, G., and Scott, V. D. (1980), *J. Phys. D: Appl. Phys.*, **13**, 995.

Love, G., and Scott, V. D. (1981), *Scanning*, **4**, 111.

Love, G., Cox, M. G. C., and Scott, V. D. (1974a), *J. Phys. D: Appl. Phys.*, **7**, 2131.

Love, G., Cox, M. G. C., and Scott, V. D. (1974b), *J. Phys. D: Appl. Phys.*, **7**, 2142.

Love, G., Cox, M. G. C., and Scott, V. D. (1975), *J. Phys. D: Appl. Phys.*, **8**, 1686.

Love, G., Cox, M. G. C., and Scott, V. D. (1976), *J. Phys. D: Appl. Phys.*, **9**, 7.

Love, G., Cox, M. G. C., and Scott, V. D. (1977a), *J. Phys. D: Appl. Phys.*, **10**, 7.

Love, G., Cox, M. G. C., and Scott, V. D. (1977b), *Developments in Electron Microscopy and Analysis 1977*, ed. D. L. Misell, (Bristol: Inst. Physics), p. 347.

Love, G., Cox, M. G. C., and Scott, V. D. (1978a), *J. Phys. D: Appl. Phys.*, **11**, 7.

Love, G., Cox, M. G. C., and Scott, V. D. (1978b), *J. Phys. D: Appl. Phys.*, **11**, 23.

Love, G., Rae, D. A., and Scott, V. D. (1980), *Electron Microscopy Vol. I Physics*, eds. P. Brederoo, and G. Boom, (Leiden: Electron Microscopy Foundation), p. 208.

Love, G., Scott, V. D., Dennis, N. M. T., and Laurenson, L. (1981), *Scanning*, **4**, 32.

Lukirskii, A. P., Savinov, E. P., Ershov, O. A., and Shepelev, Yu. F. (1964), *Optics and Spectrosc. USA*, **16**, 168.

Lurio, A., and Reuter, W. (1977), *J. Phys. D: Appl. Phys.*, **10**, 2127.

Lurio, A., Reuter, W., and Keller, J. (1977), *Advances in X-ray Analysis*, **20**, eds. H. F. McMurdie, C. S. Barrett, J. B. Newkirk, and C. O. Rudd, (New York: Plenum Press), p. 481.

Lyman, C. E., Manning, P. E., Duguette, D. J., and Hall, E. (1978), *Scanning Electron Microscopy/1978*, ed. O. Johari, (Chicago: SEM Inc.), **1**, p. 213.

McFarlane, A. A. (1972), *Micron.*, **3**, 506.

MacGillavry, C. H. (1940), *Physica*, **7**, 329.

Manzione, A. V., and Fornwalt, D. E. (1965), *Norelco Reporter*, **12**, 3.

Marshall, D. J., and Hall. T. A. (1968), *J. Phys. D: Appl: Phys.*, **1**, 1651.

Martin, P. M., and Poole, D. M. (1971), *Metallurgical Rev.*, **5**, 19.

Maur, D., and Rosner, B. (1978), *J. Phys. E: Sci. Instr.*, **11**, 1141.

Miller, W. S., and Scott, V. D. (1978), *Metal Sci.*, **12**, 95.

Moll, S. H., and Bruno, G. W., (1967), Proc. 2nd Nat. Conf. Electron Probe Analysis Soc., Boston, Mass., Paper 57.

Morris, P. L., Davies, N. C., and Treverton, J. A. (1977), *Developments in Electron Microscopy and Analysis 1977*, ed. D. L. Misell, (Bristol: Inst. Physics), p. 377.

Morris, P. L., Ball, M. D., and Statham, P. J. (1980), *Electron Microscopy and Analysis 1979*, ed. T. Mulvey, (Bristol: Inst. Physics), p. 413.

Moseley, H. G. J. (1913), *Phil. Mag.*, **26**, 1024.

Moseley, H. G. J. (1914), *Phil. Mag.*, **27**, 703.

Mulvey, T. (1964), *J. Sci. Instr.*, **41**, 61.

Murata, K. (1974), *J. Appl. Phys.*, **45**, 4110.

Murata, K., Matsukawa, T., and Shimizu, R. (1971), *Jap. J. Appl. Phys.*, **10**, 678.

Murata, K., Matsukawa, T., and Shimizu, R. (1972), *X-ray Optics and Microanalysis*, eds. G. Shinoda, K. Kohra, and T. Ichinokawa, (Tokyo: Tokyo Univ. Press), p. 105.

Musket, R. G. (1981), *Energy-Dispersive X-ray Spectrometry*, eds. K. F. J. Heinrich, D. E. Newbury, R. L. Myklebust, and C. E. Fiori, Nat. Bur. Stands. Spec. Publ. 604, (Washington: US Dept. Commerce), p. 97.

Myklebust, R. L., Newbury, D. E., and Yakowitz, H. (1976), *Use of Monte Carlo Calculations in Electron-Probe Microanalysis and Scanning Electron Microscopy*, eds. K. F. J. Heinrich, D. E. Newbury, and H. Yakowitz, Nat. Bur. Stands. Spec. Publ. 460, (Washington: US Dept. Commerce), p. 105.

Myklebust, R. L., Fiori, C. E., and Heinrich, K. F. J. (1979), *FRAME C: A Compact Procedure for Quantitative Energy-Dispersive Electron Probe X-ray Analysis*, Nat. Bur. Stands. Tech. Note 1106, (Washington: US Dept. Commerce).

Nagel, D. J. (1970), *Advances in X-ray Analysis*, **13**, eds. B. L. Henke, J. B. Newkirk, and G. Mallett, (New York: Plenum Press), p. 182.

Nasir, M. J. (1976), *J. Microsc.*, **108**, 79.

Newbury, D. E., and Myklebust, R. L. (1979), *Ultramicrosc.*, **3**, 391.

Newbury, D. E., and Myklebust, R. L. (1981), *Microbeam Analysis 1981*, ed. R. H. Geiss, (San Francisco: San Francisco Press), p. 175.

Newbury, D. E., and Yakowitz, H. (1976), *Use of Monte Carlo Calculations in Electron-Probe Microanalysis and Scanning Electron Microscopy*, eds. K. F. J. Heinrich, D. E. Newbury, and H. Yakowitz, Nat. Bur. Stands. Spec. Publ. 460, (Washington: US Dept. Commerce), p. 15.

Newbury, D. E., Myklebust, R. L., Heinrich, K. F. J., and Small, J. A. (1980), *Characterisation of Particles*, ed. K. F. J. Heinrich, Nat. Bur. Stands. Spec. Publ. 533, (Washington: US Dept. Commerce), p. 39.

Nicholas, W. W. (1929), *Bur. Stands. J. Res.*, **2**, 837.

Nicholson, J. B., and Hasler, M. F. (1966), *Advances in X-ray Analysis*, **9**, eds. G. Mallett, M. Fay, and W. W. Mueller, (New York: Plenum Press), p. 420.

Niedrig, H. (1978), *Scanning Electron Microscopy/1978*, ed. O. Johari, (Chicago: SEM Inc.), **1**, p. 841.

Nockolds. C., Nasir, M. J., Cliff, G., and Lorimer, G. W. (1980), *Electron Microscopy and Analysis 1979*, ed. T. Mulvey, (Bristol: Inst. Physics), p. 417.

Nowlin, C. H., and Blankenship, J. L. (1965), *Rev. Sci. Instr.*, **36**, 1830.

Nullens, H., Van Espen, P., and Adams, F. (1979), *X-ray Spectrom.*, **8**, 104.

Oda, Y., and Nakajima, K. (1973), *J. Jap. Inst. Met.*, **37**, 673.

Ogilvie, R. E. (1977), Proc. 12th Ann. Conf. Microbeam Analysis Soc., Boston, Mass., Paper 37.

Ong, P. S. (1966), *X-ray Optics and Microanalysis*, eds. R. Castaing, P. Deschamps, and J. Philibert, (Paris: Hermann), p. 181.

Packwood, R. H., and Brown, J. D. (1981), *X-ray Spectrom.*, **10**, 138.

Panayi, P. N., Cheshire, D. C., and Echlin, P. (1977), *SEM 1977*, ed. O. Johari, (Chicago: IITRI), **1**, p. 463.

Parrish, W., and Kohler, T. R. (1956), *Rev. Sci. Inst.*, **27**, 295.

Parobek, L., and Brown, J. D. (1974), *Advances in X-ray Analysis*, **17**, eds. C. L. Grant, C. S. Barrett, J. B. Newkirk, and C. O. Rudd, (New York: Plenum Press), p. 479.

Parobek, L., and Brown, J. D. (1978), *X-ray Spectrom.*, **7**, 26.

Pell, E. M., (1960), *J. Appl. Phys.*, **31**, 291.

Philibert, J. (1963), *X-ray Optics and X-ray Microanalysis*, eds. H. H. Pattee, V. E. Cosslett, and A. Engstrom, (New York: Academic Press), p. 379.

Philibert, J. (1969), *X-ray Optics and Microanalysis*, eds. G. Mollenstedt, and K. H. Gaukler, (Berlin: Springer), p. 114.

Philibert, J., and Tixier, R. (1968a), *Quantitative Electron Probe Microanalysis* ed. K. F. J. Heinrich, Nat. Bur. Stands., Spec. Publ. 298, (Washington: US Dept. Commerce), p. 13.

Philibert, J., and Tixier, R. (1968b), *J. Phys. D: Appl. Phys.*, **1**, 685.

Philibert, J., and Tixier, R. (1975), *Physical Aspects of Electron Microscopy and Microbeam Analysis*, eds. B. M. Siegel, and D. R. Beaman, (New York: Wiley), p. 333.

Philibert, J. Rivery, J., Bryckaert, D., and Tixier, R. (1970), *J. Phys. D: Appl. Phys.*, **3**, 70.

Poole, D. M., (1968), *Quantitative Electron-Probe Microanalysis*, ed. K. F. J. Heinrich, Nat. Bur. Stands. Spec. Publ. 298, (Washington: US Dept. Commerce), p. 93.

Poole, D. M., and Thomas, P. M. (1961–2), *J. Inst. Metals*, **90**, 228.

Powell, C. J. (1976a), *Rev. Mod. Phys.*, **48**, 33.

Powell, C. J. (1976b), *Use of Monte Carlo Calculations in Electron Probe Microanalysis and Scanning Electron Microscopy*, eds. K. F. J. Heinrich, D. E. Newbury, and H. Yakowitz, Nat. Bur. Stands. Spec. Publ. 460, (Washington: US Dept. Commerce), p. 97.

Rae, D. A., Scott, V. D., and Love, G. (1981), *Quantitative Microanalysis with High Spatial Resolution*, eds. G. W. Lorimer, M. H. Jacobs, and P. Doig, (London: Metals Society), p. 57.

Ranzetta, G. V. T., and Scott, V. D. (1963), *J. Nucl. Mater.*, **9**, 277.

Ranzetta, G. V. T., and Scott, V. D. (1966), *J. Sci. Inst.*, **43**, 816.

Ranzetta, G. V. T., and Scott, V. D. (1968), *Metals Mater.*, **1**, 146.

Rao, P., and Lifshin, E. (1977), Proc. 12th Ann. Conf. Microbeam Analysis Soc., Boston, Mass., Paper 118.

Rao-Sahib, T. S., and Wittry, D. B. (1974), *J. Appl. Phys.*, **45**, 5060.

Reed, S. J. B. (1964), Ph.D. Thesis, University of Cambridge.

Reed, S. J. B. (1965), *Brit. J. Appl. Phys.*, **16**, 913.

Reed, S. J. B. (1966), *X-ray Optics and Microanalysis*, eds. R. Castaing, P. Deschamps, and J. Philibert, (Paris: Hermann), p. 339.

Reed, S. J. B. (1972), *J. Phys. E: Sci. Instr.*, **5**, 997.

Reed, S. J. B. (1975a), *X-ray Spectrom.*, **4**, 14.

Reed, S. J. B. (1975b), *Electron Microprobe Analysis*, (Cambridge: Cambridge Univ. Press).

Reed, S. J. B. (1982), *Ultramicrosc.*, **7**, 405.

Reed, S. J. B., and Long, J. V. P. (1963), *X-ray Optics and Microanalysis*, eds. H. H. Pattee, V. E. Cosslett, and A. Engstrom, (New York: Academic Press), p. 317.

Reed, S. J. B., and Mason, P. K. (1967), Proc. 2nd Nat. Conf. Electron Probe Analysis Soc., Boston, Mass., Paper 12.

Reed, S. J. B., and Ware, N. G. (1972), *J. Phys. E: Sci. Instr.*, **5**, 582.

Reed, S. J. B., and Ware, N. G. (1973), *X-ray Spectrom.*, **2**, 69.

Reuter, W. (1972), *X-ray Optics and Microanalysis*, eds. G. Shinoda, K. Kohra, and T. Ichinokawa, (Tokyo: Tokyo Univ. Press), p. 121.

Robertson, A., Prestwich, W. V., and Kennett, T. J. (1972), *Nucl. Instr. Meth.*, **100**, 317.

Robinson, W., and Brown, J. D. (1978), Proc. 13th Ann. Conf. Microbeam Analysis Soc., Ann Arbor., Michigan, Paper 42.
Romig, A. D., and Goldstein, J. I. (1979), *Microbeam Analysis 1979*, ed. D. E. Newbury, (San Francisco: San Francisco Press), p. 124.
Ruark, A., and Brammer, F. E. (1937), *Phys. Rev.*, **52**, 322.
Russ, J. C. (1972), Proc. 7th Nat. Conf. Electron Probe Analysis Soc., San Francisco, California, Paper 76.
Russ, J. C. (1973), Proc. 8th Nat. Conf. Electron Probe Analysis Soc., New Orleans, Louisiana, Paper 30.
Russ, J. C. (1974), Proc. 9th Ann. Conf. Microbeam Analysis Soc., Ottawa, Ontario, Paper 22.
Russ, J. C. (1976), Proc. 11th Ann. Conf. Microbeam Analysis Soc., Miami, Florida: Paper 19.
Russ, J. C. (1978), Proc. 13th Ann. Conf. Microbeam Analysis Soc., Ann. Arbor., Michigan, Paper 46.
Russ, J. C., and Sandborg, A. D. (1981), *Energy Dispersive X-ray Spectrometry*, ed. K. F. J. Heinrich, D. E. Newbury, R. L. Myklebust, and C. E. Fiori, Nat. Bur. Stands. Spec. Publ. No. 604, (Washington: US Dept. Commerce), p. 71.
Russ, J. C., Sandborg, A. O., Barnhart, M. W., Soderquist, C. E., Lichtinger, R. W., and Walsh, C. J. (1973), *Advances in X-ray Analysis*, **16**, eds. L. S. Birks, C. S. Barrett, J. B. Newkirk, and C. O. Rudd. (New York: Plenum Press), p. 284.
Ruste, J. (1976), Ph.D. Thesis, University of Nancy.
Ruste, J., and Zeller, C. (1977), *C. R. Acad. Sci. Paris*, **284B**, 507.
Ruste, J., Bouchacourt, M., and Thevenot, F. (1978), *J. Less-Common Met.*, **59**, 131.
Ruthermann, G. (1941), *Naturwiss*, **29**, 648.
Ruthermann, G. (1942), *Naturwiss*, **30**, 145.
Salter, W. J. M. (1970), *Quantitative Electron Probe Microanalysis*, (London: Structural Publ. Ltd.).
Sandstrom, A. E. (1952), *Ark. Fys.*, **4**, 519.
Sayce, L. A., and Franks, A. (1964), *Proc. Roy, Soc.*, **A282**, 353.
Schamber, F. (1973), Proc. 8th Nat. Conf. Electron Probe Analysis Soc., New Orleans, Louisiana, Paper 85.
Schamber, F. (1977), *X-ray Fluorescence Analysis of Environmental Samples*, ed. T. G. Dzubay, (Michigan: Ann. Arbor. Science Publ.), p. 241.
Schamber, F. H. (1981), *Energy-Dispersive X-ray Spectrometry*, eds. K. F. J. Heinrich, D. E. Newbury, R. L. Myklebust, and C. E. Fiori, Nat. Bur. Stands. Spec. Publ. 604, (Washington: US Dept. Commerce), p. 193.
Schmitz, U., Ryder, P. L., and Pitsch, W. (1969), *X-ray Optics and Microanalysis*, eds. G. Mollenstedt, and K. H. Gaukler, (Berlin: Springer), p. 104.
Scott, V. D., and Ranzetta, G. V. T. (1961–2), *J. Inst. Met.*, **90**, 160.
Shimizu, R., and Murata, K. (1971), *J. Appl. Phys.*, **42**, 387.

Shimizu, R., Kishimoto, H., Shirai, T., Murata, K., Shinoda, G., and Miura, M. (1966), *Technol. Rep. Osaka Univ.*, **16**, No. 716–747, p. 415.

Shimizu, R., Murata, K., and Shinoda, G., (1966), *X-ray Optics and Microanalysis*, eds. R. Castaing, P. Deschamps, and J. Philibert, (Paris: Hermann), p. 127.

Shimizu, R., Kataka, Y., Matsukawa, T., Ikuta, T., Murata, K., and Hashimoto, H. (1975), *J. Phys. D: Appl. Phys.*, **8**, 820.

Shinoda, G. (1966), *X-ray Optics and Microanalysis*, eds. R. Castaing, P., Deschamps, and J. Philibert, (Paris: Hermann), p. 97.

Shinoda, G., Murata, K., and Shimizu, R. (1968), *Quantitative Electron Probe Microanalysis*, ed. K. F. J. Heinrich, Nat. Bur. Stands. Spec. Publ. 298, (Washington: US Dept. Commerce), p. 155.

Small, J. A., Heinrich, K. F. J., Fiori, C. E., Myklebust, R. L., Newbury, D. E., and Dilmore, M. F. (1978), *Scanning Electron Microscopy/1978*, ed. O. Johari, (Chicago: SEM Inc.), p. 445.

Small, J. A., Heinrich, K. F. J., Fiori, C. E., Newbury, D. E., and Myklebust, R. L. (1978), Proc. 13th Ann. Conf. Microbeam Analysis Soc., Ann. Arbor., Michigan, Paper 56.

Small, J. A., Newbury, D. E., and Myklebust, R. L. (1979), *Microbeam Analysis 1979*, ed. D. E. Newbury, (San Francisco: San Francisco Press), p. 243.

Small, J. A., Heinrich, K. F. J., Newbury, D. E., Myklebust, R. L., and Fiori, C. E. (1980), *Characterization of Particles*, ed. K. F. J. Heinrich, Nat. Bur. Stands. Spec. Publ. 533, (Washington: US Dept. Commerce), p. 29.

Smith, D. G. W. (1981), *X-ray Spectrom.*, **10**, 78.

Smith, D. G. W., and Gold, C. M. (1978), *Advances in X-ray Analysis*, **19**, eds. R. W. Gould, C. S., Barrett, J. B. Newkirk, C. O. Rudd, (Dubuque, Iowa: Kendall/Hunt Publishing Co.), p. 191.

Smith, D. G. W., and Gold, C. M. (1979), *Microbeam Analysis 1979*, ed. D. E. Newbury, (San Francisco: San Francisco Press), p. 273.

Smith, D. G. W., and Reed, S. J. B. (1981), *X-ray Spectrom.*, **10**, 198.

Smith, D. G. W., Gold, C. M., and Tomlinson, D. A. (1975), *X-ray Spectrom.*, **4**, 149.

Spendley, W., Hext, G. R., and Himsworth, F. R. (1962), *Technometrics*, **4**, 441.

Springer, G. (1966), *Mikrochim. Acta*, **3**, 587.

Springer, G. (1967), *Neues. Jahrb. Mineral Abhandl.*, **106**, 241.

Springer, G. (1972), *X-ray Optics and Microanalysis*, eds. G. Shinoda, K. Kohra, and T. Ichinokawa, (Tokyo: Tokyo Univ. Press), p. 141.

Springer, G. (1976), *X-ray Spectrom.*, **5**, 88.

Springer, G., and Nolan, B., (1976), *Canad. J. Spectrosc.*, **21**(5), 134.

Sprys, J. W., and Short, M. A. (1976), Proc. 11th Ann. Conf. Microbeam Analysis Soc., Miami, Florida, Paper 9.

Starke, H. (1898), *Ann. Phys.*, **66**, 49.

Statham, P. J. (1975), Ph.D. Thesis, University of Cambridge.

Statham, P. J. (1976a), *J. Phys. E: Sci. Intr.*, **9**, 1023.
Statham, P. J. (1976b), *X-ray Spectrom.*, **5**, 154.
Statham, P. J. (1976c), *X-ray Spectrom.*, **5**, 16.
Statham, P. J. (1977a), *X-ray Spectrom.*, **6**, 94.
Statham, P. J. (1977b), *Anal. Chem.*, **49**, 2149.
Statham, P. J. (1978), *X-ray Spectrom.*, **7**, 132.
Statham, P. J. (1979), *Mikrochim. Acta Suppl.*, **8**, 231.
Statham, P. J. (1980), *Electron Microscopy 1980, Vol. 3, Analysis*, eds. P. Brederoo, and V. E. Cosslett, (Leiden: Electron Microscopy Foundation), p. 30.
Statham, P. J. (1981), *Energy-Dispersive X-ray Spectrometry*, eds. K. F. J. Heinrich, D. E. Newbury, R. L. Myklebust, and C. E. Fiori, Nat. Bur. Stands. Spec. Publ. 604, (Washington: US Dept. Commerce), p. 127.
Statham, P. J., and Ball, M. D. (1980), *Microbeam Analysis 1980*, ed. D. B. Wittry, (San Francisco: San Francisco Press), p. 165.
Statham, P. J., and Pawley, J. B. (1978), *Scanning Electron Microscopy/1978*, ed. O. Johari, (Chicago: SEM Inc.), **1**, p. 469.
Statham, P. J., Long, J. V. P. White, G., and Kandiah, K. (1974), *X-ray Spectrom.*, **3**, 153.
Sutfin, L. V., and Ogilvie, R. E. (1971), *Amer. Soc. for Testing and Materials Spec. Tech. Publ. 485.*
Sweatman, T. R., and Long, J. V. P. (1969), *J. Petrol.*, **10**, 332.
Sweeney, W. E., Seebold, R. E., and Birks, L. S. (1960), *J. Appl. Phys.*, **31**, 1061.
Thinh, T. P., and Leroux, J. (1979), *X-ray Spectrom.*, **8**, 85.
Thomas, L. E. (1980), Proc. 38th Meeting Electron Microsc. Soc. of Amer., ed. G. W. Bailey, (Baton Rouge: Claitor), p. 90.
Thomas, P. M. (1963), *Brit. J. Appl. Phys.*, **14**, 397.
Thomas, R. N. (1961), Ph.D. Thesis, University of Cambridge.
Thompson, M. N., Doig, P., Eddington, J. W., and Flewitt, P. E. J. (1977), *Phil. Mag.*, **35**, 1537.
Tixier, R. (1979), *Microanalysis and Scanning Electron Microscopy*, eds. F. Maurice, L. Meny, and R. Tixier, (Orsay: Les Editions de Physique), p. 429.
Tixier, R., and Philibert, J. (1969), *X-ray Optics and Microanalysis*, eds. G. Mollenstedt, and K. H. Gaukler, (Berlin: Springer), p. 180.
Veigele, W. J. (1973), *Atomic Data*, **5**(1), 51.
Vesely, D., and Woodise, S. (1982), *Proc. Roy. Microsc. Soc.*, **17**, 137.
Vignes, A., and Dez, G. (1968), *J. Phys. D: Appl. Phys.*, **1**, 1309.
Wagner, E. (1917), *Physikal Zeit.*, **18**, 405.
Walitski, P. J., and Colby, J. W. (1969), Proc. 5th Nat. Conf. Electron Probe Analysis Soc., New York, Paper 19.
Walter, F. J., Stone, R., Blackburn, D. H., and Pella, P. A. (1981), *Energy-Dispersive X-ray Spectrometry*, eds. K. F. J. Heinrich, D. E. Newbury, R. L. Myklebust, and C. E. Fiori, Nat. Bur. Stands. Spec. Publ. 604, (Washington: US Dept. Commerce), p. 61.

Wardell, I. M., and Cosslett, V. E. (1966), *The Electron Microprobe*, eds. T. D. McKinley, K. F. J. Heinrich, and D. B. Wittry, (New York: Wiley), p. 23.
Ware, N. G., and Reed, S. J. B. (1973), *J. Phys. E: Sci. Instr.*, **6**, 286.
Warner, R. R., and Coleman, J. R. (1973), *Micron*, **4**, 61.
Weber, K., and Marschal, J. (1964), *J. Sci. Instr.*, **41**, 15.
Webster, D. L., Hansen, W. W., and Duveneck, F. B. (1933), *Phys. Rev.*, **43**, 839.
Wegstein, (1958), *Commun. ACM*, **1**, 9.
Weisweiler, W. (1970), *Mikrochim. Acta*, 744.
Weisweiler, W. (1975), *Mikrochim. Acta*, 365.
Wentzel, G. (1927), *Zeit. Phys.*, **43**, 524.
Whiddington, R. (1912), *Proc. Roy. Soc.*, A**86**, 360.
White, E. W., and Johnson, G. G. (1970), *X-ray Emission and Absorption Wavelengths and Two-Theta Tables*, ASTM Data Series DS 37A, Philadelphia (Amer. Soc. for Testing and Materials: Philadelphia).
Wilkes, M. V. (1966), *Introduction to Numerical Analysis*, (Cambridge: Cambridge Univ. Press), p. 55.
Williams, C. W. (1968), *Trans. IEEE Nucl. Sci.*, **15**, 297.
Williams, E. J. (1932), *Proc. Roy. Soc.*, A**130**, 320.
Wittry, D. B. (1958), *J. Appl. Phys.*, **29**, 1543.
Wood, J., Williams, D. B., and Goldstein, J. I., (1981), *Quantitative Microanalysis with High Spatial Resolution*, eds. G. W. Lorimer, M. H. Jacobs, and P. Doig, (London: Metals Society), p. 24.
Yakowitz, H. (1975), *Practical Scanning Electron Microscopy*, eds. J. I. Goldstein, and H. Yakowitz, (New York: Plenum Press), p. 388.
Yakowitz, H., and Heinrich, K. F. J. (1968), *Mikrochim. Acta*, 182.
Yakowitz, H., Myklebust, R. L., and Heinrich, K. F. J. (1973), *FRAME: An On-line Correction Procedure for Quantitative Electron Probe Microanalysis*, Nat. Bur. Stands. Tech. Note 786, (Washington: US Dept. Commerce).
Zaluzec, N. J. (1978), Ph.D. Thesis, University of Illinois; Oak Ridge Nat. Lab. Rep. ORNL/TM-6705.
Zaluzec, N. J. (1979), *Introduction to Analytical Electron Microscopy*, ed. J. J. Hren, J. I. Goldstein, and D. C. Joy, (New York and London: Plenum Press), p. 121.
Zaluzec, N. J., and Fraser, H. L. (1976), Proc. 34th Meeting Electron Microsc. Soc. of Amer., ed. G. W. Bailey, (Baton Rouge: Claitor), p. 420.
Ziebold, T. O. (1967), *Anal. Chem.*, **39**, 858.
Ziebold, T. O., and Ogilvie, R. E. (1964), *Anal. Chem.*, **36**, 322.
Ziebold, T. O., and Ogilvie, R. E. (1966), *The Electron Microprobe*, eds. T. D. McKinley, K. F. J. Heinrich, and D. B. Wittry, (New York: Wiley), p. 378.

Index

C

D

E

F

Q

R

S

Z